An Interactive Introduction to Mathematical Analysis

This book provides a rigorous course in the calculus of functions of a rea
Its gentle approach, particularly in its early chapters, makes it especially su
students who are not headed for graduate school but, for those who are, this b
provides the opportunity to engage in a penetrating study of real analysis.

 The companion onscreen version of this text contains hundreds of links to altern
approaches, more complete explanations and solutions to exercises; links that ma
more friendly than any printed book could be. In addition, there are links to a wea
of optional material that an instructor can select for a more advanced course, and th
students can use as a reference long after their first course has ended. The onscreen
version also provides exercises that can be worked interactively with the help of the
computer algebra systems that are bundled with *Scientific Notebook*.

Jonathan Lewin is professor of mathematics at Kennesaw State University in Georgia.
He is the author and co-author of three books, *Precalculus with Scientific Notebook*,
An Introduction to Mathematical Analysis, and *Exploring Mathematics with Scientific
Notebook.*

An Interactive Introduction to Mathematical Analysis

Jonathan Lewin
Kennesaw State University

CAMBRIDGE
UNIVERSITY PRESS

University Printing House, Cambridge CB2 8BS, United Kingdom

Published in the United States of America by Cambridge University Press, New York

Cambridge University Press is part of the University of Cambridge.

It furthers the University's mission by disseminating knowledge in the pursuit of education, learning and research at the highest international levels of excellence.

www.cambridge.org
Information on this title: www.cambridge.org/9781107694040

© Jonathan Lewin 2003, 2014
Scientific Notebook and *Scientific Viewer* are Registered Trademarks of MacKichan Software, Inc. © 2000-2002 MacKichan Software, Inc. All Rights Reserved.

First published 2003
Revised paperback edition 2014

A catalogue record for this publication is available from the British Library

ISBN 978-1-107-69404-0 Paperback

Contents

Preface

The Purpose of This Book

This book provides a rigorous course in the calculus of functions of a real variable. It is intended for students who have previously studied calculus at the elementary level and are possibly entering their first upper-level mathematics course. In many undergraduate programs, the first course in analysis is expected to provide students with their first solid training in mathematical thinking and writing and their first real appreciation of the nature and role of mathematical proof. Therefore, a beginning analysis text needs to be much more than just a sequence of rigorous definitions and proofs. The book must shoulder the responsibility of introducing its readers to a new culture, and it must encourage them to develop an aesthetic appreciation of this culture.

This book is meant to serve two functions (and two audiences): On the one hand, it is intended to be a gateway to analysis for students of mathematics and for certain students majoring in the sciences or technology. It is also intended, however, for other groups of students, such as prospective high school teachers, who will probably see their course in analysis as the hardest course that they have ever taken and for whom the most important role of the course will be as an introduction to mathematical thinking.

At the same time, this book is meant to be a recruiting agent. It is my desire to motivate talented students to develop their interest in mathematics and to provide them with an incentive to continue their studies after the present course has ended. Each topic is presented in a way that extends naturally to more advanced levels of study, and it should not be necessary for students to "unlearn" any of the material of this book when they enter more advanced courses in analysis and topology.

The approach in this book is particularly gentle in its first few chapters, but it gradually becomes more demanding. By the time one reaches the last few chapters, both the pace and depth have been increased. Those who reach the later chapters are probably in the second term of a two-term sequence in mathematical analysis, and I expect students who use this book for a second course (the survivors of a first course) to be generally stronger than those who take analysis for one term only. These later chapters cover quite a lot of ground and contain a number of innovative sections on topics that are not usually covered in a book at this level. A third level of coverage is provided by the optional chapters that appear only in the on-screen version of this book. These are the chapters that are marked in green in the on-screen Contents documents. However, even in the more demanding chapters, I have preserved my commitment to strong motivation and clean, well-explained proofs.

Global Structure of the Book

The book is divided into two main parts: Part I and Part II.

Part I

Part I introduces the notion of mathematical rigor and consists of Chapters 1, 2, 3, and 4 as illustrated in the following figure. The lighter boxes in this figure represent chapters in the main body of the book and the darker boxes represent chapters that can be reached only in the on-screen version.

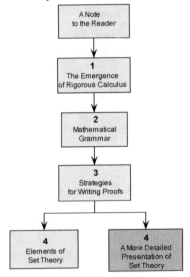

Chapter 1, *The Emergence of Rigorous Calculus*, presents a very brief view of the history of rigorous calculus and of the notion of rigor in mathematics.

Chapter 2, *Mathematical Grammar*, provides an introduction to the reading and writing of mathematical sentences and to some of the special words that we use in a mathematical argument.

Chapter 3, *Strategies for Writing Proofs*, is a sequel to the chapter on mathematical grammar. The message of this chapter is that the nature of an assertion that one wishes to prove can often suggest a strategy for the proof. Students who study this chapter will find it easier to solve problems later in the text.

Chapter 4, *Elements of Set Theory*, presents a brief elementary review of the set-theoretic and function concepts that are used throughout the text. The on-screen text offers an alternative chapter, *Set Theory*, which covers the material much more carefully and goes on to present some advanced topics.

An instructor may choose to base a significant part of the course on Part I of the text or to let the course skim through this part. Some instructors may choose to skip Part I altogether and proceed directly to Part II, where the study of mathematical analysis begins.

Part II

Part II presents an introductory course in mathematical analysis as illustrated in the following figure. The lighter boxes in this figure represent chapters in the main body of the book and the darker boxes represent chapters that can be reached only in the on-screen version. Notice that some of the interdependence arrows in the figure are dotted to remind us that we have a choice of using either a printed chapter or its alternative on-screen partner as the prerequisite to later material. For example, we can use either of the two Chapters 11 as the prerequisite to Chapter 13.

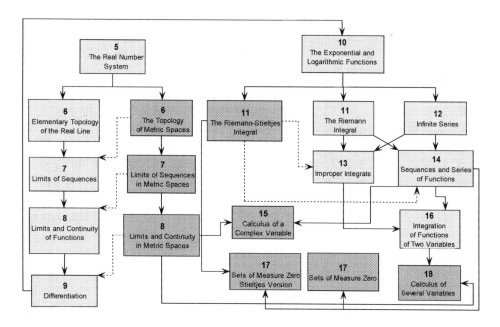

Chapter 5, *The Real Number System*, introduces the real number system and the notion of completeness that plays a prominent role throughout the succeeding chapters.

Chapter 6, *Elementary Topology of the Real Line*, introduces some of the simple topological properties of the number system $\mathbf{R}$. These properties are used extensively in the subsequent chapters on limits, continuity, differentiation, and integration. The on-screen text offers an alternative chapter, *The Topology of Metric Spaces*, for those who prefer a more advanced approach to the material.

Chapter 7, *Limits of Sequences,* presents an introduction to the theory of limits. I believe that it is easier to study limits of sequences before one studies limits of functions of a real variable. The on-screen text offers an alternative chapter, *Limits of Sequences in Metric Spaces*, for those who read the more general version of Chapter 6.

Chapter 8, *Limits and Continuity of Functions*, presents an introduction to the theory of limits and continuity of functions of a real variable. The on-screen text offers a more general chapter, *Limits and Continuity in Metric Spaces*, for those who desire the more general approach.

Chapter 9, *Differentiation,* presents an introduction to differentiation of functions of a real variable. The core of this chapter is the mean value theorem, whose proof rests heavily on the properties of continuous functions developed in Chapter 8.

Chapter 10, *The Exponential and Logarithmic Functions*, presents a rigorous definition of exponents and logarithms and derives their principal properties.

Chapter 11, *The Riemann Integral,* begins with a discussion of the integration of step functions. Although this discussion lengthens the chapter a little, I believe that the increase in length is worthwhile. The discussion of step functions and elementary sets is easy to read and takes little classroom time. Moreover, it provides simple, clean, and precise notation in which to present the mainstream of Riemann integration. This notation allows us to give short clean proofs of several theorems later in the chapter that look quite formidable in many other texts. The composition theorem for Riemann integrability is a case in point. Finally, the efficient notation developed in this chapter facilitates the proof of the Arzela bounded convergence theorem in Chapter 14. The on-screen text offers an alternative chapter, *The Riemann-Stieltjes Integral*, for those who elect to read Chapter 12 before Chapter 11 and who prefer a Stieltjes integral.

Chapter 12, *Infinite Series,* presents an introduction to the theory of infinite series. Those who wish to study this chapter before the chapter on Riemann integration may do so.

Chapter 13, *Improper Integrals,* presents a brief theory of improper integrals that runs parallel to some of the material in Chapter 12.

In Chapter 14, *Sequences and Series of Functions,* I have departed a little from some of the traditions that are canonized in most other books. This departure stems from the availability of the Arzela bounded convergence theorem. Because this theorem is presented here with a fairly simple proof, I have been able to place the concept of uniform convergence in a slightly different perspective, still important but not quite as important as it is in most texts. The fact that we do not require uniform convergence of a sequence (f_n) of functions to guarantee an equality of the form

$$\int_a^b \lim_{n \to \infty} f_n = \lim_{n \to \infty} \int_a^b f_n$$

opens the door to some beautiful and powerful theorems that appear, in Chapter 16, in a more powerful form than one would find in most texts. Among these are the theorems on differentiating under an integral sign and interchange of iterated Riemann integrals.

Chapter 15, *Calculus of a Complex Variable*, is optional and is available in the on-screen version of the book.

Chapter 16, *Integration of Functions of Two Variables*, is concerned with iterated Riemann integrals of functions defined on a rectangle in $\mathbf{R}^2$. The key theorems of the chapter are the theorem on differentiation under an integral sign and the beautiful and elegant Fichtenholz theorem on the interchange of iterated Riemann integrals. The latter theorem, which has been much neglected in analysis texts, is truly a theorem about Riemann integration. It has no analog for Lebesgue integrals and is not a special case of Fubini's theorem.

Chapter 17, *Sets of Measure Zero*, is optional and is available in the on-screen version of the book. This chapter takes the reader to the doorstep of the modern theory of integration and presents a number of interesting theorems about Riemann integration that are beyond our reach in Chapter 11. Among these are the sharp form of the composition theorem for Riemann integrability and the most natural form of the change of variable theorem. The on-screen text offers an alternative chapter, giving a version for Stieltjes integrals of the measure zero concept.

Chapter 18, *Calculus of Several Variables*, is optional and is available in the on-screen version of the book. This chapter develops many of the central topics in the calculus of several variables, including partial differentiation, integration on curves, the inverse and implicit function theorems, and the change of variable theorem for multiple integrals.

The On-Screen Version of This Book

This book is supplied both as a traditionally printed and bound textbook and also in a version that is designed for on-screen reading using the software product *Sci-*

entific Notebook[1] Version 5.5 or its free viewer version, *Scientific Viewer*. Both of these products are supplied with the on-screen book that you can download from

> http://www.math-movies.com/Update-Files/Analysis-Textbook/
> Lewin_Analysis_Interactive.zip

Scientific Notebook as supplied is a 30 day trial version that will run for 30 days and will then be converted automatically into *Scientific Viewer* unless you buy your own license for it.

Scientific Viewer is all you need for the reading of the on-screen version of this book. However, if you want to be able to revise a document and save the changes, and if you want to be able to read the interactive topics in this book using the computer algebra system with which *Scientific Notebook* comes bundled, then you need a licensed *Scientific Notebook*. For information about purchasing a permanent license for *Scientific Notebook*, go to the MacKichan Software website, http://www.mackichan.com, or telephone them at (360) 394-6033.

Disclaimer

The author of this text has no business connection with MacKichan Software Inc. and does not represent that company in any opinions or perspectives of the software products that are presented in this text.

I am, nevertheless, greatly indebted to MacKichan Software Inc. for the wonderful job that they have done, for the unique opportunities that they have provided the mathematical community with their truly unique products, and for the help that they have provided to me personally in the writing of my books.

Why an On-Screen Version?

Both the printed version and the on-screen version can play an important role in the reading of this book, and readers are encouraged to make use of both of them. The importance of a traditionally printed and bound text speaks for itself. There is, of course, nothing quite like thumbing through a printed text and there is no doubt that, at times, the reader's best course of action is to read the printed text. However, the on-screen version of the text presents special features that could never be found in a printed text, and there are times when the reader would gain by taking advantage of these special features.

The basic core of the book is contained in the printed version. The on-screen version contains all of the material in the printed version, but, taking advantage

---

[1]   *Scientific Notebook* is a product of MacKichan Software Inc.

of the powerful communication and word processing features of *Scientific Note-book*, the on-screen version also contains a large number of hyperlinks that can whisk the reader to many other items that do not appear in the printed text.

The on-screen version of the text contains the following features:

1. It includes a large number of alternative approaches and extensions of the text that can be accessed by clicking on hyperlinks. I have also provided hyperlinks that lead to hints or solutions to exercises, and, on many of the occasions on which the reader is told in the text that an assertion is "easy to see", I have provided a link to a full explanation of that assertion. In this way I have managed to make the on-screen text more friendly than any printed text could be. Some of the hyperlinks in this book lead to files located on the reader's hard drive, but others will take advantage of the Internet communication features of *Scientific Notebook* and will take the reader to targets in the World Wide Web. Links of the latter type allow me to be responsive to feedback received from students and instructors.

2. Hyperlinks in the on-screen version also allow for instant cross referencing. A single click takes us to the target of a cross reference and one more click returns us to the point at which the reference was made.

3. The on-screen version of the text contains a very useful set of Contents documents that will enable you to reach any chapter, section, or subsection that you want to see. Every header in the book also contains a link back to the appropriate position in the Contents documents, thus providing you with an instant road-map of the book at any time.

4. The navigation bar in *Scientific Notebook* provides a list of headers in the document you are reading, and you can reach any item in this list by clicking on it.

5. The on-screen version of the text includes a useful table of contents that allows you to zoom in to show more detail or zoom out to get a bird's eye view.. By clicking on any entry in the table of contents, you can jump to any desired position in the text and, from inside the text, you can jump back to the correct point in the table of contents.

6. The on-screen version of the text will contain a variety of links to sound messages and videos. Many of the videos are presented as mini-lectures that show part or all of a proof as it is being written and explained by the author. In this way, they simulate the lecture room experience with the added advantage that you can fast forward them or drag the cursor back to repeat any portions of the movie that you want to hear again.

Additional Material Provided in the On-Screen Version

One of the important roles of the on-screen version is to allow me to include a wide variety of exercises, proofs, and topics without compromising the ease with which a first course in real analysis can be selected from the printed book. The printed book remains uncluttered and contains the basic bill of fare of an introductory course, but the on-screen version provides the option to read much more, even in those chapters that also appear in the printed version.

For example, there are links to extra blocks of exercises that are not included in the printed text, there are links to proofs that were omitted in the printed book, and there are some extra optional sections to which reference may be made in later optional sections. Furthermore, the on-screen version of the text contains several additional chapters that do not appear in the printed version:

1. The chapter on set theory in the printed text contains only a minimum of material. However, at the beginning of that chapter, the on-screen version provides a link to a more extensive presentation that includes the concept of countability, the equivalence theorem, and some more advanced topics such as the axiom of choice, Zorn's lemma, the well ordering principle, and the continuum hypothesis.

2. At the beginning of the chapter on topology of the real line, the on-screen version provides a link to the topology of metric spaces.

3. At the beginning of the chapter on limits of sequences in $\mathbf{R}$, the on-screen version provides a link to the theory of limits of sequences in metric spaces.

4. At the beginning of the chapter on Riemann integration, the on-screen version provides a link to an alternative chapter on Riemann-Stieltjes integration.

5. The on-screen version provides a link to an optional chapter that introduces the calculus of a complex variable. Topics included in this optional chapter include the fundamental theorem of algebra, some elementary properties of power series, and the exponential and trigonometric functions of a complex variable. However, this chapter makes no attempt to reach any of the theorems that form the basic bill of fare in a first course in complex analysis.

6. The on-screen version provides a link to an optional chapter that introduces the concept of a set of measure zero in the line and that makes it possible to prove a variety of theorems about Riemann integrals that could not be reached in the main integration chapter. Another link offers an alternative to this chapter that presents the measure zero concept for Riemann-Stieltjes integrals.

7. The on-screen version provides a link to a chapter that develops many of the central topics in the calculus of several variables, including partial

differentiation, integration on curves, the inverse and implicit function theorems, and the change of variable theorem for multiple integrals.

My intention is to make this book serve as a reference long after the first course in analysis has ended.

Interaction with Readers

An important advantage of an on-screen text is that is can be a constantly changing product that allows me to be responsive to requests that may be made by my readers. I encourage my readers to write to me at

lewinjonathan@att.net

with comments and requests.

1. You may point out errors or omissions in the text that I can repair in the periodic update files that I shall be providing on my website. Updates for the on-screen version of the book can be obtained by clicking on a link in the table of contents.
2. You may request additional video versions of proofs.
3. If your instructor permits, then you will be given a version of the text in which the links that appear at the beginning of each set of exercises lead to documents that contain more extensive sets of solutions than those that exist in the standard solutions documents.

Based on my experiences with my own students, I am hopeful that some of my readers will feel encouraged to learn to write their mathematics in *Scientific Notebook* documents. By acquiring this skill they will become more organized in their study and will keep better records of their work. Such readers will be able to contact me by E-mail enclosing *Scientific Notebook* documents as attachments. If you wish to send me a document to which you have added pictures, please remember to save it as a rap file before you send it. Please compress any large file into ZIP form before you attach it.

Interactive Reading with *Scientific Notebook*

Readers who have a registered installation of *Scientific Notebook*, rather than *Scientific Viewer*, also have a copy of the powerful computer algebra system MuPAD[2] with which *Scientific Notebook* comes bundled. The operation of this computer algebra system within *Scientific Notebook* is particularly simple and does not require the reader to be familiar with any special syntax. Access to

---

[2] MuPAD is a trademark of Sciface Software, GmbH and Co.

the computing features allows the reader to experiment and to increase his/her understanding of the work. Items that are meant to be read interactively with *Scientific Notebook* are marked by the *Scientific Notebook* logo (**N**) .

In spite of the obvious value of interactive reading that makes use of the computing features of *Scientific Notebook*, this text does not go out of its way to present interactive reading on every page. Certainly, there are topics in this text for which the use of computing features is relevant and useful, but there are even more topics in which an attempt to use such computing features would be artificial and counterproductive.

The philosophy of this book is that, where the nature of the material being studied makes the computing features useful, these features should be exploited. However, where the material would not benefit from these computing features, the features have no place. Under no circumstances is the material of this book specifically chosen in order to provide opportunities to use the computing features of *Scientific Notebook*. In this sense, my book is not a "reform" text.

Instructor's Manual

The instructor's manual for this book is provided as a PDF file suitable for printing, and also in a form that is designed for reading on-screen. The manual elaborates on the material of the book, contains suggestions for alternative approaches, and contains solutions to most of the exercises for which a solution is not already provided in the text. Bona fide instructors may obtain instructions for downloading and using the instructor's manual by writing to

solutions@cambridge.org

The on-screen version of the instructor's manual is contained in a single executable file, the running of which will upgrade an existing installation of the book to an instructor version rather than a student version. Thus, the book must already be in place when the instructor's manual is installed. Installation of the instructor's manual will require a password that will be supplied to instructors by Cambridge University Press at the address solutions@cambridge.org.

As explained earlier, those exercises for which a solution or a hint is provided in the text are marked with the icon **H** or **?** . When one clicks on the link **?** that appears at the beginning of a block of exercises, one sees those solutions written with a cyan background. The solutions to the other exercises appear in a grey background and are visible only in an instructor installation of the book. My intention is to make it easy for instructors to supply their students with those green solutions that they wish to make available.

Preparation of This Book

This book is a sequel to the text *An Introduction to Mathematical Analysis* by Jonathan Lewin and Myrtle Lewin, whose first edition was published by Random House Inc. in 1988 and second edition was published by McGraw-Hill Inc. in 1993. I am deeply indebted to Myrtle Lewin for the many important contributions that she made to that book; they were contributions whose reach extends to the present work.

Both the printed version and the on-screen version of this work were prepared with *Scientific WorkPlace* 5.5, which is MacKichan Software's flagship product. *Scientific WorkPlace* can be thought of as *Scientific Notebook* together with support for LATEX typesetting. The page structure of the printed text is derived from an adaptation that I made of a document style supplied by MacKichan for use with the *Scientific WorkPlace* style editor. I would like to express my appreciation to the folks at MacKichan Software and, in particular, to Jon Stenerson and Jeanie Olivas, for the patient help that they gave me when I was working on this document style. I would also like to express my profound appreciation to Roger Hunter, Patti Kearney, Barry MacKichan, George Pearson, Jon Stenerson, and Steve Swanson of MacKichan Software for the valuable help and advice that they have given me in the preparation of the on-screen version of this book.

I would like to express my equally profound appreciation to my very dear friend, Natalie Kehr, who has worked through much of the manuscript, both hard copy and on-screen, and has sent me hundreds of error corrections and thoughtful suggestions. Time and time again, she pushed me, ever so tactfully, to undertake tasks that I had declared to be impossible. The result is a manuscript of a far higher quality than I would have been motivated to create alone.

I would like to express my appreciation to Eric Kehr (son of Natalie Kehr) for the valuable assistance that he provided me in order to produce an animated graph of the ruler function.

Jonathan Lewin

Reading This Book On-Screen

What do I Need to Read This Book On-Screen?

The on-screen version of this book is designed to be read with Version 5.5 of the software product *Scientific Notebook* that is made by MacKichan Software Inc. or with its free reader version, *Scientific Viewer*.[3] When you download the on-screen book from

> http://www.math-movies.com/Update-Files/Analysis-Textbook/
> Lewin_Analysis_Interactive.zip

you will see links for installing these MacKichan Software products. Choose one of them. You need to be running Microsoft Windows.

If you choose *Scientific Notebook* and register it with the serial number provided in the welcome screen, *Scientific Notebook* will run with all of its features for 30 days. After that trial period, you have two options:

1. You can register your installation of *Scientific Notebook* a second time, using a permanent serial number purchased from MacKichan Software. *Scientific Notebook* will then run permanently with all of its features on your computer. For details go to the MacKichan Software website

> http://www.mackichan.com

or telephone them at (360) 394-6033.

2. You can let the temporary registration of *Scientific Notebook* expire, in which case your installation will be converted automatically into *Scientific Viewer* that will allow you to read this book on-screen and will work permanently in your computer. Occasionally this method leads to annoying messages that tell you that Scientific Notebook is unlicensed and you can avoid these by uninstalling Scientific Notebook and installing Scientific Viewer. Note that *Scientific Viewer* will not allow you to save any changes that you make in documents and will not provide you access to the computer algebra system that come bundled with *Scientific Notebook*.

If you feel that you will not need the interactive features of *Scientific Notebook*, you can make things simple by choosing to install *Scientific Viewer* from the welcome screen instead of *Scientific Notebook*.

---

[3] When *Scientific Notebook* 6 is released, you may upgrade but do not remove Version 5.5. This book requires Version 5.5.

What Is *Scientific Notebook*?

Scientific Notebook is a combination word processor and computer algebra system. It combines the features of a powerful and friendly scientific word processor with the computing features of the powerful computing engine $MuPAD^4$ that is bundled with *Scientific Notebook*. This link to a computer algebra system makes it possible for a *Scientific Notebook* user to work directly with mathematical expressions that have been written into a document, expressions that have exactly the same form as those that one would write with pencil and paper.

The purpose of this chapter is to give you a quick training in the use of those features of *Scientific Notebook* that you will need to read this book and a brief overview of some of the computing features.

Getting Started

Using the Installation Menu for the On-Screen Book

The Welcome Screen

The on-screen book is provided in a ZIP file that you need to download and save in your local hard drive. Make an empty new folder somewhere convenient and extract the contents of the ZIP file to that folder. In that folder, you will see the executable file Lewin-Analysis.exe and a subfolder. Run the executable file and the welcome screen will come up. That welcome screen will enable you to install whichever of *Scientific Notebook* or *Scientific Viewer* you prefer, it will contain a link for installing the on-screen book itself, and it will also contain links to other important information that may be useful to you in case some troubleshooting is needed.

Installing *Scientific Notebook*

Unless you already have a MacKichan Software product Version 5.5 installed in your computer, you should begin by clicking on the option to install either *Scientific Notebook* or *Scientific Viewer*. You will be taken to an installation page that provides you with the option of installing a 30-day timelocked copy of *Scientific Notebook* 5.5. This page also shows the serial number that you should use when installing *Scientific Notebook*. Alternatively, if you choose to install *Scientific Viewer*, you will not require a serial number and no registration will be necessary.

When installing *Scientific Notebook* or *Scientific Viewer,* please say "yes" if you are asked for permission to associate the program with *.tex files. After the

---

4   *MuPAD* is a product of SciFace Software, Gmb H and Co.

installation of *Scientific Notebook* is complete, follow any instructions you see for registering it. If you have chosen *Scientific Viewer*, no registration is needed.

Installing the Book

Once *Scientific Notebook* or *Scientific Viewer* is up and running, you should run the installation menu again and, this time, you should click on the menu item in the welcome screen that installs the book.

I strongly recommend that you allow the installation program to place a short-cut to the on-screen book on your desktop. If you select that option, then you will be able to begin each reading session by clicking on this shortcut.

Occasionally, after installing the book you may have a problem getting it to show up when you click on its shortcut. If you see this problem, look at the installation welcome screen again for a link to instructions for associating *.tex files to *Scientific Notebook* or *Scientific Viewer*. You may also call the author for help at (USA) 770-973-5931, or send email to lewinjonathan@att.net, or you may use Skype. The author's Skype name is jonathan.lewin3.

Setting Your Screen View

Your enjoyment of the on-screen version of this book will be increased if you optimize your screen view. This section contains some suggestions that you may use for this purpose.

Setting Your Zoom Factor (Screen Font Size)

The ideal size for the screen fonts in your particular installation of *Scientific Notebook* depends on the screen resolution at which you run your session of Windows. The default sizes have been chosen to suit a screen resolution of 1024 by 768. If your screen resolution is higher, then you may want a larger screen font size; and if your resolution is lower, then you may want smaller screen fonts. Please note that adjusting your screen fonts is done only for your comfort in reading documents on the computer screen. In no way will a change in the size of your screen fonts affect the way in which your printed documents appear.

There are two main ways in which you can adjust the zoom factor in any given document:

- If you have chosen to show the **Standard** toolbar, then you will see the zoom factor button 150% . By clicking on the little down arrow on the right of this button you can bring up a menu of screen font sizes. You can change to any of these by pointing at them and clicking the mouse, and you can also type in your own size selection.

- At the top of your computer screen you will see several menus, one of which is labelled View. Click on this menu and it opens showing

You can choose quickly between any of the three screen font sizes shown in this menu by pointing and clicking. If you choose **Custom**, you will be presented with the opportunity of typing in your size selection.

Appearance of Graphics in Your Documents

The default view of pictures in *Scientific Notebook* displays a frame around each picture. While such a frame does no harm, this book is designed to look its best when those frames are not shown. Click on your **Tools** menu

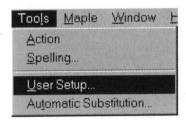

and then click on **User Setup** and select the page in the user setup labeled **Graphics**.

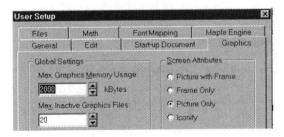

In the **Screen Attributes** part of this page, select **Picture Only**.

Navigating in the On-Screen Book

In order to read this book efficiently on the computer screen you have to be able to find your way around and move from point to point quickly and painlessly. As in most Windows products, you can use arrow keys to scroll through your documents line by line, the **Page Up** and **Page Down** keys to move up and down by one screen at a time, and the vertical scroll bar at the right of your screen to make larger jumps. However, your most valuable tool for navigating through this book is the system of hypertext links (hyperlinks) that it contains. The thousands of hyperlinks in this book will whisk you instantly to the material you want to see.

Using the On-Screen Contents Document

With the help of the hyperlinks that it contains, the on-screen Contents document provides you with the ability to jump, instantly, from anywhere in the book to any topic that you want to see, and then to jump back again. Not only does the Contents document provide you with links to the beginnings of all chapters, sections, and subsections, but the headers of all those items in the text also contain links back to the appropriate positions in the Contents document. Thus, the Contents document provides you with an instant road-map of the text, and it also allows you to thumb through the on-screen text in much the same way that one can flip through the pages of a printed book.

If you have placed a shortcut to the Contents document on your desktop, you may want to begin each session by opening the Contents document and jumping from there to the item you wish to read.

Appearance and Use of Hyperlinks

Each hyperlink appears either with a yellow background or a green background. I have made an effort to use yellow hyperlinks for the utility links between the various book features and for links to material that is covered in the printed version of the book. The green links lead to material that appears only in the on-screen version of the book.

This color coding applies to the vast majority of hyperlinks, but you will find a few exceptions. In some cases, where the target of a link is closely related to material in the printed text, it is not easy to decide what color the link should be. Furthermore, there are some technical limitations to the choice of hyperlink colors, and so a few links do have the wrong color.

In some documents you may find it necessary to **hold down your control key** when you click on a link.

After operating a hyperlink, you can return to where you had been by clicking on the button  or  in your link or history toolbar.

Using the On-Screen Index

The index to the on-screen version of this book is a single document that contains the index entries in alphabetical order. Each item appears with a brief description and one or more hyperlinks to positions in the text where that item appears. You will also find a variety of links to biographical information that is located on the World Wide Web. To reach the index you can click any of the many links to it that appear in the book.

Using the Navigation Toolbar

Another useful way of navigating in any large document is to use the **Navigate** and **Link** toolbars.

If these toolbars do not appear in your screen, click on the **View** menu at the top of your screen. When it opens you will see

Click on **Toolbars** and you will see the toolbar menu. Open it and make sure that the toolbars that you want to see are checked.

If you click on the down arrow in your **Navigate** toolbar, it will show you a list of document headers and you can click on any of these headers to jump to it.

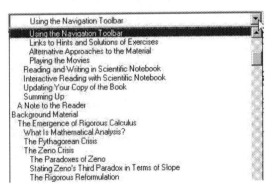

Links to Hints and Solutions of Exercises

At the beginning of each set of exercises you will see a hyperlink ⊘ on which you can click to jump to the hints and solutions. Those exercises for which either a hint or a complete solution is provided in the solutions document are marked with the icon ▉ or ⊘ . You will also be invited to send questions, comments, and suggestions to the author. You may write to the author at

lewinjonathan@att.net

Alternative Approaches to the Material

Some of the links take you to documents that extend the material as it appears in the text or provide an alternative approach to it. Such links can also take you to a variety of interesting exercises that do not appear in the main body of the book.

A yellow colored link takes you to material that also appears in the printed text. A green colored link usually takes you to special material that can be seen only in the on-screen text. Sometimes, you may see links that show with special icons such as ▣ or ◈ . A link to an item that contains material that is likely to be significantly more difficult than the material in the text will appear with the icon ☠ .[5]

A link to a sound message will appear with the icon ▣ .

Playing the Movies

This book contains about 50 sound movie "mini-lectures" that let you view a portion of the text in lecture form with some of the atmosphere that exists with an instructor in a classroom. At any time while a movie is playing, you can press your spacebar to make it pause, and you can drag the cursor back and forth to make the movie play at any point that you wish to see. Each movie appears with the icon ▣ or with the icon ▣ .

---

[5] Why a skull and crossbones? Well, I think that even a mathematics book should be permitted a bit of humour from time to time.

Reading and Writing in *Scientific Notebook*

The text editing component of *Scientific Notebook* is a word processing system that has the ability to distinguish between **text** and **mathematics**. At any time, *Scientific Notebook* is either in **text mode** or it is in **mathematics mode**. You can see which mode is active at any given time by looking at your standard tool bar

If the button **T** is showing there, then *Scientific Notebook* is in text mode and any symbols that you type will be treated as text. Such symbols will usually be black but, depending on their position in the document and the typeface being used, they may also appear in other colors. If the mathematics tool bar shows the symbol **M** , then *Scientific Notebook* is in mathematics mode and any symbols you type will be typeset according to mathematical conventions and also recognized as mathematics for the purposes of mathematical operations. There are many ways to change the mode of *Scientific Notebook* from text to mathematics and back again. One way is to point the mouse at the symbol **T** and click it to change to mathematics mode and to point at the symbol **M** and click to change to text mode.

Interactive Reading with *Scientific Notebook*

If you are reading this text with the full version of *Scientific Notebook*, rather than the free viewer version, or if you are reading this book with the MacKichan Software flagship product *Scientific WorkPlace*, then your software comes bundled with the computer algebra system *MuPAD*. The operation of the computer algebra system within *Scientific Notebook* is particularly simple and does not require you to be familiar with any special syntax. If you have set up your copy of *Scientific Notebook* to display the Computing toolbar,

then you can perform some of the most common computing operations with a single mouse click. Alternatively, you may click on the **Compute** item at the top of your screen to open the pull-down menu, which gives you a brief listing of the operations that you can perform. As you can see, many of the items have submenus that can be opened by clicking on the arrow to the right. For example, if you move the cursor down to **Calculus**, then you see

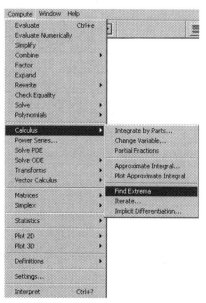

Each of the computing operations can be performed by the simple operation of pointing at one of these menu items and clicking. Thus, for example, if you place your cursor in the expression $\int_0^1 \sqrt{1 - x^2}\,dx$ and then click on the item **Evaluate** in the Compute menu, you will see

$$\int_0^1 \sqrt{1 - x^2}\,dx = \frac{1}{4}\pi.$$

Updating Your Copy of the Book

The table of contents of the on-screen book will include the array

Welcome   **Preface**   **Reading on-Screen**   **Troubleshooting**
Interdependence   **Movies**   **About This Document**   **Check for Updates**

that shows eight important links. By clicking on **Check for Updates**, you will be able to see the build date of your version of the file. Then, you will see a second link that takes you to the update information where you can see announcements of the kinds of updated patches that are available and that might be appropriate for you. Where possible, incremental upgrade patches will be provided so that you can avoid the hefty download of the entire on-screen text.

Summing Up

Both the printed bound copy and the on-screen version of this book will have a role to play as you study the material that is contained in this book. Each version has its own advantages and disadvantages. Try to preserve a balance between the two versions as you read this book so that you can benefit from the features that each of them provides.

A Note to the Reader

Mathematical analysis is the critical and careful study of calculus that rests upon some of the ground-breaking discoveries that were made during the nineteenth century, discoveries that made it possible, for the first time, to appreciate the nature of our number system and the concepts of limit, continuity, derivative, and integral.

If you are a typical reader of this book, then you have already completed some courses in a "first calculus" sequence, and the words *limit, continuity, derivative,* and *integral* are already familiar to you. But there is an important difference between the way you will see these concepts in this book and the way they are presented in most elementary calculus courses. There, the purpose was to get on with the material as quickly as possible so as to give you a bird's-eye view of the subject and to allow you to see some of its applications. Here, on the other hand, we shall strive for *understanding*. Now that you already have your bird's-eye view, it is time to go back and make a careful and critical study of the *central ideas* of calculus.

Although you will encounter many exciting new ideas in this book, it is not our purpose to study the complete spectrum of calculus topics. In fact, your elementary calculus courses covered many more topics than we shall study here. And yet you will be working hard. Very hard! Do not be discouraged by the prospect of the hard work that lies before you. The fruits of your labors as you study this book will more than repay you for the efforts that you will make. Your reward will be the thrill of genuine understanding and, as your understanding of mathematics increases, so will your appreciation of its beauty. You will experience in mathematics the kind of stimulation and pleasure that we associate with the great masterpieces in art, literature, and music. Perhaps you will come to feel that mathematical analysis is the greatest masterpiece of them all.

PART I

Background Material

This part of the text contains a brief history of the emergence of rigorous calculus, an introduction to mathematical grammar, an introduction to the art of reading and writing mathematical proofs, and a brief introduction to the theory of sets and functions. Use as little or as much of this material as you need.

Please insert a blank page here.

Chapter 1
The Emergence of Rigorous Calculus

1.1   What Is Mathematical Analysis?

Mathematical analysis[8] is the critical and careful study of calculus with an emphasis on understanding of its basic principles. As opposed to *discrete* mathematics or *finite* mathematics, mathematical analysis can be thought of as being a form of *infinite* mathematics. As such, it must rank as one of the greatest, most powerful, and most profound creations of the human mind.

> The infinite! No other question has ever moved so profoundly the spirit of man — David Hilbert (1921).

Now, as you may expect, great, profound, and powerful thoughts do not often appear overnight. In fact, it took the best part of 2500 years from the time the first calculus-like problems tormented Pythagoras, until the first really solid foundations of mathematical analysis were laid in the nineteenth century. During the seventeenth and eighteenth centuries calculus blossomed, becoming an important branch of mathematics and, at the same time, a powerful tool, able to describe such physical phenomena as the motion of the planets, the stability of a spinning top, the behavior of a wave, and the laws of electrodynamics. This period saw the emergence of almost all of the concepts that one might expect to see in an elementary calculus course today.

But if the blossoms of calculus were formed during the seventeenth and eighteenth centuries, then its roots were formed during the nineteenth. Calculus underwent a revolution during the nineteenth century, a revolution in which its fundamental ideas were revealed and in which its underlying theory was properly understood for the first time. In this revolution, calculus was rewritten from its foundations by a small band of pioneers, among whom were Bernhard Bolzano, Augustin Cauchy, Karl Weierstrass, Richard Dedekind, and Georg Cantor. You will see their names repeatedly in this book, for it was largely as a result of their efforts that the subject that we know today as *mathematical analysis* was born. Their work enabled us to appreciate the nature of our number system and gave us our first solid understanding of the concepts of limit, continuity, derivative, and

---

[8]   Note to instructors: This chapter is not designed for in-class teaching. It is intended to be a reading assignment, possibly in conjunction with other material that the student can find in the library.

integral. This is the great and profound theory to which you, the reader of this book, are heir.

In this chapter we shall focus on three earth-shaking events that have taken place during the past 2500 years and which helped to pave the way for the emergence of rigorous mathematics as we know it today. These events are sometimes known as the **Pythagorean crisis**, the **Zeno crisis**, and the **set theory crisis**.

1.2   The Pythagorean Crisis

In about 500 B.C.E. an individual in the Pythagorean school noticed that, according to the Greek concepts of number and length, it is impossible to compare the length of a side of a square with the length of its diagonal. The Greek concept of length required that, in order to compare two line segments AB and CD, we need to be able to find a measuring rod that fits exactly a whole number of times into each of them. If, for example, the measuring rod fits 6 times into AB and 10 times into CD, as shown in Figure 1.1, then we have

$$\frac{AB}{CD} = \frac{6}{10}.$$

More generally, if the measuring rod fits exactly m times into AB and exactly n

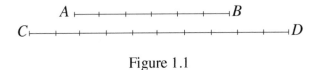

Figure 1.1

times into CD, then we have

$$\frac{AB}{CD} = \frac{m}{n}.$$

Note that this kind of comparison requires that the ratio of any two lengths must be a rational number.

The crisis came when the young Pythagorean drew a square with a side of one unit as shown in Figure 1.2 and applied the theorem of Pythagoras to find the length of the diagonal. As we know, the length of this diagonal is $\sqrt{2}$ units. From the fact that the number $\sqrt{2}$ is irrational he concluded that the equation

$$\frac{\sqrt{2}}{1} = \frac{m}{n}$$

is impossible if m and n are integers and that, consequently, it is impossible to compare the side of this square with its diagonal.

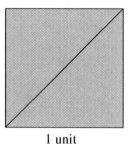

1 unit

Figure 1.2

From our standpoint today, we can see that this discovery reveals the inadequacy of the rational number system and of the Greek concept of length; but to them, the discovery was a real shocker. Just how much of a shock it was can be gauged from the writings of the Greek philosopher Proclus, who tells us that the Pythagorean who made this terrible discovery suffered death by shipwreck as a punishment for it.

1.3   The Zeno Crisis

1.3.1   The Paradoxes of Zeno

In the fifth century B.C.E., Zeno of Elea came up with four innocent-sounding statements that plagued the philosophers all the way up to the time of Bolzano and Cauchy early in the nineteenth century. These four statements are known as the **paradoxes of Zeno**, and the first three of these appear in Bell [4] as follows:

1. *Motion is impossible, because whatever moves must reach the middle of its course before it reaches the end; but before it has reached the middle, it must have reached the quarter mark, and so on, indefinitely. Hence the motion can never start.*

2. *Achilles running to overtake a crawling tortoise ahead of him can never overtake it, because he must first reach the place from which the tortoise started; when Achilles reaches that place, the tortoise has departed and so is still ahead. Repeating the argument, we easily see that the tortoise will always be ahead.*

3. *A moving arrow at any instant is either at rest or not at rest, that is, moving. If the instant is indivisible, the arrow cannot move, for if it did, the instant would immediately be divided. But time is made up of instants. As the arrow cannot move in any one instant, it cannot move in any time. Hence it always remains at rest.*

Much has been said about these paradoxes, and, quite obviously, we are not going to do them justice here. But let's talk about the third paradox for a moment. At any one instant of time, the arrow does not move. Does that really mean that the arrow will not find its target? Would Zeno have been prepared to stand in front of the arrow? We think not. Then what was Zeno trying to tell us? Zeno's statement warns us that velocity can be meaningful in any physical sense only as an *average velocity over a period of time*. If an arrow covers a distance of 60 feet during the course of a second, we can say that the arrow has an average velocity of 60 feet per second. But Zeno's statement warns us that our senses can make nothing out of a notion of *velocity of the arrow at any one instant*.

1.3.2   Stating Zeno's Third Paradox in Terms of Slope

To state Zeno's third paradox in terms of slope, we shall suppose that A is the point $(x_1, f(x_1))$ on the graph of a function f, and that B is some other point $(x_1 + \Delta x, f(x_1 + \Delta x))$, as shown in Figure 1.3. As usual, the slope of the line

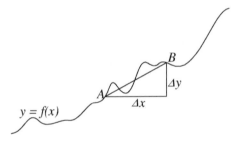

Figure 1.3

segment AB is defined to be to be the ratio $\Delta y / \Delta x$, where

$$\Delta y = f(x_1 + \Delta x) - f(x_1).$$

This ratio $\Delta y / \Delta x$ is the average slope of the graph of f between the points A and B. However, Zeno's third paradox serves as a warning that there is no obvious physical meaning to the notion of *slope of the graph at the point A*.

"But" you may ask, "isn't this what calculus is all about? Are the paradoxes of Zeno trying to tell us to abandon the idea of a derivative?" They are not. But what we should learn from these paradoxes is that if we want to *define* the derivative of the function f at the point A to be

$$\lim_{h \to 0} \frac{f(x + h) - f(x)}{h},$$

then that's just fine with Zeno. Only we can't blame Zeno if this derivative that we have *defined* doesn't measure how the function f increases at A, because, as Zeno quite rightly tells us, the function f can't change its value at any one point. We may therefore think of Zeno's paradoxes as telling us that (referring to Figure 1.3) even though we may speak of the slope $\frac{\Delta y}{\Delta x}$ of the line segment AB, and even though we may *define* the derivative of f at A and call it $\frac{dy}{dx}$ and have

$$\frac{\Delta y}{\Delta x} \to \frac{dy}{dx} \quad \text{as} \quad \Delta x \to 0,$$

we may not think of $\frac{dy}{dx}$ as the ratio of two quantities dy and dx, the amounts by which y and x increase at the point A, because, as Zeno quite rightly tells us, there are *no* increases in y and x at the point A.

1.3.3 The Rigorous Reformulation

Mathematics prior to the dawn of the nineteenth century was much less precise than mathematics as we know it today. The core of pre-nineteenth century mathematics was the calculus that had been developed by Newton, Leibniz, and others during the seventeenth century. That calculus represented a magnificent contribution. It gave us the notation for derivatives and integrals that we still use today and provided a mathematical basis for the understanding of such physical phenomena as the motion of the planets, the motion of a spinning top and the vibration of a violin string. But the calculus of Newton and Leibniz did not rest on a solid foundation.

The problem with Newtonian calculus is that it was not based on an adequate theory of limits. In fact, prior to the nineteenth century, there was not much understanding that calculus needs to be based on a theory of limits at all. Nor was there much understanding of the nature of the number system $\mathbf{R}$ and the role of what we call today the *completeness* of the number system $\mathbf{R}$. In a sense, the calculus of Newton and Leibniz did not pay sufficient heed to the paradoxes of Zeno. Although Newton and Leibniz themselves may have had some appreciation of the fundamental ideas upon which the concepts of derivative and integral depend, many of those who followed them did not. Until the end of the eighteenth century the majority of mathematicians based their work upon an impossible mythology. During this time, proofs of theorems in calculus commonly depended on a notion of "infinitely small" numbers, numbers that were zero for some purposes yet not for others. These were known as *evanescent numbers*, *differentials*, or *infinitesimals*, and, undeniably, their use provided a beautiful, revealing, and elegant way of looking at many of the important theorems of calculus. Even today we like to use the notion of an infinitesimal to motivate some of the theorems in calculus, and scientists use them even more frequently than mathematicians. But it is one thing to use the idea of an infinitesimal to *motivate* a theory, and it is quite another matter to base virtually the entire theory upon them. Today, the concept of an infinitesimal can actually be made precise in a modern mathematical theory that is known as *nonstandard analysis*, but there was no precision in the way infinitesimals were used in the eighteenth century.

During the eighteenth century, the voices of critics began to be heard. In 1733, Voltaire [32] described calculus as

> *The art of numbering and measuring exactly a thing whose existence cannot be conceived.*

Then, in 1734, Bishop George Berkeley, the philosopher, wrote an essay, Berkeley [5], in which he rebuked the mathematicians for the weak foundations upon which their calculus had been based, and he no doubt took great pleasure in asking

> *Whether the object, principles, and inferences of the modern analysis are more distinctly conceived, or more evidently deduced than religious mysteries and points of faith.*

Some mathematicians composed weak answers to Berkeley's criticism, and others tried vainly to make sense of the idea of infinitely small numbers, but it was not until the early nineteenth century that any real progress was made. The turning point came with the work of Bernhard Bolzano, who gave us the first coherent definition of limits and continuity and the first understanding of the need for a complete number system. Then came the work of Cauchy, Weierstrass, Dedekind, and Cantor that placed calculus on a rigorous foundation and settled many important questions about the nature of the number system $\mathbf{R}$. The work of these pioneers has made possible the understanding that we have promised you.

1.4   The Set Theory Crisis

Following the work of the nineteenth-century pioneers, the mathematical community began to believe that true understanding was at last within its grasp. All of the fundamental concepts seemed to be rooted solidly in Cantor's theory of sets. But the collective sigh of relief had hardly died away when a new kind of paradox burst upon the scene. In 1897, the Italian mathematician Burali-Forti discovered what is known today as the **Burali-Forti paradox**, which shows that there are serious flaws in Cantor's theory of sets, upon which our understanding of the real number system had been based. Then, a few years later, Bertrand Russell discovered his famous paradox. Like Burali-Forti's paradox, Russell's paradox demonstrates the presence of flaws in Cantor's set theory.

To see just how much these paradoxes stunned the mathematical community, one might want to look at Frege [9], *Grundgesetze der Arithmetik* (*The Fundamental Laws of Arithmetic*), which was written by the German philosopher Gottlob Frege and published in two volumes, the first in 1893 and the second in 1903. This book was Frege's life work, and it was his pride and joy. He had bestowed upon the mathematical community the first sound analysis of the meaning of number and the laws of arithmetic and, although the book is quite technical in places, it is worth skimming through, if only to see the sarcastic way in which Frege speaks of the "stupidity" of those who had come before him. An example of this sarcasm is Frege's description of his attempt to induce other mathematicians to tell him what the number *one* means. "One object," would be the reply. "Very well," answered Frege, "I choose the *moon*! Now I ask you please to tell me: *Is one plus one still equal to two?*" As things turned out, the second volume of Frege's book came out just after Russell had sent Frege his famous paradox. There was just enough space at the end of Frege's book for the following acknowledgment:

> *A scientist can hardly encounter anything more undesirable than to have the foundation collapse just as the work is finished. I was put in this position by a letter from Mr. Bertrand Russell when the work was almost through the press.*

As Frege said, the foundation collapsed. It would not be stretching the truth too much to say that all of mathematics perished in the fire storm that was ignited by the paradoxes of Russell and Burali-Forti. The mathematics that we know today is what emerged from that storm like a phoenix from the ashes, and it depends upon a new theory of sets that is known as **Zermelo-Fraenkel set theory** which was developed in the first few decades of the twentieth century. Within the framework of Zermelo-Fraenkel set theory, we can once again make use of Frege's important work.

One question that remains is whether we are now safe from new paradoxes that might ignite a new fire storm, and the answer is that we don't know. A theorem of Gödel guarantees that, unless someone actually discovers a new paradox that destroys Zermelo-Fraenkel set theory, we shall never know whether such a paradox exists. Thus it is entirely possible that you, the reader of this book, may stumble upon a snag that shows that mathematics as we know it does not work. But don't hold your breath. The chances of your encountering a new paradox are very remote.

Chapter 2
Mathematical Grammar

In this chapter you will learn how to read and understand the language in which mathematical proofs are written. Just like any other language, the language of mathematics contains its own special words and rules of grammar that govern the way in which these words may be combined into sentences. This chapter will acquaint you with many of these words, and it will teach you how the rules of grammar apply to them.

2.1   The Quantifiers *For Every* and *There Exists*

2.1.1   Unknowns in a Mathematical Statement

Some mathematical statements refer only to predefined mathematical objects. For example, the statement $1 + 1 = 2$ refers only to the numbers 1 and 2 and the operation $+$. However, mathematical statements can also refer to objects that have not already been introduced. We shall call such objects **unknowns** in the statement. To understand how unknowns can appear, look at the following simple question about elementary algebra:

Is it true that

$$(x + y)^2 = x^2 + y^2? \tag{2.1}$$

Perhaps the answer "no" is hovering on your lips. If so, you are being a little hasty, for is it not true that

$$(3 + 0)^2 = 3^2 + 0^2?$$

As you can see, the truth or falsity of Equation (2.1) depends on the values of x and y that we are talking about. Until we receive more information about x and y, we cannot say whether or not Equation (2.1) is true. Here are some analogs of the above question that can be answered specifically:

- Is it true that, if $x = 2$ and $y = 3$, then Equation (2.1) is true? No!
- Is it true that there exist numbers x and y such that Equation (2.1) holds? Yes!
- Is it true that for every number x there exists a number y such that Equation (2.1) holds? Yes!

12

- Is it true that there exists a number x such that for every number y Equation (2.1) holds? Yes!
- Is it true that for all numbers x and y Equation (2.1) holds? No!

If you look at any one of these five questions, you will see that each of the symbols x and y was introduced either by saying "for every" or by saying "there exists". The symbols in these questions have been what we call **quantified**. The statements that follow exhibit some more examples of quantified symbols.

1. *Every tall man in this theater is wearing a hat.* In this statement, "tall man" is introduced with *for every*.
2. *No tall man in this theater is wearing a hat.* This statement can be interpreted as saying that every man wearing a hat in this theater fails to be tall.
3. *Some tall men in this theater are wearing hats.* This statement introduces "tall man" with *there exists*. It says that there exists at least one tall man in this theater who is wearing a hat.
4. *Not all of the tall men in this theater are wearing hats.* This statement can be interpreted as saying that there exists at least one tall man in this theater who is not wearing a hat. Interpreted this way, the statement introduces "tall man" with *there exists*.
5. *For every positive integer n there is a prime number p such that $p > n$.* This statement contains two unknowns, n and p. The unknown n is introduced with *for every* and then, after n has been introduced, the unknown p is introduced with *there exists*.

2.1.2   The Quantifiers

The phrase *for every* is called the **universal quantifier** and, depending on the context, it appears sometimes simply as *every*, sometimes as *all*, and sometimes as the symbol $\forall$. The phrase *there exists* is called the **existential quantifier**, and, depending on the context, it sometimes appears as *there is, we can find, it is possible to find, there must be, there is at least one, some*, or as the symbol $\exists$.

2.1.3   Exercises on the Use of Quantifiers

Except in Exercise 2, decide whether the sentence that appears in the exercise is meaningful or meaningless. If the sentence is meaningful, say whether what it says is true or false.

1. (a) ▣ $\sqrt{x^2} = x$.
   (b) ▣ For every real number x we have $\sqrt{x^2} = x$.
   (c) ▣ For every positive number x we have $\sqrt{x^2} = x$.
2. (a) Ⓝ Point at the expression $\sqrt{x^2}$ and click on the Evaluate button ▣.

(b) Ⓝ Point at the expression $\sqrt{x^2}$ and click on the Simplify button ⊞ .

(c) Ⓝ Point at the equation $\sqrt{x^2} = x$, open the Compute menu, and click on Check Equality.

(d) Ⓝ Point at the equation $x = -2$ and click on the button ⊞ to supply the definition $x = -2$ to *Scientific Notebook*. Then try a Check Equality on the equation $\sqrt{x^2} = x$.

3. ▣ For every number x and every number y there is a number z such that $z = x + y$.

4. ▣ For every number x there is a number z such that for every number y we have $z = x + y$.

5. ▣ For every number x and every number z there is a number y such that $z = x + y$.

6. ▣ $\sin^2 x + \cos^2 x = 1$.

7. ▣ For every number x we have $\sin^2 x + \cos^2 x = 1$.

8. ▣ For every integer $n > 1$, if $n^2 \leq 3$, then the number 57 is prime.

2.1.4   Order of Appearance of Unknowns in a Statement

As we have seen, if a statement contains some unknowns, then its truth or falsity depends on the way in which these unknowns have been quantified. If a statement contains more than one unknown, then the order in which these unknowns appear can also be very important. The following examples illustrate how we can change the meaning of a statement by changing the order in which two unknowns are introduced.

1. If two unknowns are introduced with $\forall$, one directly after the other, then the order in which they are introduced is not important. For example, the following two statements say exactly the same thing.

    (a) *For every positive number x and every negative number y, the number xy is negative.*

    (b) *For every negative number y and every positive number x, the number xy is negative.*

2. If two unknowns are introduced with $\exists$, one directly after the other, then the order in which they are introduced is not important. For example, the following two statements say exactly the same thing.

    (a) *There exists a positive integer m and a positive integer n such that $\sqrt{m^2 + n^3}$ is an integer.*

    (b) *There exists a positive integer n and a positive integer m such that*

$\sqrt{m^2 + n^3}$ *is an integer.*

3. If one unknown in a sentence is introduced with $\forall$ and another is introduced with $\exists$, then the order in which they appear is very important. For example, compare the following two sentences:

    (a) *For every positive number x there exists a positive number y such that $y < x$.*

    (b) *There exists a positive number y such that for every positive number x we have $y < x$.*

    Although these statements may look similar, they do not say the same thing. As a matter of fact, the first one is true and the second one is false.

4. In this example we look at some sentences that contain three unknowns.

    (a) *For every number x and every number z, there exists a number y such that $x + y = z$.*

    (b) *For every number x there exists a number y such that for every number z we have $x + y = z$.*

    (c) *There exists a number y such that for every number x and every number z we have $x + y = z$.*

    If you look at these statements carefully, you will see that the first one is true and the other two are false.

5. In this example we look at two more statements with three unknowns. Try to decide whether they are true or false. We shall provide the answers in Subsection 3.8.1.

    (a) *For every number x there exists a positive number δ such that for every number t satisfying the inequality $|t - x| < \delta$ we have $|t^2 - x^2| < 1$.*

    (b) *There exists a positive number δ such that for every number x and for every number t satisfying the inequality $|t - x| < \delta$ we have $|t^2 - x^2| < 1$.*

6. In this example we look at some statements that contain four unknowns. Try to decide whether they are true or false. We shall provide some of the answers in Section 3.8.

    (a) *For every number $\varepsilon > 0$ and for every number $x \in [0, 1]$ there exists a positive integer N such that for every integer $n \geq N$ we have*

    $$\frac{nx}{1 + n^2 x^2} < \varepsilon.$$

    (b) *For every number $x \in [0, 1]$ and for every number $\varepsilon > 0$ there exists a* *positive integer N such that for every integer $n \geq N$ we have*

$$\frac{nx}{1 + n^2 x^2} < \varepsilon.$$

(c) *For every number $\varepsilon > 0$ there exists a positive integer N such that for every number $x \in [0, 1]$ and for every integer $n \geq N$ we have*

$$\frac{nx}{1 + n^2 x^2} < \varepsilon.$$

2.1.5   Exercises on Order of Appearance of Unknowns

For each of the following pairs of statements, decide whether or not the statements are saying the same thing. Except in the first two exercises, say whether or not the given statements are true.

1. (a) Every person in this room has seen a good movie that has started playing this week.
   (b) **H** A good movie that has started playing this week has been seen by every person in this room.

2. (a) Only men wearing top hats may enter this hall.
   (b) Only men may enter this hall wearing top hats.
   (c) Men wearing top hats only may enter this hall.
   (d) Men wearing only top hats may enter this hall.
   (e) **H** Men wearing top hats may enter this hall only.

3. (a) **H** For every nonzero number x there is a number y such that $xy = 1$.
   (b) **H** There is a number y for which the equation $xy = 1$ is true for every nonzero number x.

4. (a) **H** For every number $x \in [0, 1)$ there exists a number $y \in [0, 1)$ such that $x < y$.
   (b) **H** There is a number $y \in [0, 1)$ satisfying $x < y$ for every number $x \in [0, 1)$.

5. (a) For every number $x \in [0, 1]$ there exists a number $y \in [0, 1]$ such that $x < y$.
   (b) **H** There is a number $y \in [0, 1]$ satisfying $x < y$ for every number $x \in [0, 1]$.

6. (a) For every number $x \in [0, 1)$ there exists a number $y \in [0, 1)$ such that $x \leq y$.
   (b) There is a number $y \in [0, 1)$ satisfying $x \leq y$ for every number $x \in [0, 1)$.

7. (a) For every number $x \in [0, 1]$ there exists a number $y \in [0, 1]$ such that $x \leq y$.

   (b) There is a number $y \in [0, 1]$ satisfying $x \leq y$ for every number $x \in [0, 1]$.

8. (a) For every odd integer m it is possible to find an integer n such that mn is even.

   (b) ■ It is possible to find an integer n such that for every odd integer m the number mn is even.

9. (a) For every number x it is possible to find a number y such that $xy = 0$.

   (b) ■ It is possible to find a number y such that for every number x we have $xy = 0$.

10. (a) For every number x it is possible to find a number y such that $xy \neq 0$.

   (b) ■ It is possible to find a number y such that for every number x we have $xy \neq 0$.

11. (a) ■ For every number a and every number b there exists a number c such that $ab = c$.

   (b) ■ For every number a there exists a number c such that for every number b we have $ab = c$.

12. (a) For every number a and every number c there exists a number b such that $ab = c$.

   (b) ■ For every number a there exists a number b such that for every number c we have $ab = c$.

13. (a) ■ For every nonzero number a and every number c there exists a number b such that $ab = c$.

   (b) ■ For every nonzero number a there exists a number b such that for every number c we have $ab = c$.

2.2   Negating a Mathematical Sentence

When you are trying to read and understand a mathematical statement P, you will often find it useful to ask yourself what it would mean to say that the statement P is false. The assertion that the given statement P is false is called the **negation** or **denial** of the statement P and is written as $\neg P$.

Thus, for example, if P is the assertion that the number 4037177 is prime, then the statement $\neg P$ is the assertion that the number 4037177 is composite. As a matter of fact, since

$$4037177 = 17 \times 19 \times 29 \times 431,$$

we know that the statement P is false and that the statement $\neg P$ is true. In general, if P is any mathematical statement, then one of the statements P and $\neg P$ will be true and the other will be false.

2.2.1   Negations and the Quantifiers

Suppose that a given statement P asserts that *Everyone in this room can speak French*. The denial of P does not claim that no one in the room can speak French. To deny the assertion that everyone in the room can speak French, all we have to do is find one person in the room who cannot speak French.

More generally, if a statement P contains the quantifier "for every", saying that every object x of a certain type has a certain property, then the denial of P says that there exists at least one object of this type that fails to have the required property.

On the other hand, if a statement P says that at least one person in this room has a dirty face, then the denial of P says that every person in the room has a clean face. Note how the quantifiers "for every" and "there exists" change places as we move from a statement to its denial.

2.2.2   Some Exercises on Negations and the Quantifiers

Write a negation for each of the following statements or say that the statement is meaningless.

1. All roses are red.
2. In Sam's flower shop there is at least one rose that is not red.
3. In every flower shop there is at least one rose that is not red.
4. I believe that all roses are red.
5. There is at least one person in this room who thinks that all roses are red.
6. Every person in this room believes that all roses are red.
7. At least half of the people in this room believe that all roses are red.
8. Every man believes that all women believe that all roses are red.
9. You were at least an hour late for work every day last week.
10. It has never rained on a day on which you have remembered to take your umbrella.
11. You told me that it has never rained on a day on which you have remembered to take your umbrella.
12. You lied when you told me that it has never rained on a day on which you have remembered to take your umbrella.

13. ⊞ I was joking when I said that you lied when you told me that it has never rained on a day on which you have remembered to take your umbrella.

14. ⊞ This, Watson, if I mistake not, is our client now.[9]

15. (a) ⊞ For every real number x there exists a real number y such that

$$\frac{2x^2 + xy - y^2}{x^3 - y^3} = \frac{2}{3(x - y)} + \frac{5y + 4x}{3(x^2 + xy + y^2)}.$$

Is this statement true?

(b) ⊞ There exists a real number x such that for every real number y we have

$$\frac{2x^2 + xy - y^2}{x^3 - y^3} = \frac{2}{3(x - y)} + \frac{5y + 4x}{3(x^2 + xy + y^2)}.$$

Is this statement true?

(c) ⊞ For every real number x and every real number $y \neq x$ we have

$$\frac{2x^2 + xy - y^2}{x^3 - y^3} = \frac{2}{3(x - y)} + \frac{5y + 4x}{3(x^2 + xy + y^2)}.$$

Is this statement true?

2.3 Combining Two or More Statements

Two or more given statements can be combined into a single statement using one or more of the conjunctions and conditionals

and	or	if	only if	$\Rightarrow$	$\Leftarrow$	$\Longleftrightarrow$

In this section we study the way in which these combined statements may be formed and some of the relationships between statements of this type.

2.3.1 The Conjunction *and*

If P and Q are given statements, then the assertion

$$P \text{ and } Q$$

says that both of the statements P and Q are true. This assertion is sometimes written as $P \wedge Q$.

---

[9] The purpose of this exercise is to invite you to discuss a rather strange statement that was made by Sherlock Holmes in one of the stories by Sir Arthur Conan Doyle. Is Holmes' statement meaningful?

For example, if P is the statement

> *Every tall man in this theater is wearing a hat.*

and Q is the statement

> *Every man wearing a hat in this theater is inconsiderate.*

then the statement $P \wedge Q$ says that

> *Every tall man in this theater is wearing a hat and every man wearing a hat in this theater is inconsiderate.*

From the statement $P \wedge Q$ we can deduce that every tall man in this theater is inconsiderate.

2.3.2   The Conjunction *or*

If P and Q are given statements, then the assertion

$$P \text{ or } Q$$

says that at least one of the statements P and Q is true. This assertion is often written as $P \vee Q$.

For example, if P is the statement

> *You have a cracked radiator.*

and Q is the statement

> *Your water pump needs replacing.*

then the sentence P and Q says

> *Either you have a cracked radiator or your water pump needs replacing.*

Note that the condition $P \vee Q$ includes the possibility that both P and Q are true. Thus, if you hear the words

> *Either you have a cracked radiator or your water pump needs replacing.*

then you are going to have at least one job performed on your car but there is also the possibility that both your radiator and your water pump need to be replaced.

2.3.3   Some Examples on the Use of *and* and *or*

1. The equation $x^2 - 3x - 4 = 0$ is equivalent to the condition

$$x = -1 \quad \text{or} \quad x = 4.$$

It would be quite wrong to write the solution of the equation as $x = -1$ *and* $x = 4$ because the equation does not require x to be equal to both of the numbers -1 and 4. If x is equal to either one of the numbers -1 and 4, the equation will hold.

2. The inequality $x^2 - 3x - 4 \leq 0$ is equivalent to the condition

$$-1 \leq x \quad \text{and} \quad x \leq 4.$$

In order for this quadratic inequality to hold we have to have *both* of the conditions $-1 \leq x$ and $x \leq 4$.

3. The inequality $x^2 - 3x - 4 \geq 0$ is equivalent to the condition

$$x \leq -1 \quad \text{or} \quad x \geq 4.$$

4. In this example we assume that m and n are integers. The condition for the number mn to be even is equivalent to the condition

$$m \text{ is even} \quad \text{or} \quad n \text{ is even.}$$

Note that mn will certainly be even in the event that both of the integers m and n are even, but the condition that mn is even does not *require* that both of m and n are even. All it requires is that at least one of the integers m and n is even.

5. Again in this example we assume that m and n are integers. The condition for the number mn to be odd is equivalent to the condition

$$m \text{ is odd} \quad \text{and} \quad n \text{ is odd.}$$

6. In this example we assume that x and y are nonzero numbers. The condition for the equation $|x + y| = |x| + |y|$ to hold is that

$$\text{either } x < 0 \text{ and } y < 0, \text{ or } x > 0 \text{ and } y > 0.$$

2.3.4 The Conditional *if*

If P and Q are given statements, then the sentence

$$\text{If } P, \text{ then } Q$$

says that, in the event that P is true, the sentence Q must also be true. One of the most important facts about the sentence "If P, then Q" is that it places no demands on Q when P is false. The sentence "If P, then Q" can be thought of as saying that

I don't know whether or not P is true and I don't care. However, if the

statement P does happen to be true, then the statement Q must also be true.

The condition "If P, then Q" can be written in several equivalent ways. One of these is

$$Q \text{ if } P,$$

because the latter statement also says that Q must hold true in the event that P is true. We can also write the condition "If P, then Q" in the form

$$P \text{ only if } Q,$$

because the latter condition says that the only way that P can be true is that Q is also true. The condition "If P, then Q" can also be written without using the word "if" because it says that

$$\text{either } P \text{ is false or } Q \text{ is true.}$$

The following table lists some of the many equivalent ways in which we can write the condition "If P, then Q":

If P, then Q	P implies Q
Q if P	$P \Rightarrow Q$
P only if Q	Q is implied by P
P is a sufficient condition for Q	$Q \Leftarrow P$
Q is a necessary condition for P	Either $\neg P$ or Q
$(\neg Q) \Rightarrow (\neg P)$	$(\neg P) \vee Q$

Finally, the sentence P if and only if Q can be written in any of the following equivalent forms:

P is necessary and sufficient for Q	P iff Q
$(P \Rightarrow Q) \wedge (Q \Rightarrow P)$	$P \Leftrightarrow Q$

The assertion $P \Leftrightarrow Q$ tells us that either the statements P and Q are both true or they are both false.

2.3.5 Negations and the Conjunctions and Conditionals

Suppose that P and Q are given mathematical statements.

1. Since the condition $P \wedge Q$ asserts that both of the statements P and Q must be true, the denial of this condition says that one (or both) of the statements P and Q is false. Thus the denial of the condition $P \wedge Q$ says that either P is false or Q is false.
2. Since the condition $P \vee Q$ says that at least one of the statements P and Q must be true, the denial of this condition says that neither of the statements P and Q is true. In other words, the denial of the condition $P \vee Q$ says that P is false and Q is false.
3. Since the condition $P \Rightarrow Q$ says that either P is false or Q is true, the denial of the condition $P \Rightarrow Q$ says that P is true and Q is false. Thus, for example, the denial of the assertion

    If you eat that grape, you will die.

    says that

    You will eat that grape and you will not die.

2.3.6   Contrapositives and Converses

As we know, if P and Q are mathematical statements, then the assertion $P \Rightarrow Q$ says that either P is false or Q is true. We shall now make the observation that the assertion $(\neg Q) \Rightarrow (\neg P)$ says exactly the same thing. In fact, the assertion $(\neg Q) \Rightarrow (\neg P)$ says that either $\neg Q$ is false or $\neg P$ is true; in other words, it says that either Q is true or P is false.

Thus, if P and Q are mathematical statements, then the assertions $P \Rightarrow Q$ and $(\neg Q) \Rightarrow (\neg P)$ are logically equivalent to each other. We are therefore at liberty to look at whichever of these two assertions looks easier to understand. The assertion $(\neg Q) \Rightarrow (\neg P)$ is called the **contrapositive** form of the assertion $P \Rightarrow Q$.

Of course, the statements $P \Rightarrow Q$ and $Q \Rightarrow P$ do not say the same thing. We call the statement $Q \Rightarrow P$ the **converse** of the statement $P \Rightarrow Q$.

Look, for example, at the statement: *All roses are red*. This statement can be thought of as saying that if an object is a rose, then it must be red. The contrapositive form of this statement says that all things that are not red must fail to be roses. The converse of this statement says that all red things are roses.

2.3.7   A Word of Warning

The statement "If P, then Q" is often confused with the slightly more complex sentence "P, and therefore Q". The meaning of the latter sentence is as follows:

The statement P is known to be true and we also know that P implies Q.

Therefore we can assert that the statement Q must be true.

To help us understand the distinction between "If P, then Q" and "P, and therefore Q" we shall take P to be the statement: *It is raining outside* and Q to be the statement: *You will get wet*. The sentence "If P, then Q" says:

> *If it is raining outside, then you will get wet.*

This sentence does not assert that it is raining. It says only that in the event that it is raining, you will get wet. This sentence is always true on a dry day. On the other hand, the sentence "P, and therefore Q" says:

> *It is raining outside and therefore you will get wet.*

This sentence *does* assert that it is raining and also that you will get wet. The latter sentence is always false on a dry day.

2.3.8   Some Exercises on the Use of Conditionals

In the exercises that follow you should assume that P, Q, R, and S are given statements that may be either true or false.

1. ▛ Write down the denial, the converse, and the contrapositive form of each of the following statements:

   (a) *All cats scratch.*

   (b) *If what you said yesterday is correct, then Jim has red hair.*

   (c) *If a triangle $\triangle ABC$ has a right angle at C, then*

$$(AB)^2 = (AC)^2 + (BC)^2.$$

   (d) *If some cats scratch, then all dogs bite.*

   (e) *It is with regret that I inform you that someone in this room is smoking.*

   (f) *If a function is differentiable at a given number, then it must be continuous at that number.*

   (g) *Every boy or girl alive is either a little liberal or else a conservative.*

2. ▛ Write down a denial of each of the following statements.

   (a) *All cats scratch and some dogs bite.*

   (b) *Either some cats scratch or, if all dogs bite, then some birds sing.*

   (c) *He walked into my office this morning, told me a pack of lies and punched me on the nose.*

   (d) *No one has ever seen an Englishman who is not carrying an umbrella.*

   (e) *For every number x there exists a number y such that $y > x$.*

3. ▣ In each of the following exercises we assume that f and g are given functions. Write down a denial of each of the following statements:

   (a) *Whenever $x > 50$, we have $f(x) = g(x)$.*

   (b) *There exists a number w such that $f(x) = g(x)$ for all numbers $x > w$.*

   (c) *For every number x there exists a number $\delta > 0$ such that for every number t satisfying the condition $|x - t| < \delta$ we have $|f(x) - f(t)| < 1$.*

   (d) *There exists a number $\delta > 0$ such that for every pair of numbers x and t satisfying the condition $|x - t| < \delta$ we have $|f(x) - f(t)| < 1$.*

   (e) *For every number $\varepsilon > 0$ and for every number x, there exists a number $\delta > 0$ such that for every number t satisfying $|x - t| < \delta$, we have $|f(t) - f(x)| < \varepsilon$.*

   (f) *For every positive number ε there exists a positive number δ such that for every pair of numbers x and t satisfying the condition $|x - t| < \delta$ we have $|f(x) - f(t)| < \varepsilon$.*

4. Explain why the statement $\neg (P \Rightarrow Q)$ is equivalent to the statement $P \wedge (\neg Q)$.

5. Explain why the statement $\neg (P \Leftrightarrow Q)$ is equivalent to the assertion that either (P is true and Q is false) or (P is false and Q is true).

6. Explain why the statement $\neg (P \vee Q)$ is equivalent to the statement $(\neg P) \wedge (\neg Q)$.

7. Explain why the statement $\neg (P \Rightarrow (Q \vee R))$ is equivalent to the assertion that P is true and that both of the statements Q and R are false.

8. Explain why the converse of the statement $P \Rightarrow (Q \vee R)$ is equivalent to the condition $(R \Rightarrow P) \wedge (Q \Rightarrow P)$.

9. Write the assertion $P \Rightarrow (Q \vee R)$ as simply as you can in its contrapositive form.

10. Write the assertion $(P \wedge Q) \Rightarrow (R \vee S)$ as simply as you can in its contrapositive form.

Chapter 3
Strategies for Writing Proofs

3.1 Introduction

The purpose of this chapter is to teach you how to study mathematical proofs that have been provided for you and how to think up proofs of your own. As you will see, these two tasks are almost the same because, in the process of reading a proof, you are constantly filling in details in order to justify the assertions that the author has made. Thus the key to learning a proof is *understanding*. A time honored practice that students have employed in order to "learn" a proof without having to go to the trouble of actually understanding it is to commit it to memory. Don't make that mistake! Memorizing a proof that you do not understand makes about as much sense as studying a great musical composition by memorizing the notes and musical symbols such as ♭, ♮, and ♯ that appear on a piece of paper, without having the slightest idea of what these symbols mean or what melody is written there.

Nor should you be content, when you are studying a proof, with the knowledge of how each individual step follows from the one before it. Every proof has a *theme*, a master plan, that suggests to us what the individual steps should be. You have understood a proof only when you have looked into it deeply enough to perceive that theme. Sometimes, when you are reading a proof, it will take some careful digging to unearth its underlying theme, especially when the proof is very polished. One of the difficulties that faces you is that the proof you are reading is a completed article. It works. It is valid. But it doesn't always reveal all of the thoughts that led to its discovery and the thoughts that guided your teacher to write it in the form that lies before you.

To help yourself to discover the underlying theme of a proof and to anticipate the way in which it may be laid out, you may want to ask yourself the following two kinds of question:

- What are we trying to prove here? What does this statement mean to me? How else can I write this statement? What would it mean to say that this statement is not true? What other proofs do I know that are used to prove statements like this one?
- What is the given information? What does the given information tell me? How does one usually go about using this kind of given information? What other proofs do I know that use this kind of information.

Sometimes, when a proof is difficult, you will not be able to anticipate it. Read the proof one step at a time and, when you understand the individual steps, go back to the job of trying to anticipate it. You will gradually come to understand the bridge between the given information and the required statement that the proof provides. As your understanding of the proof solidifies, make sure that you understand where *all* of the given information is used. If any of this information wasn't used, then either the theorem can be improved or (more likely) you are misunderstanding something.

When you really understand a proof, you will feel able to explain it to others. In fact, you will *want* to do so; in much the same way that you might look around for someone to whom you could tell a good joke that you have just heard. One of the best ways to learn a proof is to imagine that you are going to teach it to someone else. As you study it, write it down on a blank piece of paper and imagine that you are, in fact, explaining it to another person. You have understood the proof if and only if you have the feeling that you did a really fine job of explaining it.

The way we approach the task of proving a particular mathematical statement depends on the nature of that statement and, in particular, on the way the grammatical symbols *if, and, or,* $\Rightarrow$, $\forall$, and $\exists$ appear. Exactly where these words appear and how they appear plays a major role when we devise our proof writing strategy.

3.2 Statements that Contain the Word *and*

If P and Q are mathematical statements, then, as you know, the statement $P \wedge Q$ asserts that both of the statements P and Q are true. In this section we shall discuss some strategies for proving a statement of the form $P \wedge Q$, and we shall also discuss some strategies for using information that is phrased in this form.

3.2.1 Proving a Statement of the Form $P \wedge Q$

If P and Q are mathematical statements, then, in order to prove the statement $P \wedge Q$, we have two tasks to perform: We need to show that P is true, and we also need to show that Q is true. This kind of proof can be broken down into two steps.

- Step 1: Show that the statement P is true.
- Step 2: Show that the statement Q is true.

Suppose, for example, that you want to prove that neither of the integers 1037173 nor 4087312111 is prime. We have two tasks to perform. We need to show that the number 1037173 is not prime, and then we need to show that the integer 4087312113 is not prime. To accomplish these two tasks we can observe first that

$$1037173 = 223 \times 4651,$$

and then we can observe that

$$4087312111 = 587 \times 6963053.$$

Watch for the word *and* in statements that you want to prove. The appearance of *and* is often an indication that your problem can be broken down into several simpler parts. Even if you fail to solve the problem as a whole, you may still succeed with some of these parts. Remember: A partial solution of a mathematical problem is better than no solution at all.

3.2.2   Using Information of the Form $P \wedge Q$

Suppose that we are given information of the form $P \wedge Q$ and that we want to prove a statement R. We cannot easily break this kind of proof into several parts. We have been given the truth of each of the statements P and Q, and we need to use this information to obtain the statement R.

3.2.3   Example of a Proof Using Information Containing *and*

To illustrate this kind of proof we shall prove an elementary fact about integers. As you know, if an integer a is even, then a has the form $2n$ for some integer n and if a is odd, then a has the form $2n + 1$. We shall now prove the following statement:

If a and b are odd integers, then their product ab is also odd.

Proof. Suppose that a and b are odd integers. We begin by using the fact that a is odd to write a in the form $2m + 1$ for some integer m. Next we use the fact that b is odd to write b in the form $2n + 1$ for some integer n. We observe that

$$ab = (2m + 1)(2n + 1) = 2(2mn + m + n) + 1,$$

and we conclude that ab is odd. ∎

3.2.4   Some Exercises on Statements Containing *and*

1. Prove the following assertions:
    (a) ⊞ If $a, b, c, x,$ and y are positive numbers and $a^x = b$ and $b^y = c$, then $a^{xy} = c$.
    (b) ⊞ If two integers m and n are both even, then mn has a factor 4.
    (c) If an integer m is even and an integer n has a factor 3, then mn has a factor 6.
2. "Ladies and gentlemen of the jury" said the prosecutor, "We shall

demonstrate beyond a shadow of doubt that, on the night of June 13, 1997, the accused, Slippery Sam Carlisle, did willfully, unlawfully and maliciously kill and murder the deceased, Archibald Bott, by striking him on the head with a blunt instrument". Outline a strategy that the prosecutor might use to prove this charge. How many separate assertions must the prosecutor prove in order to carry out his promise to the jury?

3. ▉ One of the basic laws of arithmetic tells us that if a and b are any two numbers satisfying the condition $a < b$, and if $x > 0$, then $ax < bx$. Show how this law may be used to show that if $0 < u < 1$ and $0 < v < 1$, then $0 < uv < 1$.

4. ▉ In this exercise, if we are given three nonnegative integers a, b, and c, then the integer that consists of a hundreds, b tens and c units will be written as $\lceil a, b, c \rceil$. Given nonnegative integers a, b, and c, prove the assertion $P \wedge Q \wedge R \wedge S$, where P, Q, R and S are, respectively, the following assertions:

P. If the number $\lceil a, b, c \rceil$ is divisible by 3, then $a + b + c$ is also divisible by 3.

Q. If the number $a + b + c$ is divisible by 3, then $\lceil a, b, c \rceil$ is also divisible by 3.

R. If the number $\lceil a, b, c \rceil$ is divisible by 9, then $a + b + c$ is also divisible by 9.

S. If the number $a + b + c$ is divisible by 9, then $\lceil a, b, c \rceil$ is also divisible by 9.

3.3   Statements that Contain the Word *or*

Suppose that P and Q are mathematical statements. As you know, the statement $P \vee Q$ asserts that at least one of the statements P and Q must be true. The statement $P \vee Q$ does not guarantee that both of the statements P and Q are true, although they might be. In this section we shall discuss some strategies for proving a statement of the form $P \vee Q$, and we shall also discuss some strategies for using information that is phrased in this form.

3.3.1   Proving a Statement of the Form $P \vee Q$

If P and Q are mathematical statements, then, in order to prove the statement $P \vee Q$, we need to show that at least one of the statements P and Q is true. In other words, we need to show that if either of the statements P and Q happens to be false, then the other one must be true. Among the ways in which one may approach a proof of this type, the following two are worth mentioning:

1. Try to prove that the statement P is true. If you succeed, you are done. If you can't see why P is true, or if you suspect that P may be false, try to prove that Q is true.
2. This approach is a slight refinement of the first approach. If you know that the statement P is false, then add the information that P is false to the information that you already have and try to prove that Q is true.

3.3.2   Example of a Proof of a Statement Containing *or*

We now return to the example that we saw in Subsection 3.2.3. Suppose that m and n are two given integers. The following statements all say the same thing:

1. If both m and n are odd, then mn must also be odd.
2. If mn is even, then either m is even or n is even.
3. If mn is even and m is odd, then n must be even.

In Subsection 3.2.3, we proved this statement in its first form. Now we shall prove it in its second form and you can decide which of the two proofs you prefer.

Suppose then that mn is even. We need to explain why either m or n must be even. For this purpose we shall show that if m happens to be odd, then the number n has to be even. (In other words, we are really proving the statement in its third form.) Suppose that m is odd. Using this assumption, choose an integer a such that $m = 2a + 1$ and, using the fact that mn is even, choose an integer b such that $mn = 2b$. Thus

$$(2a + 1)\, n = 2b,$$

from which we deduce that

$$n = 2\,(b - an)\,,$$

and we have shown that n must be even.

3.3.3   Using Information of the Form $P \vee Q$

Suppose that we want to prove a statement R using given information of the form $P \vee Q$, where P and Q are mathematical statements. In order to construct such a proof you need to accomplish two tasks:

(a) Show that if P is true, then R is true. You can do this by assuming that P is true and showing that R is true.
(b) Show that if Q is true, then R is true. You can do this by assuming that Q is true and showing that R is true.

A proof of this type can therefore be separated into different cases.

3.3.4   Example of a Proof Using Information Containing *or*
If $x = \cos 20°$ *or* $x = -\cos 40°$ *or* $x = -\cos 80°$, *then*

$$8x^3 - 6x - 1 = 0. \tag{3.1}$$

Proof. In order to prove this statement we need to perform three tasks. We need to show that $x = \cos 20°$ is a solution of Equation (3.1). Then we need to show that $x = -\cos 40°$ is a solution of Equation (3.1). Finally we need to show that $x = -\cos 80°$ is a solution of Equation (3.1). To perform the first of these three tasks we deduce from the trigonometric identity

$$\cos 3\theta = 4\cos^3 \theta - 3\cos \theta$$

that, if $x = \cos 20°$, then

$$
\begin{aligned}
8x^3 - 6x - 1 &= 2\left(4\cos^3 20° - 3\cos 20°\right) - 1 \\
&= 2\cos 60° - 1 = 0.
\end{aligned}
$$

We leave the second and third of these three tasks as exercises. ∎

3.3.5   Some Exercises on Statements Containing *or*

1. ▦ Given that m and n are integers and that the number mn is not divisible by 4, prove that either m is odd or n is odd.
2. ▦ Given that m and n are integers, that neither m nor n is divisible by 4 and that at least one of the numbers m and n is odd, prove that the number mn is not divisible by 4.
3. ▦ Prove that if $x = -\cos 40°$ or $x = -\cos 80°$, then

$$8x^3 - 6x - 1 = 0.$$

4. ▦ A theorem in elementary calculus, known as Fermat's theorem, says that if a function f defined on an interval has either a maximum or a minimum value at a number c inside that interval, then either $f'(c) = 0$ or $f'(c)$ does not exist. Give a brief outline of a strategy for approaching the proof of this theorem.
5. A well-known theorem on differential calculus that is known as **L'Hôpital's rule** may be stated as follows:

    Suppose that f and g are given functions, that c is a given number, that

$$\lim_{x \to c} \frac{f'(x)}{g'(x)} = L,$$

and that one or the other of the following two conditions holds

(a) *Both $f(x)$ and $g(x)$ approach 0 as $x \to c$.*
(b) *$g(x) \to \infty$ as $x \to c$.*
Then

$$\lim_{x \to c} \frac{f(x)}{g(x)} = L.$$

Describe how the proof of L'Hôpital's rule can be broken down into two parts. For each part of the proof, say what is being assumed and what is being proved.

3.4   Statements of the Form $P \Rightarrow Q$

As you know, if P and Q are mathematical statements, then $P \Rightarrow Q$ says that in the event that P is true, the statement Q must also be true. The assertion $P \Rightarrow Q$ can also be looked upon as a statement containing the word *or* because it says that either P is false or Q is true. In this section we shall discuss the strategy for proving a statement of the form $P \Rightarrow Q$ and the strategy for using information that is phrased in the form $P \Rightarrow Q$.

3.4.1   Proving a Statement of the Form $P \Rightarrow Q$

If P and Q are mathematical statements, then to prove the statement $P \Rightarrow Q$ we need to show that either P is false or that Q is true. We can therefore follow the procedure given in Subsection 3.3.1 and use one of the following methods:

- Assume that P is true and use this to show that Q must be true.
- Assume that Q is false and use this to show that P must be false. In other words, prove the contrapositive form of the statement $P \Rightarrow Q$.

3.4.2   Using Information of the Form $P \Rightarrow Q$

Suppose that we want to prove a statement R using given information of the form $P \Rightarrow Q$, where P and Q are mathematical statements. Since the given information $P \Rightarrow Q$ can be interpreted as saying that either P is false or Q is true, we can use the method described in Subsection 3.3.3. Another approach to this sort of proof is to reason that the condition $P \Rightarrow Q$ is interesting only when P is true. If you know (or can prove) that P is true, then you can deduce from the condition $P \Rightarrow Q$ that Q is true and use the fact that both P and Q are true to show that R is true.

On the other hand, if you know that P is false, then the statement Q is irrelevant. In this case, use the fact that P is false to show that R is true.

3.4.3   Example of a Proof Using Information Containing $\Rightarrow$

We shall take P to be the statement that $x \neq 1$ and Q to be the statement that $x = 3$. Suppose that we are given that $P \Rightarrow Q$ and that we want to prove that the equation

$$x^2 - 4x + 3 = 0$$

holds. In the event that the statement P is true we know that $x = 3$, and we can verify that the equation holds in this case. In the event that the statement P is false we have $x = 1$, and, once again, we can verify that the equation holds.

3.4.4   Exercises on the Symbol $\Rightarrow$

1. (a) ▉ Outline a strategy for proving an assertion that has the form
    $P \Rightarrow (Q \wedge R)$.
   (b) ▉ Write down the assertion $P \Rightarrow (Q \wedge R)$ in its contrapositive form and outline a strategy for proving it in this form.
2. (a) Outline a strategy for proving an assertion that has the form
    $P \Rightarrow (Q \vee R)$.
   (b) Write down the assertion $P \Rightarrow (Q \vee R)$ in its contrapositive form and outline a strategy for proving it in this form.
3. (a) ▉ Outline a strategy for proving an assertion that has the form
    $(P \wedge Q) \Rightarrow R$.
   (b) ▉ Write down the assertion $(P \wedge Q) \Rightarrow R$ in its contrapositive form and outline a strategy for proving it in this form.
4. (a) Outline a strategy for proving an assertion that has the form
    $(P \vee Q) \Rightarrow R$.
   (b) Write down the assertion $(P \vee Q) \Rightarrow R$ in its contrapositive form and outline a strategy for proving it in this form.
5. (a) ▉ Outline a strategy for proving an assertion that has the form
    $(P \vee (Q \Rightarrow P)) \Rightarrow R$.
   (b) ▉ Write down the assertion $(P \vee (Q \Rightarrow P)) \Rightarrow R$ in its contrapositive form and outline a strategy for proving it in this form.

3.5   Statements of the Form $\exists x\,(P(x))$

As you know, if $P(x)$ is a statement about an unknown x, then the assertion $\exists x\,(P(x))$ says that there is at least one object x for which the statement $P(x)$ is true. In this section we shall discuss the strategy for proving a statement of the form $\exists x\,(P(x))$ and the strategy for using information that is phrased in the form $\exists x\,(P(x))$.

3.5.1   Proving A Statement of the Form $\exists x\,(P(x))$

How does one prove that something exists? Legend has it that the Greek philosopher Diogenes traveled from place to place with a lantern in his hand looking for one honest man. By finding one he could certainly have proved that such a man exists. Diogenes' attempt to find an honest man demonstrates one very good way of proving that an object exists: Find one. We use this kind of proof very commonly in mathematics. Suppose, for example, that we want to prove that there exists a prime number between 80 and 90. We can achieve this proof by demonstrating that the number 83 is prime.

Although the method of giving an example is the most common way of proving existence, it has the drawback that we have to be able to find that example. This may be very hard to do. It may even be impossible. Fortunately we also have other ways of proving existence. Even though we may not know of any examples of a certain kind of object, sometimes there will be indirect evidence that the object exists. For example, if you are sitting in a crowded restaurant, you do not have to be able to see and identify a smoker in order to know that someone is smoking. All you have to do is try to breathe.

There are many important theorems in mathematics that assert the existence of an object even when it is hard or impossible to lay our hands on one. We look at two examples:

1. The well-known theorem in differential calculus that we call the **mean value theorem**, and which is illustrated in Figure 3.1, can be stated as follows:

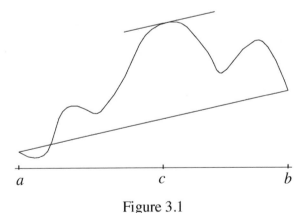

Figure 3.1

Suppose that a function f is continuous on an interval [a, b] (where a < b) and that the derivative f'(x) exists for every x ∈ (a, b). Then there is at least one number c ∈ (a, b) such that

$$f'(c) = \frac{f(b) - f(a)}{b - a}. \qquad\qquad (3.2)$$

The importance of the mean value theorem lies in the fact that, although there must exist a number c that makes Equation (3.2) hold, we usually cannot lay our hands on such a number. Had we been able to do so, the mean value theorem would not have been nearly as interesting. As it happens, the mean value theorem makes a very deep statement about the nature of our number system. It tells us that, even though we may not be able to find the number c explicitly, it exists anyway.

2.   A real number x is said to be **algebraic** if it is a solution of an equation of the form

$$a_n x^n + a_{n-1} x^{n-1} + \cdots a_2 x^2 + a_1 x + a_0 = 0,$$

where n is a positive integer, and the numbers $a_0, \ldots a_n$ are integers, and $a_n \neq 0$. A number that is not algebraic is said to be **transcendental**. Although there are many common examples of transcendental numbers, including such numbers as $2^{\sqrt{2}}$, $\log_2 3$, e, π, and e^{π}, it is by no means easy to prove that any given number is transcendental. It is therefore hard to prove that transcendental numbers exist if the method of proof is to exhibit an example of one. In the latter part of the nineteenth century, Georg Cantor came up with a different kind of proof of the existence of transcendental numbers that is based upon his theory of sets. What Cantor showed is that the set $\mathbf{R}$ of all real numbers has a property that is called *uncountability* but that the set A of algebraic numbers does not have this property. Therefore, he reasoned, the sets $\mathbf{R}$ and A cannot be equal to each other and so there must be real numbers that are not algebraic. Thus we can use Cantor's proof to guarantee the existence of transcendental numbers even if we have never seen one.

To sum up, if we can lay our hands on an example of a certain kind of object, then we know that the object exists. This method is the most satisfying way of proving existence, but it isn't always possible. Sometimes we have to use an indirect method to prove an existence theorem and, as you may expect, the existence theorems that require indirect proofs are usually the most interesting ones.

3.5.2   Using Information of the Form $\exists x\,(P(x))$

In the last subsection we discussed proofs of statements of the type $\exists x\,(P(x))$. Now we shall assume that a statement of this type has been given and we shall discuss the possible ways of using this information. What we have assumed is that there exists at least one object x with certain properties. For this fact to be of any interest to us we must have a need for an object of this type. Our procedure is to *choose* an object x with the desired properties and to use it for whatever purpose it is needed. If the statement $\exists x\,(P(x))$ has been given, and if we need an object x satisfying the condition $P(x)$, then we need to write

choose x such that the condition $P(x)$ is true.

To illustrate this kind of proof we shall prove the following fact about differential calculus:

Suppose that f is a continuous function on an interval $[a, b]$ and that $f'(x) = 0$ for every number $x \in (a, b)$. Then $f(a) = f(b)$.

Proof. The mean value theorem guarantees that there exists at least one number $c \in (a, b)$ such that

$$f'(c) = \frac{f(b) - f(a)}{b - a}.$$

and so we *choose* a number $c \in (a, b)$ such that this equation holds. From the fact that $f'(x) = 0$ for every $x \in (a, b)$ we certainly have $f'(c) = 0$. Therefore

$$\frac{f(b) - f(a)}{b - a} = 0,$$

and we conclude that $f(a) = f(b)$. ∎

3.5.3   A Note of Caution

When you are writing a mathematical proof, do not make the common mistake of thinking that, just because a condition of the form $\exists x\,(P(x))$ is given, such an object x has been chosen for you automatically. You do not have such an object x in your hands until you have written:

Choose x such that the condition $P(x)$ is true.

Consider the following pair of sentences:

There exists a number x such that $0 < x < 1$.
Clearly, $x^2 < x$.

The second of these two sentences is meaningless. If you were to write these sentences in an examination, the examiner would respond with something like: "*What is x?*" You cannot answer that x is "the number" introduced in your first sentence because all the first sentence says is that there is a number between 0 and 1.

On the other hand it, would be perfectly legitimate to say:

There exists a number x *such that* $0 < x < 1$.
Furthermore, for every $x \in (0, 1)$ *we have* $x^2 < x$.

Alternatively, you could say:

Using the fact that $(0, 1)$ *is nonempty, choose* $x \in (0, 1)$.
Note that $x^2 < x$.

3.6   Statements of the Form $\forall x \, (P(x))$

As you know, if $P(x)$ is a statement about an unknown x, then the assertion $\forall x \, (P(x))$ says that the condition $P(x)$ is true for *every* object x. In this section we shall discuss the strategy for proving a statement of the form $\forall x \, (P(x))$ and the strategy for using information that is phrased in the form $\forall x \, (P(x))$.

3.6.1   Inductive and Deductive Reasoning

A statement of the type $\forall x \, (P(x))$ can *never* be proved by giving an example. All an example can tell us is that there is one object x for which $P(x)$ is true. It says nothing to guarantee that $P(x)$ is true for *every* x. Thus, for example, you cannot conclude that the identity

$$\sqrt[3]{\frac{3\sqrt{3}x + (2x^2 + 1)\sqrt{4 - x^2}}{6\sqrt{3}}} + \sqrt[3]{\frac{3\sqrt{3}x - (2x^2 + 1)\sqrt{4 - x^2}}{6\sqrt{3}}} = x$$

holds for all numbers $x \in [-2, 2]$ merely by observing that it holds when $x = 0$. Even if we were to find thousands of numbers x in $[-2, 2]$ for which this identity is true, in the absence of a general proof we would still not know for sure that the identity holds for *all* $x \in [-2, 2]$.

Outside of mathematics there is a type of reasoning called **inductive reasoning**, which is used to validate statements of the type $\forall x \, (P(x))$. The idea of inductive reasoning is that $P(x)$ probably holds for all x if it is known to hold for a representative selection of objects x. For example, if every time you have used a certain laundry detergent you have broken out in a rash, you could reasonably conclude that this detergent is causing your problem. Every time you use the detergent and see the rash appear you feel a little more confident that your theory is correct.

Inductive reasoning plays an important role in such walks of life as science, medicine, and even law. Doctors must use inductive reasoning to make decisions upon which our very lives depend. Courts must render decisions based on inductive reasoning. There is, in fact, a branch of mathematics, known as *statistics,* that is devoted to the problem of measuring how confidently one may assert a statement that was obtained using inductive reasoning. But inductive reasoning can never be used to make a conclusion in mathematics.

3.6.2   Proving a Statement of the Form $\forall x \, (P(x))$

In order to verify an assertion of the type $\forall x \, (P(x))$ we need to show that the condition $P(x)$ holds for every member x of some specified set. If that set is infinite, then we can't live long enough to test the conditions $P(x)$ one at a time, but, fortunately, there is a more efficient approach. We prove that the condition $P(x)$ holds for what we call an *arbitrary* value of x that we introduce using the word *suppose*.

3.6.3   Example of a Proof of a Statement Containing *for every*

For every number $x \in [-2, 2]$ we have

$$\sqrt[3]{\frac{3\sqrt{3}x + (2x^2 + 1)\sqrt{4 - x^2}}{6\sqrt{3}}} + \sqrt[3]{\frac{3\sqrt{3}x - (2x^2 + 1)\sqrt{4 - x^2}}{6\sqrt{3}}} = x.$$

Proof. Because we want to prove that a condition is true for every number x in the interval $[-2, 2]$, we begin our proof by writing

$$\text{Suppose } x \in [-2, 2].$$

In writing this opening sentence we do not have to know that there are any numbers in the interval $[-2, 2]$. What we are saying is that, in case there are any, suppose that x is an arbitrary one of them, coming without any restrictions, so that anything we might be able to prove about x would apply just as well to any other number in the interval. In other words, let x be an arbitrary number in the interval $[-2, 2]$ that has come to challenge us to prove that

$$\sqrt[3]{\frac{3\sqrt{3}x + (2x^2 + 1)\sqrt{4 - x^2}}{6\sqrt{3}}} + \sqrt[3]{\frac{3\sqrt{3}x - (2x^2 + 1)\sqrt{4 - x^2}}{6\sqrt{3}}} = x.$$

Now that we have this challenger in our hands we can begin our proof. To obtain the desired result we shall first show that

$$\sqrt[3]{\frac{3\sqrt{3}x + (2x^2 + 1)\sqrt{4 - x^2}}{6\sqrt{3}}} = \frac{x}{2} + \frac{1}{2}\sqrt{\frac{4 - x^2}{3}}.$$

and

$$\sqrt[3]{\frac{3\sqrt{3}x - (2x^2 + 1)\sqrt{4 - x^2}}{6\sqrt{3}}} = \frac{x}{2} - \frac{1}{2}\sqrt{\frac{4 - x^2}{3}}.$$

Once these two equations have been established, the desired result will follow at once. To establish the first of these two equations we expand and simplify the expression

$$\left(\frac{x}{2} + \frac{1}{2}\sqrt{\frac{4 - x^2}{3}} \right)^3$$

to obtain

$$\left(\frac{x}{2} + \frac{1}{2}\sqrt{\frac{4 - x^2}{3}} \right)^3 = \frac{3\sqrt{3}x + (2x^2 + 1)\sqrt{4 - x^2}}{6\sqrt{3}}.$$

The second equation can be obtained similarly by showing that

$$\left(\frac{x}{2} - \frac{1}{2}\sqrt{\frac{4 - x^2}{3}} \right)^3 = \frac{3\sqrt{3}x - (2x^2 + 1)\sqrt{4 - x^2}}{6\sqrt{3}}.$$

Before leaving this topic we should notice that, although the proof we have just given is valid, it is not very satisfying. It guarantees that the equation

$$\sqrt[3]{\frac{3\sqrt{3}x + (2x^2 + 1)\sqrt{4 - x^2}}{6\sqrt{3}}} + \sqrt[3]{\frac{3\sqrt{3}x - (2x^2 + 1)\sqrt{4 - x^2}}{6\sqrt{3}}} = x$$

holds for every $x \in [-2, 2]$, but it doesn't tell us how we could have anticipated this equation. It doesn't motivate it. If you would like to see a better approach, go to the on-screen text and click on the icon [icon], which will take you to some exercises in the optional document on cubic equations. Of course, those exercises should not be attempted unless you decide to read that optional document.

3.6.4   Using Information of the Form $\forall x \, (P(x))$

Information of the type $\forall x \, (P(x))$ can be useful to us only when there are certain objects x that are of particular interest to us and for which we would like to know that the condition $P(x)$ is true. We can then say that because the condition $P(x)$ is true for *every* x, certainly $P(x)$ will be true for those objects x that are of interest to us.

Suppose, for example, that I am standing in a used car lot and I want to buy a used car. The knowledge that all used car sales personnel are honest is very comforting to me. It tells me that I can place my implicit trust in the gentleman who is assisting me.

3.6.5   Example of a Proof Using Information Containing *for every*

We shall revisit the proof that appears in Subsection 3.5.2. There, we were given a function f that is continuous on an interval $[a, b]$ and for which $f'(x) = 0$ for every $x \in (a, b)$, and we proved that $f(a) = f(b)$. We began by choosing a number $c \in (a, b)$ such that

$$f'(c) = \frac{f(b) - f(a)}{b - a}.$$

Then we made use of the fact that $f'(x) = 0$ for *every* $x \in (a, b)$. Among all the numbers $x \in (a, b)$, the number c is of particular interest to us. Since $f'(x) = 0$ for every $x \in (a, b)$, we know that $f'(c) = 0$. Therefore

$$\frac{f(b) - f(a)}{b - a} = 0,$$

and we deduce that $f(a) = f(b)$.

3.6.6   Exercises on Statements Containing Quantifiers

1. Physicist's proof that all odd positive integers are prime: 1 is prime.[10] 3 is prime. 5 is prime. 7 is prime. 9 is experimental error. 11 is prime. 13 is prime. We have now taken sufficiently many readings to verify the hypothesis. Comment!

2. ▉ You know that there are 1000 people in a hall. Upon inspection you determine that 999 of these people are men. What can you conclude about the 1000th person?

3. ▉ The *product rule for differentiation* says that for every number x and all functions f and g that are differentiable at x, we have

    $$(fg)'(x) = f'(x)g(x) + f(x)g'(x).$$

    Write down the opening line of a proof of the product rule. Your opening line should start: *Suppose that ...*

4. ▉ Given that $P(x)$ and $Q(x)$ are statements that contain an unknown x and that S is a set, outline a strategy for proving the assertion $P(x) \Rightarrow Q(x)$ for every $x \in S$. Write down the opening line of your proof.

---

[10]   Mathematicians are divided on the issue of whether or not the number 1 should be called prime.

5. ■ Given that $P(x)$ is a statement that contains an unknown x and that S is a set, write down an opening line of a proof of the assertion that $P(x)$ is true for every $x \in S$.

6. ■ You are given that $P(x)$ and $Q(x)$ are statements that contain an unknown x, that S is a set, that $P(x)$ is true for every $x \in S$, and that $P(x) \Rightarrow Q(x)$. Is it possible to deduce that $Q(x)$ is true for every member x of the set S?

7. ■ You are given that $P(x)$ and $Q(x)$ are statements that contain an unknown x, that S is a set, that $P(x)$ is true for every $x \in S$, and that $P(x) \Rightarrow Q(x)$. Is it possible to deduce that $Q(x)$ is true for at least one member x of the set S?

8. ■ Write down the contrapositive form of the statement that for every member x of a given set S we have $P(x) \Rightarrow Q(x)$.

9. ■ Write down the denial of the statement that for every x we have $P(x) \Rightarrow Q(x)$.

10. Prove that, for every number x in the interval $[-2, 2]$, if we define

$$u = \sqrt[3]{\frac{3\sqrt{3}x + (2x^2 + 1)\sqrt{4 - x^2}}{6\sqrt{3}}}$$

and

$$v = \sqrt[3]{\frac{3\sqrt{3}x - (2x^2 + 1)\sqrt{4 - x^2}}{6\sqrt{3}}},$$

then

$$u^2 + v^2 = 1 + uv.$$

Hint: With an eye on the proof in Subsection 3.6.3, show that $u^3 + v^3 = u + v$.

3.7 Proof by Contradiction

The idea of proof by contradiction is that if we can deduce a contradiction by assuming that a certain statement P is false, then P must be true. Proof by contradiction is usually most useful when the statement P that we are considering seems rather intangible but its denial $\neg P$ is nice and concrete. If P is this kind of statement, it is sometimes hard to find a direct proof that P is true. However, the assumption that P is false may place some very concrete information in our hands. If we can show that this information leads to a contradiction, then we have proved indirectly that P is true.

For an example of a statement P of this type, suppose that x is a given number and take P to be the statement that x is irrational.[11] Since P asserts that it is impossible to find integers m and n such that $x = \frac{m}{n}$, a direct proof of the statement P must tell us why it is impossible to find such integers m and n. How does one prove that it is impossible to do something? Now consider the statement $\neg P$. The statement $\neg P$ says that it *is* possible to find integers m and n such that $x = \frac{m}{n}$. If we are assuming the statement $\neg P$, then at least we know how to begin our proof:

Choose integers m and n such that $x = \dfrac{m}{n}$.

As you may expect from this discussion, we often use proof by contradiction to prove that a given real number is irrational.

3.7.1   Some Examples of Proof by Contradiction

1. In this example we shall show that the number $\log_2 6$ is irrational. To prove this assertion by the method of proof by contradiction we begin:

   To obtain a contradiction, assume that $\log_2 6$ is rational.

   Using the assumption that $\log_2 6$ is a positive rational number, choose positive integers m and n such that

   $$\log_2 6 = \frac{m}{n}.$$

   We observe that $2^{m/n} = 6$ and therefore

   $$2^m = 6^n.$$

   But the right side of the latter identity has a factor 3 while the left side does not, and so the two sides cannot be equal. Since the assumption that $\log_2 6$ is rational led us to a contradiction, we conclude that $\log_2 6$ must be irrational.

2. In this example we again concern ourselves with the irrationality of a number. This time we show that the number $\sqrt{2}$ is irrational. To obtain a contradiction we shall assume that $\sqrt{2}$ is rational. Choose positive integers m and n such that

   $$\sqrt{2} = \frac{m}{n}.$$

   We now cancel out any factors that are common to the numerator and

---

[11]   Recall that a number is rational if it can be expressed in the form of a ratio $\frac{m}{n}$, where m and n are integers.

denominator of the fraction m/n yielding a fraction p/q in which p and q are positive integers that have no common factor. Since $\sqrt{2} = p/q$, we have

$$p^2 = 2q^2.$$

Now, since q has no common factor with p we see that q has no common factor with p^2. But, on the other hand, since q is a factor of $2q^2$, we know that q is a factor of p^2. Therefore $q = 1$ and we see that $p^2 = 2$. The latter condition is certainly impossible because there is no integer whose square is 2. Since the assumption that $\sqrt{2}$ is rational leads to a contradiction, we conclude that $\sqrt{2}$ must be irrational.

3.7.2   Drawbacks of Proof by Contradiction

Although the method of proof by contradiction can be very useful when we are proving certain kinds of statements, some members of the mathematical community are less than enthusiastic about using it. They argue that it is better to give a positive reason why a statement P must be true rather than to conclude that P must be true because the statement $\neg P$ leads to a contradiction. Therefore, they feel that whenever we can see a direct way of showing that a statement P is true, we should use it.

Then there are the logical purists who point out that, if we can deduce a contradiction by assuming that a certain statement P is false, then this leaves open the remote possibility that we may also be able to deduce a contradiction by assuming that P is true. In this event, we would conclude that there is an inherent contradiction in mathematics. Therefore, the logical purists point out, if we can deduce a contradiction by assuming that a statement P is false, we need to say that, as long as there is no inherent contradiction in mathematics, P must be true.

3.7.3   Exercises on Proof by Contradiction

1. Prove that the following numbers are irrational:
   (a) $\log_{10} 5$.
   (b) ⬛ $\log_{12} 24$.
   (c) $\sqrt[3]{4}$.
   (d) ⬛ Any solution of the equation $8x^3 - 6x - 1 = 0$.
2. Given that m and n are integers and that mn does not have a factor 3, prove that neither m nor n can have a factor 3.
3. Suppose that we know that $x^2 - 2x < 0$ and that we wish to prove that $0 < x < 2$. Write down the first line of a proof of this assertion that uses the method of proof by contradiction. Do this in such a way that your proof

splits into two cases and complete the proof in each of these cases.

4. █ Suppose that f is a given function defined on the interval $[0, 1]$ and suppose that we wish to prove that this function f has the property that there exists a number $\delta > 0$ such that, whenever t and x belong to the interval $[0, 1]$ and $|t - x| < \delta$, we have $|f(t) - f(x)| < 1$. Write down the first line of a proof of this assertion that uses the method of proof by contradiction.

5. Suppose that $\{x_1, x_2, \cdots x_n\}$ is a subset of a vector space[12] V and that we wish to prove that the set $\{x_1, x_2, \cdots x_n\}$ is linearly independent. Write down the first line of a proof of this assertion that uses the method of proof by contradiction. (Try to be specific. Don't just suppose that the set is linearly dependent.)

3.8   Some Further Examples

This section contains a few extra examples in which we prove or disprove statements of the form $\forall x\, (P(x))$. The proofs contained here will help you to develop skills that will be useful to you when you read some of the later chapters.

3.8.1   A Fact About Inequalities

In this subsection we discuss the two statements that appeared in Example 5 of Subsection 2.1.4. We begin by proving that the first of the statements is true. This statement is as follows:

For every number x there exists a positive number δ such that for every number t satisfying the inequality $|t - x| < \delta$ we have $|t^2 - x^2| < 1$.

Proof. Since we want to prove that a condition holds for every number x, we begin this proof by writing:

Suppose that x is any real number.

Now that we have a challenger x in our hands we need to prove that there exists a positive number δ such that for every number t satisfying $|t - x| < \delta$ we have $|t^2 - x^2| < 1$. We shall demonstrate the existence of such a number δ by finding one.

To help us find this number, we observe that if $\delta > 0$, then for any number t satisfying $|x - t| < \delta$ we have

$$
\begin{aligned}
\left|x^2 - t^2\right| &= |x - t|\,|x + t| = |x - t|\,|(t - x) + 2x| \\
&\leq |x - t|\,(|x - t| + 2\,|x|) < \delta\,(\delta + 2\,|x|)\,.
\end{aligned}
$$

---

12   Skip this exercise if you have not had a course in linear algebra.

So, to guarantee the inequality $|t^2 - x^2| < 1$, all we have to do is find a positive number δ for which

$$\delta\left(\delta + 2\,|x|\right) \leq 1,$$

and, with this thought in mind, we define

$$\delta = \frac{1}{1 + 2\,|x|}.$$

Note that $\delta \leq 1$ and so

$$\delta\left(\delta + 2\,|x|\right) \leq \frac{1}{1 + 2\,|x|}\left(1 + 2\,|x|\right) = 1.$$

Now that both x and δ have been introduced, we want to show that for every number t satisfying the inequality $|t - x| < \delta$ we have $|t^2 - x^2| < 1$. We therefore continue the proof by writing:

Suppose that t is any number satisfying $|t - x| < \delta$.

And we complete the proof by observing that

$$\left|t^2 - x^2\right| < \delta\left(1 + 2\,|x|\right) \leq 1.$$

We shall now show that the second of the two statements that appeared in Example 5 of Subsection 2.1.4 is false. This statement is as follows:

There exists a positive number δ such that for every number x and for every number t satisfying the inequality $|t - x| < \delta$ we have $|t^2 - x^2| < 1$.

Proof. To see that this statement is false, suppose that δ is any positive number. Define $x = 1/\delta$ and

$$t = x + \frac{\delta}{2}.$$

Then, although $|t - x| = \delta/2 < \delta$, we have

$$t^2 - x^2 = x\delta + \frac{\delta^2}{4} > 1. \blacksquare$$

3.8.2   Another Fact About Inequalities

In this subsection we prove Statement 6a that appeared in Example 6 in Subsection 2.1.4:

For every number $\varepsilon > 0$ and for every number $x \in [0, 1]$ there exists a positive integer N such that for every integer $n \geq N$ we have

$$\frac{nx}{1 + n^2 x^2} < \varepsilon.$$

Proof. Since we want to prove that a condition holds for every $\varepsilon > 0$, we begin the proof by writing:

$$\text{Suppose that } \varepsilon > 0.$$

Now that we have this challenger ε in our hands we need to prove that for every number $x \in [0, 1]$ there exists a positive integer N such that for every integer $n \geq N$ we have

$$\frac{nx}{1 + n^2 x^2} < \varepsilon.$$

We therefore continue the proof by writing:

$$\text{Suppose that } x \in [0, 1].$$

Now that we have this challenger x in our hands we need to show that there exists a positive integer N such that for every integer $n \geq N$ we have

$$\frac{nx}{1 + n^2 x^2} < \varepsilon,$$

and we shall demonstrate the existence of such a number N by finding one. In the event that $x = 0$, the desired inequality holds for all values of n and we simply define $N = 1$. We see that

$$\frac{nx}{1 + n^2 x^2} = 0 < \varepsilon$$

for all $n \geq N$. In the event that $0 < x \leq 1$, we choose an integer $N > \frac{1}{\varepsilon x}$ and we observe that for all $n \geq N$,

$$\frac{nx}{1 + n^2 x^2} \leq \frac{nx}{0 + n^2 x^2} = \frac{1}{nx} \leq \frac{1}{Nx} < \varepsilon. \blacksquare$$

3.8.3   Yet Another Fact About Inequalities

In this subsection we show that the assertion 6c that we saw in Example 6 in Subsection 2.1.4 is false. This statement says:

For every number $\varepsilon > 0$ there exists a positive integer N such that for every *number $x \in [0, 1]$ and for every integer $n \geq N$ we have*

$$\frac{nx}{1 + n^2x^2} < \varepsilon.$$

To prove that this statement is false we shall prove that its denial is true. We therefore need to prove the following statement:

There exists a number $\varepsilon > 0$ such that for every positive integer N there exists a number $x \in [0, 1]$ and there exists an integer $n \geq N$ such that

$$\frac{nx}{1 + n^2x^2} \geq \varepsilon.$$

We shall demonstrate the existence of such a number ε by finding one. As a matter of fact, we shall show that the number $1/2$ is an example of such a number ε. We begin by writing:

Define $\varepsilon = 1/2$.

Now we must show that for every positive integer N there exists a number $x \in [0, 1]$ and there exists an integer $n \geq N$ such that

$$\frac{nx}{1 + n^2x^2} \geq \varepsilon,$$

and so we continue by writing:

Suppose that N is any positive integer.

To complete the proof we need to demonstrate the existence of a number $x \in [0, 1]$ and an integer $n \geq N$ such that

$$\frac{nx}{1 + n^2x^2} \geq \varepsilon,$$

and we shall do so by finding two such numbers. We define $n = N$ and $x = 1/N$, and we observe that

$$\frac{nx}{1 + n^2x^2} = \frac{1}{1 + 1} = \varepsilon. \blacksquare$$

3.8.4   Some Additional Exercises

In each of the following exercises, decide whether the statement is true or false and then write a carefully worded proof to justify your assertion.

1. ▦ For every number $x \in [0, 1]$ there exists a positive integer N such that for every number $\varepsilon > 0$ and every integer $n \geq N$ we have

$$\frac{nx}{1 + n^2 x^2} < \varepsilon.$$

2. ▦ For every number $x \in [0, 1]$ and every number $\varepsilon > 0$ there exists a number $\delta > 0$ such that for every number $t \in [0, 1]$ satisfying $|t - x| < \delta$ we have $|t^2 - x^2| < \varepsilon$.

3. ▦ For every number $\varepsilon > 0$ and every number $x \in [0, 1]$ there exists a number $\delta > 0$ such that for every number $t \in [0, 1]$ satisfying $|t - x| < \delta$ we have $|t^2 - x^2| < \varepsilon$.

4. ▦ For every number $\varepsilon > 0$ there exists a number $\delta > 0$ such that for every number $x \in [0, 1]$ and every number $t \in [0, 1]$ satisfying $|t - x| < \delta$ we have $|t^2 - x^2| < \varepsilon$.

5. ▦ For every number $\varepsilon > 0$ and every number x there exists a number $\delta > 0$ such that for every number t satisfying $|t - x| < \delta$ we have $|t^2 - x^2| < \varepsilon$.

6. ▦ For every number $\varepsilon > 0$ there exists a number $\delta > 0$ such that for every number x and every number t satisfying $|t - x| < \delta$ we have $|t^2 - x^2| < \varepsilon$.

7. ▦ For every number $x \in (0, 1]$ there exists a number $\delta > 0$ such that for every number $t \in (0, 1]$ satisfying $|t - x| < \delta$ we have

$$\left| \frac{1}{t} - \frac{1}{x} \right| < 1.$$

8. ▦ There exists a number $\delta > 0$ such that for every number $x \in (0, 1]$ and every number $t \in (0, 1]$ satisfying $|t - x| < \delta$ we have

$$\left| \frac{1}{t} - \frac{1}{x} \right| < 1.$$

9. ▦ For every number $p \in (0, 1]$ there exists a number $\delta > 0$ such that for every number $x \in [p, 1]$ and every number $t \in [p, 1]$ satisfying $|t - x| < \delta$ we have

$$\left| \frac{1}{t} - \frac{1}{x} \right| < 1.$$

10. (a) If either $0 < \theta < \pi$ or $\pi < \theta < 2\pi$, then

$$\arctan\left(\tan\left(\theta/2\right)\right) + \arctan\left(\tan\left(\pi/2 - \theta\right)\right) = \frac{\pi - \theta}{2}.$$

(b) ■ N Ask *Scientific Notebook* to solve the equation

$$\arctan\left(\tan\left(\theta/2\right)\right) + \arctan\left(\tan\left(\pi/2 - \theta\right)\right) = \frac{\pi - \theta}{2}.$$

Are you satisfied with the answer that it gives?

11. If x is any rational number, then

$$\lim_{n \to \infty} \left(\lim_{m \to \infty} \left(\cos\left(n!\pi x\right)\right)^m \right) = 1.$$

12. If x is any irrational number, then

$$\lim_{n \to \infty} \left(\lim_{m \to \infty} \left(\cos\left(n!\pi x\right)\right)^m \right) = 0.$$

Chapter 4
Elements of Set Theory

This chapter provides a brief introduction to those concepts from elementary set theory that we shall need in the rest of the book. Since many of these concepts may already be familiar to you, you should read as much or as little of this chapter as you need. If you are reading the on-screen version of this book, you can find a more extensive presentation of the set theory (that includes the concepts of countability and set equivalence, the Schröder-Bernstein equivalence theorem, and some more advanced topics) by clicking on the icon ▨ .

4.1   Introduction

In the latter part of the nineteenth century Georg Cantor revolutionized mathematics by showing that all mathematical ideas can trace their origins to a single concept: the concept of a *set*. On the face of it, Cantor's notion of a set was very simple: A set is a collection of objects. But the theory of sets that he created out of this idea was so profound that it has become the foundation stone of every branch of mathematics. Using Cantor's theory, mathematicians were able at last to provide precise definitions of important mathematical concepts such as real numbers, points in space, and continuous functions. In addition, Cantor's theory of sets provided a startling insight into the nature of infinity. Perhaps the most significant of his discoveries was the fact that some infinite magnitudes are larger than others.

Our notion of a set in this book will be the same one that Cantor used. We shall think of a set as being a collection of objects, and, when it suits us, we shall replace the word *set* by *class, family,* or *collection*. All of these words will have the same meaning. So, for example, the symbol

$$\{2, -1, 7\}$$

stands for the set whose members are the numbers 2, -1, and 7. When an object x belongs to a given set S we write $x \in S$. For example, $2 \in \{2, -1, 7\}$.

However, before we go any further with this topic we are going to pause briefly to confront some of the logical flaws that lurk in Cantor's definition of a set as a collection of objects. These are the flaws that we mentioned in Section 1.4 and which forced Frege to admit that the foundation of his monumental book had collapsed just as it went to press.

4.1.1 Bertrand Russell's Paradox

The trouble with Cantor's theory of sets is that when we are collecting objects together to make a set we have to allow for the possibility that some of the objects that we are collecting might be sets themselves. For example, the set

$$\{3, 8, \{2, 3\}\}$$

contains the number 3, the number 8, and the set $\{2, 3\}$ whose members are the numbers 2 and 3. Notice how the set $\{2, 3\}$ is a member of the set $\{3, 8, \{2, 3\}\}$. It is therefore entirely possible that a set A may be one of its *own* members. Look at the following two examples:

1. Suppose that E is the set of all cows. Since E is not a cow, it is not one of its own members.
2. Suppose that E is the set of all of those things that we could ever talk about that are not cows. Since this set E is not a cow, we see that this set E is one of its own members.

Mathematicians prefer to avoid sets that are members of themselves. They feel that if we are going to collect some objects together to make a set A, then all of those objects ought to be well known to us *before* we collect them together. Now what if one of those objects is the set A itself? We would need to know what the set A is even before we have collected its members together to define it. So mathematicians dislike the idea of sets that are members of themselves. Note, however, that this dislike is not the devastating paradox of Bertrand Russell. It is merely a warning that something nasty is going on and that a paradox may be lurking somewhere.

We shall call a set A **self-possessed** if A is one of its own members. In other words, A is self-possessed if and only if $A \in A$. Bertrand Russell's idea was to consider the set S of all of those sets A that are not self-possessed. He then asked himself a question:

Is this set S self-possessed?

He discovered that, no matter how one answers this question, the answer leads to a contradiction. In other words, not only does the assumption that S *is* self-possessed lead to a contradiction, but so does the assumption that S *isn't* self-possessed. To see how the two contradictions may be obtained one may reason as follows:

- Suppose that S is self-possessed; in other words, suppose that S belongs to S. Thus S is a self-possessed member of S. But from the definition of S we see that the members of S have to be sets that are not self-possessed, and we have arrived at a contradiction.

- Suppose that S is not self-possessed. Since every set that isn't self-possessed must belong to S, the set S must belong to S, and we conclude that S is self-possessed. Once again we have arrived at a contradiction.

It is this double contradiction that propels us into the vicious circle known as Bertrand Russell's paradox, the paradox that threw the mathematical world into an uproar in the year 1901.

4.1.2   The Zermelo-Fraenkel Axioms

The message of Russell's paradox is that it is the very existence of the set S described in Subsection 4.1.1 that forces a contradiction upon us. Russell's paradox warns us that not every collection of objects should be thought of as a *set,* especially if some of its members are already sets. For this reason, the modern theory of sets rejects Cantor's definition of a set as being simply a collection of objects, and replaces this notion by an abstract idea that is based on a system of axioms. The most common of these systems of axioms is known as the **Zermelo-Fraenkel** system.

Since the modern theory of sets is an advanced topic, we shall not present it here. We shall continue to think of a set as a collection of objects, but we shall keep at the back of our minds the warning that not every such collection should be allowed. Fortunately, the Zermelo-Fraenkel axioms provide us with all of the sets that appear in algebra, analysis, topology, and other branches of mathematics. All of the sets that we shall mention from now on in this book can be verified as legitimate in a Zermelo-Fraenkel system, and we shall therefore cease to worry about paradoxes.

4.2   Sets and Subsets

To begin this section we shall extend our language a little. As we have said, a set is a collection of objects. Any object that belongs to a given set S is said to be a **member** of S, an **element** of S, or a **point** of the set S. If x is a member of a set S, then we write $x \in S$ and say that x **belongs** to S or is **contained** in S. We also say that S **contains** x. If an object x does not belong to a set S, then we write $x \notin S$. Two sets A and B are equal when they contain precisely the same objects. The **empty set** $\emptyset$ is the set that has no members at all.[13]

4.2.1   Subsets of a Given Set

If A and B are two given sets, then we say that A is a **subset** of B or, alternatively, that A is **included** in B if every member of A is also a member of B. When A

---

[13]   The symbol $\emptyset$ can be read aloud as "emptyset".

is a subset of B we write $A \subseteq B$, and we also write this condition as $B \supseteq A$ and say that B **includes** A. Figure 4.1 shows how we may picture the condition $A \subseteq B$. Note that two sets A and B are equal if and only if both of the conditions $A \subseteq B$ and $B \subseteq A$ hold. In the event that $A \subseteq B$ and the sets A and B are not equal, we say that A is **properly included** in B and that A is a **proper subset** of B.

If A and B are any two sets, then the denial of the condition $A \subseteq B$ says that there exists $x \in A$ such that $x \notin B$. In the event that $A = \emptyset$, we certainly can't find a member $x \in A$ such that $x \notin B$ because there are no members of A to find. We therefore conclude that $\emptyset \subseteq B$ for every set B.

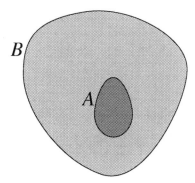

Figure 4.1

4.2.2   Equality of Sets

Two sets A and B are equal when they contain precisely the same members. Thus if A and B are two given sets, then the condition $A = B$ says that every member of A must belong to B and every member of B must belong to A. In other words, $A = B$ is equivalent to the condition that $A \subseteq B$ and also $B \subseteq A$.

4.2.3   The Power Set of a Given Set

Given any set A, the family of all subsets of A is called the power set of A and is written as $p(A)$. For a simple example, observe that

$$p(\{1,2,3\}) = \{\emptyset, \{1\}, \{2\}, \{3\}, \{1,2\}, \{1,3\}, \{2,3\}, \{1,2,3\}\}.$$

4.2.4   Set Builder Notation

We have already used notation such as $\{2, 7, -3\}$, which stands for the set whose members are the numbers 2, 7, and -3. This is a form of what is known as **set builder notation**. There is also another form of set builder notation. Suppose that $P(x)$ is a statement that contains a single unknown x. The notation $\{x \mid P(x)\}$

stands for the set of all of those objects x for which the statement $P(x)$ is true. This notation is particularly useful to describe subsets of a given set. The following examples illustrate this notation:

1. Writing the set of all real numbers[14] as $\mathbf{R}$ and taking $P(x)$ to be the condition $-3 \leq x < 2$, we have

$$\{x \in \mathbf{R} \mid P(x)\} = \{x \in \mathbf{R} \mid -3 \leq x < 2\} = [-3, 2).$$

2. Taking $P(x)$ to be the condition $8x^3 - 6x - 1 \geq 0$, we have

$$\{x \in \mathbf{R} \mid P(x)\} = \{x \in \mathbf{R} \mid 8x^3 - 6x - 1 \geq 0\}.$$

3. In this example we refer to the set $\mathbf{Q}$ of all rational numbers. The set

$$\{x \in \mathbf{Q} \mid x < 0 \text{ or } x^2 < 2\}$$

is the set of all of those rational numbers that are less than $\sqrt{2}$.

4.2.5   The Union of Two Sets

Suppose that A and B are given sets. The symbol $A \cup B$ stands for the set of all of those objects that belong to at least one of the two sets A and B and is called the **union** of A and B. Thus

$$A \cup B = \{x \mid x \in A \text{ or } x \in B\}.$$

4.2.6   The Intersection of Two Sets

Suppose that A and B are given sets. The symbol $A \cap B$ stands for the set of all of those objects that belong to *both* of the sets A and B and is called the **intersection** of the sets A and B. Thus

$$A \cap B = \{x \mid x \in A \text{ and } x \in B\}.$$

In the event that $A \cap B = \emptyset$, we say that the sets A and B are **disjoint** from each other.

4.2.7   The Difference of Two Sets

Suppose that A and B are given sets. The **difference** $A \setminus B$ of the sets is defined by

$$A \setminus B = \{x \in A \mid x \notin B\}.$$

---

[14]   In this chapter we draw freely from the set $\mathbf{R}$ of real numbers, the set $\mathbf{Q}$ of rational numbers, the set $\mathbf{Z}$ of integers, and the set $\mathbf{Z}^+$ of natural numbers for our examples. A more careful presentation of these sets can be found in the chapters that follow.

Another way of looking at $A \setminus B$ is to say that

$$A \setminus B = \{x \mid x \in A \text{ and } x \notin B\}.$$

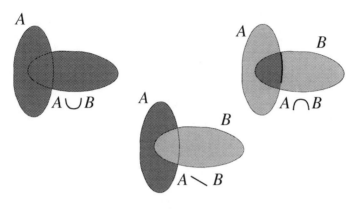

Figure 4.2

Figure 4.2 shows how we may picture the union, intersection and difference of two sets A and B.

4.2.8   The Cartesian Product of Two Sets

If A and B are any two sets, then their **Cartesian product** $A \times B$ is the set of all ordered pairs (x, y) that have $x \in A$ and $y \in B$. In other words,

$$A \times B = \{(x, y) \mid x \in A \text{ and } y \in B\}.$$

4.2.9   Some Common Sets

1. The set of all real numbers is usually written as $\mathbf{R}$ or $\mathbb{R}$.
2. The set of all rational numbers is usually written as $\mathbf{Q}$ or $\mathbb{Q}$.
3. The set of all integers is usually written as $\mathbf{Z}$ or $\mathbb{Z}$.
4. The set of all positive integers is usually written as $\mathbf{Z}^+$, or $\mathbb{Z}^+$, or $\mathbf{N}$, or $\mathbb{N}$.
5. If a and b are real numbers and $a \le b$, then we shall use the standard interval notation

$$
\begin{aligned}
[a, b] &= \{x \in \mathbf{R} \mid a \le x \le b\} \\
[a, b) &= \{x \in \mathbf{R} \mid a \le x < b\} \\
(a, b] &= \{x \in \mathbf{R} \mid a < x \le b\} \\
(a, b) &= \{x \in \mathbf{R} \mid a < x < b\}.
\end{aligned}
$$

As you may know, an interval of the form $[a, b]$ is called a **closed interval** and an interval of the form (a, b) is called an **open interval.** We can also define intervals of infinite length: If a is any real number, then we define

$$
\begin{aligned}
[a, \infty) &= \{x \in \mathbf{R} \mid a \leq x\} \\
(a, \infty) &= \{x \in \mathbf{R} \mid a < x\} \\
(-\infty, a] &= \{x \in \mathbf{R} \mid x \leq a\} \\
(-\infty, a) &= \{x \in \mathbf{R} \mid x < a\},
\end{aligned}
$$

and finally, the symbol $(-\infty, \infty)$ is the set $\mathbf{R}$ of all real numbers. Note that if a is any real number then the interval (a, a) is the empty set $\emptyset$.

6. The set $\mathbf{R} \times \mathbf{R}$ of all ordered pairs of real numbers is called the **Euclidean plane** and is also written as $\mathbf{R}^2$.

4.2.10   Exercises on Set Notation

1. ◫ Given objects a,b, and y and given that $\{a, b\} = \{a, y\}$, prove that $b = y$.
2. ◫ Prove that if a, b, x, and y are any given objects and if $\{a, b\} = \{x, y\}$, then either $a = x$ and $b = y$, or $a = y$ and $b = x$.
3. Prove that if a, b, x, and y are any given objects and if

$$
\{\{a\}, \{a, b\}\} = \{\{x\}, \{x, y\}\},
$$

then $a = x$ and $b = y$.
4. ◫ Describe the set $p(\emptyset)$.
5. ◫ Describe the set $p(p(\emptyset))$.
6. ◫ Given that $A = \{a, b, c, d\}$, list all of the members of the set $p(A)$.
7. Given that $A = \{a, b\}$, list all of the members of the set $p(p(A))$.
8. Ⓝ Use the Evaluate operation in *Scientific Notebook* to evaluate the sets $\{1, 2, 3\} \cap \{2, 3, 4\}$ and $\{1, 2, 3\} \cup \{2, 3, 4\}$.

4.2.11   The DeMorgan Laws

Suppose that A, B, and C are given sets. The following simple identities are known as the **DeMorgan laws:**

1. $A \cup (B \cap C) = (A \cup B) \cap (A \cup C)$.
2. $A \cap (B \cup C) = (A \cap B) \cup (A \cap C)$.
3. $A \setminus (B \cup C) = (A \setminus B) \cap (A \setminus C)$.
4. $A \setminus (B \cap C) = (A \setminus B) \cup (A \setminus C)$.

Proof. We shall prove law 2 and leave the proofs of the other three laws as exercises. The proof that we are about to give is highly detailed. When you prove the other three laws, write your proofs first in as much detail as is given here. Then rewrite them more briefly.

To prove law 2 we shall begin by proving that

$$A \cap (B \cup C) \subseteq (A \cap B) \cup (A \cap C).$$

We want to prove that every member of the left side must belong to the right side. Since we want to prove a "for every" statement, we begin our proof with the words

Suppose that $x \in A \cap (B \cup C)$.

We know that $x \in A$ and also that $x \in B \cup C$. In other words, we know that $x \in A$ and either $x \in B$ or $x \in C$. Thus either $x \in A$ and $x \in B$ or otherwise $x \in A$ and $x \in C$. Since the first of these two possibilities says that $x \in A \cap B$ and the second possibility says that $x \in A \cap C$, we therefore know that x belongs to at least one of the two sets $A \cap B$ and $A \cap C$; in other words,

$$x \in (A \cap B) \cup (A \cap C).$$

We have therefore shown that

$$A \cap (B \cup C) \subseteq (A \cap B) \cup (A \cap C),$$

and we shall now complete the proof of law 2 by showing that

$$(A \cap B) \cup (A \cap C) \subseteq A \cap (B \cup C).$$

We want to prove that every member of the left side must belong to the right side. Since we want to prove a "for every" statement, we begin our proof with the words

Suppose that $x \in (A \cap B) \cup (A \cap C)$.

We know that x must belong to at least one of the sets $A \cap B$ and $A \cap C$. Since the given information says that one or the other of two conditions must hold, we proceed according to the method described in Subsection 3.3.3. We have two tasks to perform:

(a) We need to show that if $x \in A \cap B$, then $x \in A \cap (B \cup C)$.
(b) We need to show that if $x \in A \cap C$, then $x \in A \cap (B \cup C)$.

To perform the first of these two tasks we suppose that $x \in A \cap B$. Since $x \in A$ and $x \in B$, we know that $x \in A$ and that x belongs to at least one of the sets B and C. So in this case we certainly have $x \in A \cap (B \cup C)$.

To perform the second of these two tasks we suppose that $x \in A \cap C$. Since $x \in A$ and $x \in C$, we know that $x \in A$ and that x belongs to at least one of the sets B and C. So in this case we certainly have $x \in A \cap (B \cup C)$.

4.2.12   Exercises on Set Operations

1. Express the set $[-2, 3] \setminus (0, 1]$ as the union of two intervals.
2. Given two sets A and B, prove that the condition $A \subseteq B$ is equivalent to the condition $A \cup B = B$.
3. Given two sets A and B, prove that the condition $A \subseteq B$ is equivalent to the condition $A \cap B = A$.
4. Given two sets A and B, prove that the condition $A \subseteq B$ is equivalent to the condition $A \setminus B = \emptyset$.
5. Illustrate the identity

$$A \setminus (B \cup C) = (A \setminus B) \cap (A \setminus C)$$

by drawing a figure. Then write out a detailed proof.
6. Illustrate the identity

$$A \setminus (B \cap C) = (A \setminus B) \cup (A \setminus C)$$

by drawing a figure. Then write out a detailed proof.
7. Given that A, B, and C are subsets of a set X, prove that the condition $A \cap B \cap C = \emptyset$ holds if and only if

$$(X \setminus A) \cup (X \setminus B) \cup (X \setminus C) = X.$$

8. ▦ Given sets A, B, and C, determine which of the following identities are true.
    (a) $A \cap (B \setminus C) = (A \cap B) \setminus (A \cap C)$.
    (b) $A \cup (B \setminus C) = (A \cup B) \setminus (A \cup C)$.
    (c) $A \cup (B \setminus C) = (A \cup B) \cap (A \setminus C)$.
    (d) $A \cup (B \setminus C) = (A \cup B) \setminus (A \cap C)$.
    (e) $A \setminus (B \setminus C) = (A \setminus B) \setminus C$.
    (f) $A \setminus (B \setminus C) = (A \setminus B) \setminus (A \setminus C)$.
    (g) $A \setminus (B \setminus C) = (A \setminus B) \cap (A \setminus C)$.
    (h) $A \setminus (B \setminus C) = (A \setminus B) \cup (A \setminus C)$.
    (i) $A \setminus (B \setminus C) = (A \setminus B) \cup (A \cap C)$.
    (j) $A \setminus (B \setminus C) = (A \setminus B) \cap (A \cup C)$.
    (k) $A = (A \cap B) \cup (A \setminus B)$.
    (l) $p(A \cup B) = p(A) \cup p(B)$.

(m)$p(A \cap B) = p(A) \cap p(B)$.

(n) $A \times (B \cup C) = (A \times B) \cup (A \times C)$.

(o) $A \times (B \cap C) = (A \times B) \cap (A \times C)$.

(p) $A \times (B \setminus C) = (A \times B) \setminus (A \times C)$.

9. 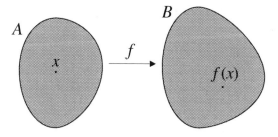 Is it true that if A and B are sets and $A = A \setminus B$, then the sets A and B are disjoint from each other?

10. Given that A is a set with 10 members, B is a set with 7 members, and that the set $A \cap B$ has 4 members, how many members does the set $A \cup B$ have?

11. ◩ Give an example of a set A that contains at least three members and that satisfies the condition $A \subseteq p(A)$.

12. ◩ For which sets A do we have $A \in p(A)$?

4.3   Functions

4.3.1   Intuitive Definition of a Function

In this brief presentation of set theory we shall be content with an intuitive view of a function f defined on a set A as a rule that assigns to each member x of the set A a unique object written as $f(x)$ and called the **value** of the function f at x.

If f is a function with domain A and if $f(x) \in B$ for every $x \in A$, then we say that f is a function **from** A **to** B and we write $f : A \rightarrow B$. When f is a function from A to B we also say that f **maps** A to B. Figure 4.3 shows how we may picture this idea.

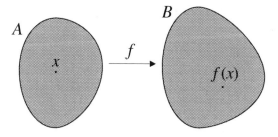

Figure 4.3

4.3.2   Domain of a Function

If f is a function defined on a set A, then the set A is said to be the **domain** of the function f. Thus the domain of a function f is the set of all objects x for which the symbol $f(x)$ is defined.

4.3.3   Range of a Function

If f is a function defined on a set A, then the set

$$\{f(x) \mid x \in A\}$$

is called the **range** of the function f. Note that if $f : A \to B$, then the range of f has to be a subset of the set B.

4.3.4   Some Examples of Functions

1. Suppose that we have defined

    $$f(x) = x^2$$

    for every number x. Then f is a function defined on the set $\mathbf{R}$ of all real numbers. The set $\mathbf{R}$ is the domain of f and the range of f is $[0, \infty)$.
2. Suppose that we have defined

    $$f(x) = x^2$$

    for every number $x \in [-1, 3]$. Then f is a function defined on $[-1, 3]$ and this interval is the domain of f. The range of f is the interval $[0, 9]$.
3. Suppose that C is a bag of jelly beans, any one of which can be yellow, green, red, or blue. Suppose that for every $x \in C$, the symbol $f(x)$ is defined to be the color of the jelly bean x. Then f is a function defined on C and the range of f is the set that contains the four words, yellow, green, red, and blue.

4.3.5   Restriction of a Function to a Set

Suppose that $f : A \to B$ and that E is any subset of A. The **restriction** of f to E is defined to be the function g from E to B defined by $g(x) = f(x)$ for every $x \in E$. Furthermore, the **image** $f[E]$ **of** E **under** f is defined by

$$f[E] = \{f(x) \mid x \in E\}.$$

Figure 4.4 shows how we may picture this idea. Note that if A is the domain of a function f, then the range of f is the set $f[A]$.

4.3.6   Preimage of A Set

Suppose that $f : A \to B$. If E is any subset of B, then the **preimage of** E **under** f is the set $f^{-1}[E]$ defined by

$$f^{-1}[E] = \{x \in A \mid f(x) \in E\}.$$

Figure 4.5 illustrates the idea of a preimage.

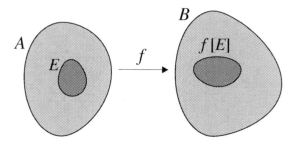

Figure 4.4

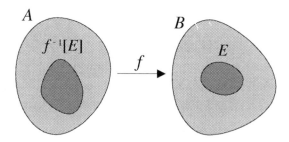

Figure 4.5

Note that the idea of a preimage does not depend on the idea of an inverse function f^{-1}, which will be defined in Subsection 4.3.12.

4.3.7   One-One Functions

A given function f is said to be **one-one** if it is impossible to find two different members x_1 and x_2 in the domain of f for which $f(x_1) = f(x_2)$. Another way of saying that a given function f is one-one is to say that whenever x_1 and x_2 belong to the domain of f and $x_1 \neq x_2$ we must have $f(x_1) \neq f(x_2)$. One-one functions are sometimes called **injective.**

Since no two different real numbers can have the same cube, we see that if $f(x) = x^3$ for every $x \in \mathbf{R}$, then the function f is one-one. However, if $f(x) = x^2$ for every $x \in \mathbf{R}$, then f is not one-one because $f(2) = f(-2)$. If f is a function from a set of real numbers into $\mathbf{R}$, then we can picture the condition that f be one-one as saying that no horizontal line can meet the graph of f more than once. Figure 4.6 illustrates the graph of a one-one function. Notice how some horizontal lines meet the graph of this function exactly once while others don't meet it at all.

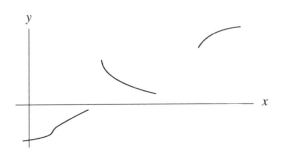

Figure 4.6

4.3.8   Monotone Functions

Suppose that S is a set of real numbers and that $f : S \to \mathbf{R}$. We say that the function f is **increasing** if whenever t and x belong to S and $t < x$ we have $f(t) \leq f(x)$. We can picture an increasing function as one whose graph never falls as we move from left to right. The graph of this type of function is illustrated in Figure 4.7. If the graph of f actually rises as we move from

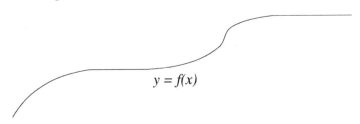

$$y = f(x)$$

Figure 4.7

left to right, then we say that the function f is **strictly increasing.** Thus the function f is strictly increasing if whenever t and x belong to S and $t < x$ we have $f(t) < f(x)$. Decreasing functions and strictly decreasing functions are defined similarly. A function that is either increasing or decreasing is said to be **monotone**, and a function that is either strictly increasing or strictly decreasing is said to be **strictly monotone.**

We see at once that a strictly monotone function is always one-one. However, a one-one function does not have to be strictly monotone. Figure 4.6 illustrates the graph of a one-one function that fails to be monotone.

4.3.9   Functions onto a Given Set

Suppose that f is a function from a set A to a set B. The range $f[A]$ of f is a subset of B. In the event that the range of f is the entire set B, we say that the function f is **onto** the set B. In other words, f is onto the set B if $f[A] = B$.

Another way of saying this is to say that for every member y of B there is at least one member x of A such that $y = f(x)$.

Suppose, for example, that $f(x) = x^2$ for every $x \in \mathbf{R}$. Although $f : \mathbf{R} \rightarrow \mathbf{R}$, the function f is not onto the set $\mathbf{R}$. On the other hand, f is onto the interval $[0, \infty)$.

4.3.10 A Remark About Terminology

In Subsection 4.3.7 we mentioned that a one-one function is sometimes called an *injective function*. When we say that a given function is injective we are saying that the function is of a certain special type. In other words, the word *injective* is an adjective. We can say that a particular function is injective in exactly the same way that we can say that a leaf is green.

However, unlike the concept of a one-one function, the concept that we studied in Subsection 4.3.9 does not describe a property of functions. When we say that a given function f is onto a given set B we are not only talking about the function f; we are talking about the function f and the set B. For example, if $f(x) = x^2$ for every real number x, then f fails to be onto the set $\mathbf{R}$ of real numbers even though it is onto the set $[0, \infty)$.

We can ask whether a given function f is onto a given set B but we can never ask whether or not a given function is *onto;* because such a question makes no sense. It makes no more sense than it would make if you were to ask me whether I am sitting *on*. I would have to answer such a question by asking: "Sitting on *what?*" The point of this remark is that the word *onto* is not an adjective. It is a preposition and we need to keep in mind that one of the fundamental rules of grammar prohibits the use of a preposition at the end of a sentence.

In some mathematical writing, the word *surjective* is used to describe the fact that a given function is "onto" some unspecified set. This word *surjective* is misleading because it looks like an adjective even though it is not one, and it will not be used in this book.

4.3.11 Composition of Functions

If f is a function from a set A to a set B and g is a function from B to a set C, then the **composition** $g \circ f$ of f and g is the function from A to C defined by

$$(g \circ f)(x) = g(f(x))$$

for every $x \in A$. Figure 4.8 illustrates this idea. For example, if $f(x) = x^2$ for every $x \in \mathbf{R}$ and $g(x) = 1 + 2x$ for every $x \in [0, \infty)$, then $(g \circ f)(x) = 1 + 2x^2$ for every $x \in \mathbf{R}$ and $(f \circ g)(x) = (1 + 2x)^2$ for every $x \in [0, \infty)$.

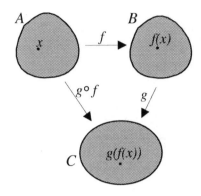

Figure 4.8

4.3.12   Inverse Functions

Suppose that f is a one-one function from a set A onto a set B. Given any member y of B there is one and only one member x of A such that $y = f(x)$. If for each member y of the set B we define $g(y)$ to be the one and only one member x of A for which $y = f(x)$, then we have defined a function $g : B \to A$. This function g is called the **inverse function** of f and is written as f^{-1}.

Note that $f^{-1}\left(f(x)\right) = x$ for all $x \in A$ and $f\left(f^{-1}(y)\right) = y$ for all $y \in B$.

Note finally that if E is any subset of B, then the expression $f^{-1}\left[E\right]$ could be interpreted both as the preimage of the set E under the function f and also as the image under the function f^{-1} of the set E. Fortunately, these two interpretations yield the same set.

4.3.13   Inverse Functions Are also One-One

Suppose that f is a one-one function from a set A onto a set B. Then the function f^{-1} is also one-one. Moreover, $\left(f^{-1}\right)^{-1} = f$.

Proof. To show that f^{-1} is one-one we need to show that if y_1 and y_2 belong to B and $f^{-1}(y_1) = f^{-1}(y_2)$, then $y_1 = y_2$. Suppose that y_1 and y_2 belong to B and $f^{-1}(y_1) = f^{-1}(y_2)$. Then we have

$$y_1 = f\left(f^{-1}(y_1)\right) = f\left(f^{-1}(y_2)\right) = y_2.$$

Finally, given $x \in A$, the symbol $\left(f^{-1}\right)^{-1}(x)$ stands for the member of B that f^{-1} sends to x. But since $f^{-1}\left(f(x)\right) = x$ we know that this member of B is $f(x)$. Therefore $f(x) = \left(f^{-1}\right)^{-1}(x)$ and we conclude that $\left(f^{-1}\right)^{-1} = f$. ■

4.3.14   Some Examples of Inverse Functions

1. If we define $f(x) = x^2$ for every $x \in [0, \infty)$, then f is a one-one function from $[0, \infty)$ onto $[0, \infty)$ and for every $y \in [0, \infty)$ we have $f^{-1}(y) = \sqrt{y}$.
2. If we define $f(x) = \tan x$ for every $x \in (-\pi/2, \pi/2)$, then f is a one-one function from $(-\pi/2, \pi/2)$ onto $\mathbf{R}$ and for every real number y we have $f^{-1}(y) = \arctan y$.
3. Suppose that $a \in \mathbf{R} \setminus \{-1, 1\}$. We begin with the observation that if x is any number unequal to $1/a$ and $y \neq -1/a$, then the equations

$$y = \frac{x - a}{1 - ax}$$

and

$$x = \frac{y + a}{1 + ay}.$$

are equivalent statements. On the other hand, it is easy to see that if either $x = 1/a$ or $y = -1/a$, then both of these equations are impossible. From this observation we deduce that the function f defined by

$$f(x) = \frac{x - a}{1 - ax}$$

for all $x \in \mathbf{R} \setminus \{1/a\}$ is a one-one function from $\mathbf{R} \setminus \{1/a\}$ onto the set $\mathbf{R} \setminus \{-1/a\}$. Furthermore, for every $y \in \mathbf{R} \setminus \{-1/a\}$, we have

$$f^{-1}(y) = \frac{y + a}{1 + ay}.$$

4. This example makes use of a little elementary calculus. Given any number a satisfying $-1 < a < 1$, we define the function f_a on the interval $[-1, 1]$ by defining

$$f_a(x) = \frac{x - a}{1 - ax}$$

for every number $x \in [-1, 1]$. From the observations that we made in Example 3 we know that each of these functions f_a is one-one. Since these functions are rational functions, they are also continuous on the interval $[-1, 1]$.

   Now suppose that $-1 < a < 1$. Because f_a is one-one and because $f_a(-1) = -1$ and $f_a(1) = 1$, the number $f_a(x)$ cannot be equal to -1 or 1 for any x in the open interval $(-1, 1)$. It therefore follows from the elementary properties of continuous functions that either $f_a(x) < -1$ for every $x \in (-1, 1)$, or $-1 < f_a(x) < 1$ for every $x \in (-1, 1)$, or $f_a(x) > 1$

for every $x \in (-1, 1)$. But $f_a(0) = -a \in (-1, 1)$, and so we conclude that

$$f_a : [-1, 1] \to [-1, 1].$$

Finally, to see that f_a is onto the interval $[-1, 1]$ we observe that if $y \in [-1, 1]$, then $f_{-a}(y) \in [-1, 1]$ and it is easy to see that $y = f_a(f_{-a}(y))$. We deduce that for each a, the function f_a is one-one from $[-1, 1]$ onto $[-1, 1]$ and that f_{-a} is the inverse function of f_a.

4.3.15   Exercises on Functions

1. Given that $f(x) = x^2$ for every real number x, simplify the following expressions:

    (a) $f[[0, 3]]$.
    (b) $f[[(-2, 3]]$.
    (c) $f^{-1}[[-3, 4]]$.

2. 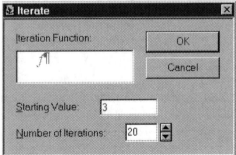 Point at the equation $f(x) = x^2$ and then click on the button $f(x)$ in your computing toolbar. Then work out the expressions in parts (a) and (b) of Exercise 1 by pointing at them and clicking on the evaluate button.

3. Supply each of the definitions $f(x) = x^2$ and $g(x) = 2 - 3x$ to *Scientific Notebook* and then ask *Scientific Notebook* to solve the equation

    $$(f \circ g)(x) = (g \circ f)(x).$$

4. Supply the definition

    $$f(x) = \frac{x - 2}{1 - 2x}$$

    to *Scientific Notebook*. In this exercise we shall see how to evaluate the composition of the function f with itself up to 20 times starting at a variety of numbers. Open the Compute menu, click on Calculus, and move to the right and select Iterate.In the iterate dialogue box fill in the function as f, the starting value as 3, and the number of iterations as 20. Repeat this process with different starting values. Can you draw a conclusion from what you see?

5. Given that $f(x) = x^2$ for all $x \in \mathbf{R}$ and $g(x) = 1 + x$ for all $x \in \mathbf{R}$, simplify the following expressions:

    (a) $(f \circ g)[[0, 1]]$.
    (b) $(g \circ f)[[0, 1]]$.
    (c) $(g \circ g)[[0, 1]]$.

6. (a) Given that $f(x) = (3x - 2)/(x + 1)$ for all $x \in \mathbf{R} \setminus \{-1\}$, determine whether or not f is one-one and find its range.

    (b) (N) Point at the equation

    $$y = \frac{3x - 2}{x + 1}$$

    and ask *Scientific Notebook* to solve for x. How many values of x are given? Is this result consistent with the answer that you gave in part (a) of the question?

7. Suppose that $f : A \to B$ and that $E \subseteq A$. Is it true that $E = f^{-1}[f[E]]$? What if f is one-one? What if f is onto B?

8. Suppose that $f : A \to B$ and that $E \subseteq B$. Is it true that $E = f[f^{-1}[E]]$? What if f is one-one? What if f is onto B?

9. (H) Suppose that $f : A \to B$ and that P and Q are subsets of B. Prove the identities

$$f^{-1}[P \cup Q] = f^{-1}[P] \cup f^{-1}[Q]$$

$$f^{-1}[P \cap Q] = f^{-1}[P] \cap f^{-1}[Q]$$

$$f^{-1}[P \setminus Q] = f^{-1}[P] \setminus f^{-1}[Q].$$

10. (H) Suppose that $f : A \to B$ and that P and Q are subsets of A. Which of the following statements are true? What if f is one-one? What if f is onto B?

$$f[P \cup Q] = f[P] \cup f[Q].$$

$$f[P \cap Q] = f[P] \cap f[Q].$$

$$f[P \setminus Q] = f[P] \setminus f[Q].$$

11. ■ Given that f is a one-one function from A to B and that g is a one-one function from B to C, prove that the function $g \circ f$ is one-one from A to C.

12. Given that f is a function from A onto B and that g is a function from B onto C, prove that the function $g \circ f$ is a function from A onto C.

13. Given that $f : A \to B$, that $g : B \to C$, and that the function $g \circ f$ is one-one, prove that f must be one-one. Give an example to show that the function g does not have to be one-one.

14. ■ Given that f is a function from A onto B, that $g : B \to C$, and that the function $g \circ f$ is one-one, prove that both of the functions f and g have to be one-one.

15. Given any set S, the **identity function** i_S on S is defined by $i_S(x) = x$ for every $x \in S$. Prove that if f is a one-one function from a set A onto a set B, then $f^{-1} \circ f = i_A$ and $f \circ f^{-1} = i_B$.

16. Suppose that $f : A \to B$.

    (a) Given that there exists a function $g : B \to A$ such that $g \circ f = i_A$, what can be said about the functions f and g?

    (b) Given that there exists a function $h : B \to A$ such that $f \circ h = i_B$, what can be said about the functions f and h?

    (c) Given that there exists a function $g : B \to A$ such that $g \circ f = i_A$ and that there exists a function $h : B \to A$ such that $f \circ h = i_B$, what can be said about the functions f, g, and h?

17. As in Example 4 of Subsection 4.3.14, we define

$$f_a(x) = \frac{x - a}{1 - ax}$$

whenever $a \in (-1, 1)$ and $x \in [-1, 1]$.

    (a) Prove that if a and b belong to $(-1, 1)$, then so does the number

$$c = \frac{a + b}{1 + ab}.$$

Hint: A quick way to do this exercise is to observe that $c = f_{-b}(a)$.

    (b) Given a and b in $(-1, 1)$ and

$$c = \frac{a + b}{1 + ab},$$

prove that $f_b \circ f_a = f_c$.

PART II

Elementary Concepts of Analysis

This part of the text introduces the basic principles of analysis and provides a careful introduction to the concepts of limit, continuity, derivative, integral, and infinite series. We begin with a chapter on the real number system upon which all of these concepts depend.

Please insert a blank page here.

Chapter 5
The Real Number System

The **real number system** or "real line" $\mathbf{R}$ is the lifeblood of calculus, and so the level at which we can understand calculus is dependent upon how well we understand the system $\mathbf{R}$. Since real numbers play a central role in every branch of mathematics, you are already familiar with the number line, but that familiarity may not be a sufficiently strong foundation upon which to build the theory of calculus. This chapter will allow you to deepen your understanding of the number line and, in so doing, will open the door to the chapters that follow.

5.1 Introduction to the System $\mathbf{R}$

5.1.1 Philosophical Introduction to the System $\mathbf{R}$

In approaching the concept of a number we are confronted by some important questions: *What are numbers? Do they exist in the physical world? Did we find them or did we have to invent them?* There are two main philosophical approaches to this question:

1. The real numbers are *there*. They were there long before our species first began to roam this planet, and they will be there long after we are gone. Don't ask what they are. Work with them. Study them. Study them in order to verify that the system $\mathbf{R}$ has the properties that we want to use in other branches of mathematics. Then get on with the job of doing mathematics.

2. Real numbers do not have any existence in the physical world. They are just figments of our imaginations. In order to use them in mathematics we must invent them and ask ourselves:

   ▶ *How are real numbers defined, and what do we need to assume in order to be able to define them?*

   ▶ *What fundamental properties of the number system $\mathbf{R}$ can we deduce from our definitions?*

   ▶ *What role is played by these fundamental properties as we develop algebra, calculus (analysis), topology, and other fields of mathematics?*

Modern mathematics usually takes the point of view that we have to *invent* the real numbers, but we cannot undertake this process of invention with empty hands. The most common starting point for the real number system and all of the mathematics that depends upon it is a fundamental system of axioms known as the **Zermelo-Fraenkel** system of axioms. Starting with these axioms, it is possible to *define* the system $\mathbf{Z}^+$ of positive integers and to extend $\mathbf{Z}^+$ to obtain the system $\mathbf{Z}$ of integers and the system $\mathbf{Q}$ of rational numbers. Once the system $\mathbf{Q}$ of rational numbers has been constructed, one may construct the system $\mathbf{R}$ of real numbers by a method known as the method of **Dedekind cuts**.

Unfortunately, all of this construction takes considerably more time than can be allocated in the courses for which this book has been written. It must be left for a more advanced course. Therefore, instead of attempting to *define* the real number system, we shall assume its existence and we shall write down a list of its fundamental properties that can be used to deduce all of its other properties.

5.1.2   Intuitive Introduction to the System $\mathbf{R}$

Most of us come to understand the idea of a real number quite gradually as we progress from childhood to adulthood. We become aware of the positive integers first. Then, as our need for numbers develops, we become aware of more complicated numbers like negative numbers and rationals. As little children we become aware of positive integers when we start counting precious commodities like candies. The process of hoarding candies teaches us about addition and multiplication. When we eat the candies we learn that subtraction can be performed under certain restricted conditions. (You can't eat what you don't have.) The process of sharing them forces us to confront the operation of division. The idea of a rational number enters our consciousness the first time we have two candy bars to be shared among three people. Somewhat more difficult is the notion of a negative number, but even these are injected into our young lives. Do negative numbers really exist, or are they merely figments of our imaginations? Ask any little kid who has already spent next week's allowance.

When we have accepted both fractions and negative numbers into our lives, our concept of a number is that of a *rational* number. Recall that a number is said to be **rational** if it can be written as a ratio m/n, where m and n are integers. The set of rational numbers is usually written as $\mathbf{Q}$. For a while, the number system $\mathbf{Q}$ seems to be perfectly adequate, but eventually its defects come to light. Not only is the system $\mathbf{Q}$ inadequate for advanced mathematics, but it doesn't even support all of the activity that makes up the basic bill of fare of middle and high school mathematics courses. For a nice description of this basic bill of fare, look at the following excerpt from the song of the "Modern Major General" in Gilbert and Sullivan's *Pirates of Penzance:*

> I am very well acquainted, too, with matters mathematical,
>
> I understand equations, both the simple and quadratical,
>
> About binomial theorem I'm teeming with a lot of news,
>
> With many cheerful facts about the square of the hypotenuse.

The moment we began to solve quadratic equations like

$$x^2 - 2x - 5 = 0, \tag{5.1}$$

irrational numbers entered our lives. And one of the cheerful facts about the square of the hypotenuse introduced the number $\sqrt{2}$ into geometry resulting in death by shipwreck (see Section 1.2). This part of the "Modern Major General's" song refers to a richer number system than the system **Q** of rational numbers. To gain some idea of what kind of numbers are needed, look at Equation (5.1) again. As you know, its solution is $x = 1 \pm \sqrt{6}$. Now consider the slightly more complicated equation

$$x^6 - 2x^3 - 5 = 0.$$

This equation is also a quadratic equation. It is quadratic in x^3, and its solution is

$$x = \sqrt[3]{1 + \sqrt{6}} \quad \text{or} \quad x = \sqrt[3]{1 - \sqrt{6}}.$$

The numbers needed by a "Modern Major General" are the numbers that we call **surds**. These are the numbers that one can assemble starting with positive integers and using the operations $+, -, \times, \div,$ and $\sqrt[n]{\ }$. Numbers of this type can be pretty complicated. For example,

$$\sqrt[17]{\frac{\left(3 + \sqrt{\frac{5 - \sqrt{3}}{2 + \sqrt[3]{7}}}\right)\left(\sqrt[7]{2} + \sqrt{9 - \sqrt[5]{\frac{\frac{1}{1 + \sqrt[5]{5}} + \sqrt{2}}{20 - \sqrt[3]{5}}}}\right)}{3 + \sqrt[7]{\frac{4 - \sqrt{3}}{8 - \sqrt[3]{3 + \sqrt{5}}}} - \sqrt[11]{\frac{3 + \sqrt{\frac{2 + \sqrt[3]{5}}{6 + 5\sqrt[3]{7}}}}{8 - \sqrt{\frac{7 + 5\sqrt{3}}{6 + 4\sqrt[3]{2}}}}}}} + \sqrt[12]{6 - \frac{1 + \sqrt[4]{3}}{4 + \sqrt[5]{4} + \sqrt{5}}}.$$

The number system of surds is a much richer number system than the system **Q** of rational numbers. In the system of surds we can add, subtract, multiply, and divide just as we can add, subtract, multiply, and divide rational numbers. But we can do one more thing: We can apply a radical sign $\sqrt[n]{\ }$ to any positive surd and to yield another surd. This extra property of the system of surds enables us to solve quadratic equations; for if $a, b,$ and c are surds and $a \neq 0$, then the solution

$$x = \frac{-b \pm \sqrt{b^2 - 4ac}}{2a}$$

of the quadratic equation

$$ax^2 + bx + c = 0$$

is a surd as long as $b^2 - 4ac \geq 0$. Surds can also be used to solve some equations that are too hard for most modern major generals. For example, it can be shown that the solution of the equation

$$x^6 - 6x^2 - 6 = 0$$

is

$$x = \pm\sqrt{\sqrt[3]{2} + \sqrt[3]{4}}.$$

From the perspective of elementary mathematics, the system of surds seems to allow us to do anything we might want to do with numbers. It seems to contain all of the numbers that could possibly exist, and, indeed, many people who have not studied advanced mathematics are unaware that there *are* any other kinds of numbers.

But there are! One such number is $\cos 20°$. The complexity of the number $\cos 20°$ may come as a surprise because it looks deceptively simple. For example, $\cos 20°$ is the length of the side AB in $\triangle ABC$ as shown in Figure 5.1. Thus,

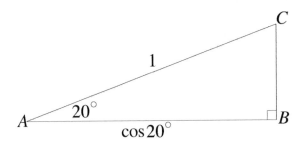

Figure 5.1

from a geometric point of view, the number $\cos 20°$ seems to be quite simple. Now if we substitute $\theta = 20°$ in the familiar trigonometric identity

$$4\cos^3 \theta - 3\cos \theta = \cos 3\theta,$$

we obtain

$$4\cos^3 20° - 3\cos 20° = \cos 60°,$$

which we can write as

$$4\cos^3 20° - 3\cos 20° = \frac{1}{2},$$

and from this fact we can see that the number $x = \cos 20°$ is one of the solutions of the equation

$$8x^3 - 6x - 1 = 0. \tag{5.2}$$

Therefore the number $\cos 20°$ also seems to possess some algebraic simplicity.

However, in spite of its apparent simplicity, the equation $8x^3 - 6x - 1 = 0$ can be shown to be very complicated, and it is known that (within the system of real numbers) its solutions are *not* surds.[15] What this means is that, within the system of real numbers, there is no formula for solving cubic equations. If we are allowed to use complex numbers, then there is a formula that gives the solutions of the cubic equation

$$ax^3 + bx^2 + cx + d = 0$$

as follows:

$$x = \sqrt[3]{\left(\frac{9cba - 27da^2 - 2b^3}{54a^3} + \frac{\sqrt{(12ac^3 - 3c^2b^2 - 54abcd + 81a^2d^2 + 12b^3d)}}{18a^2} \right)}$$

$$-\frac{3ca - b^2}{9a^2 \sqrt[3]{\left(\frac{9cba - 27da^2 - 2b^3}{54a^3} + \frac{\sqrt{(12ac^3 - 3c^2b^2 - 54abcd + 81a^2d^2 + 12b^3d)}}{18a^2} \right)}} - \frac{1}{3}\frac{b}{a}$$

or

$$x = -\frac{1}{2}\sqrt[3]{\left(\frac{9abc - 27a^2d - 2b^3}{54a^3} + \frac{\sqrt{(12ac^3 - 3c^2b^2 - 54abcd + 81a^2d^2 + 12b^3d)}}{18a^2} \right)}$$

$$+\frac{3ac - b^2}{18a^2 \sqrt[3]{\left(\frac{9abc - 27a^2d - 2b^3}{54a^3} + \frac{\sqrt{(12ac^3 - 3c^2b^2 - 54abcd + 81a^2d^2 + 12b^3d)}}{18a^2} \right)}} - \frac{1}{3}\frac{b}{a}$$

$$+\frac{i\sqrt{3}}{2}\sqrt[3]{\left(\frac{9abc - 27a^2d - 2b^3}{54a^3} + \frac{\sqrt{(12ac^3 - 3c^2b^2 - 54abcd + 81a^2d^2 + 12b^3d)}}{18a^2} \right)}$$

---

[15]   The proof of this assertion relies on some advanced topics in abstract algebra.

$$+\frac{i\sqrt{3}}{18}\frac{3ac-b^2}{a^2\sqrt[3]{\left(\frac{9abc-27a^2d-2b^3}{54a^3}+\frac{\sqrt{(12ac^3-3c^2b^2-54abcd+81a^2d^2+12b^3d)}}{18a^2}\right)}}$$

or

$$x=-\frac{1}{2}\sqrt[3]{\left(\frac{9abc-27a^2d-2b^3}{54a^3}+\frac{\sqrt{(12ac^3-3c^2b^2-54abcd+81a^2d^2+12b^3d)}}{18a^2}\right)}$$

$$+\frac{3ac-b^2}{18a^2\sqrt[3]{\left(\frac{9abc-27a^2d-2b^3}{54a^3}+\frac{\sqrt{(12ac^3-3c^2b^2-54abcd+81a^2d^2+12b^3d)}}{18a^2}\right)}}-\frac{1}{3}\frac{b}{a}$$

$$-\frac{i\sqrt{3}}{2}\sqrt[3]{\left(\frac{9abc-27a^2d-2b^3}{54a^3}+\frac{\sqrt{(12ac^3-3c^2b^2-54abcd+81a^2d^2+12b^3d)}}{18a^2}\right)}$$

$$-\frac{i\sqrt{3}}{18}\frac{3ac-b^2}{a^2\sqrt[3]{\left(\frac{9abc-27a^2d-2b^3}{54a^3}+\frac{\sqrt{(12ac^3-3c^2b^2-54abcd+81a^2d^2+12b^3d)}}{18a^2}\right)}}.$$

Looking at this formula, we can understand why the "Modern Major General" studiously avoided any mention of cubic equations in his song. If this formula is applied to the equation $8x^3-6x-1=0$, then the solutions come out as

$$x=\frac{\left(\sqrt[3]{\left(4+4i\sqrt{3}\right)}\right)^2+4}{4\sqrt[3]{\left(4+4i\sqrt{3}\right)}}$$

or

$$x=-\frac{4+\sqrt[3]{\left(4+4i\sqrt{3}\right)}+i\sqrt{3}\sqrt[3]{\left(4+4i\sqrt{3}\right)}}{2\left(\sqrt[3]{\left(4+4i\sqrt{3}\right)}\right)^2}$$

or

$$x = \frac{2 - 2i\sqrt{3} - \sqrt[3]{\left(4 + 4i\sqrt{3}\right)} + i\sqrt{3}\sqrt[3]{\left(4 + 4i\sqrt{3}\right)}}{2\left(\sqrt[3]{\left(4 + 4i\sqrt{3}\right)}\right)^2}.$$

The interesting thing about these solutions is that, although they are all *real* numbers, we need *complex* numbers to express them as surds. What we have here are three real numbers that are not surds in the system of real numbers but which *are* surds if we widen our scope to the system of complex numbers.

Thus the system of surds is inadequate, even for elementary mathematics. There is a still larger system of numbers called the system of **algebraic numbers**. These are the numbers that are solutions of equations of the form

$$a_n x^n + a_{n-1} x^{n-1} + \cdots + a_1 x + a_0 = 0,$$

where the numbers $a_n, \cdots, a_0$ are integers and $a_n \neq 0$. It can be shown that all surds are algebraic numbers but many algebraic numbers, like $\cos 20°$, are not surds. Unfortunately, even the system of algebraic numbers falls short because there are real numbers that fail to be algebraic. Such numbers are called **transcendental**. It is possible (but not easy) to show that the numbers $e, \pi, e^\pi, 2^{\sqrt{2}}$, and $\log_3 4$ are all transcendental.

The real number system **R** is therefore much richer and much more complicated than one may suspect. Modern mathematics (and almost every major theorem of calculus) depends strongly on a "completeness property" of the system **R** that guarantees that the real numbers as we know them are *all* of the real numbers that need to be. We can think of this property as saying that if the real numbers are laid out on a "number line" in the traditional way, then *every* point on that number line must be occupied by a real number. This condition is by no means satisfied by the rational number system, which is full of little punctures where the irrationals should be. In a certain sense we can say that only a very few points on the number line are occupied by rational numbers, by surds, or even by algebraic numbers. Speaking of the set of rational numbers, Cantor put it this way:

> *The rationals are spotted on the line like stars in a black sky. The dense blackness is the firmament of the irrationals.*

You can find a precise description of Cantor's spectacular statement in the more detailed presentation of set theory that is provided as an alternative to Chapter 4. We shall begin our study of the real number system in the next section,

which begins a list of the basic axioms from which all of the other properties of
the system $\mathbf{R}$ can be deduced.

5.1.3   Some Exercises on Surds

1. Given that a, b, and c are surds; that $a \neq 0$; and that $b^2 - 4ac \geq 0$, explain
    why the solutions of the equation

    $$ax^2 + bx + c = 0$$

    are surds.

2. A detailed discussion of the material of this exercise can be found in the
    document on cubic equations ▨ .

    (a) Verify by direct multiplication that, for any numbers u, v and x,

    $$(u + v + x)\left(u^2 + v^2 + x^2 - uv - ux - vx\right) = u^3 + v^3 + x^3 - 3uvx.$$

    (b) Given numbers x, a, and b, show that the expression $x^3 + ax + b$ can be
        written in the form

        $$x^3 + ax + b = x^3 + u^3 + v^3 - 3uvx$$

        by solving the simultaneous equations

        $$3uv = -a$$
        $$u^3 + v^3 = b.$$

    Now show that as long as $27b^2 + 4a^3 \geq 0$ these equations can be solved
    giving

    $$u = \sqrt[3]{\frac{3\sqrt{3}b + \sqrt{27b^2 + 4a^3}}{6\sqrt{3}}}$$

    and

    $$v = \sqrt[3]{\frac{3\sqrt{3}b - \sqrt{27b^2 + 4a^3}}{6\sqrt{3}}}.$$

    (c) Deduce that if a and b are surds and $27b^2 + 4a^3 \geq 0$, then the solutions
        of the cubic equation

        $$x^3 + ax + b = 0 \tag{5.3}$$

        are also surds. Why doesn't this fact contradict the claim made earlier
        that the solutions of Equation (5.2) are not surds?

(d) How many real solutions does Equation (5.3) have in the case
$27b^2 + 4a^3 > 0$? What if $27b^2 + 4a^3 = 0$?

3. Ⓝ Use *Scientific Notebook* to find the exact form of the solutions of the
equation

$$125x^3 - 300x^2 + 195x - 28 + 4\sqrt{7} = 0$$

and show graphically that only one of these solutions is real.

5.2   Axioms for the Real Number System

In this section we list a set of axioms from which all of the properties of the
number system **R** may be deduced.

5.2.1   The Raw Material

The raw material for the real number system consists of the following ingredients:

- The *set* **R** of real numbers.
- Two different special numbers called 0 and 1.
- The arithmetical operations $+$ and $\times$ of addition and multiplication. The
  operation $+$ associates a single number $a + b$ (called the **sum** of a and b) to
  any given ordered pair (a, b) of real numbers. The operation $\times$ associates a
  single number $a \times b$ (called the **product** of a and b and also written as ab) to
  any given ordered pair (a, b) of real numbers.
- The order relation $<$. Given any real numbers a and b, the statement $a < b$ is
  either true or it is false.

In the following subsections we list the fundamental facts about the numbers
0 and 1, the arithmetical operations, and the order relation $<$. All other properties
of the real number system can be deduced from these fundamental facts that we
call the **axioms for the real number system**.

1. **The Associative Laws**
    If a, b, and c are any numbers, then

$$a + (b + c) = (a + b) + c$$

and

$$a(bc) = (ab)c.$$

2. **The Commutative Laws**
   If a, b, and c are any numbers, then

   $$a + b = b + a \quad \text{and} \quad ab = ba.$$

3. **The Roles of 1 and 0**
   Given any number x we have

   $$1x = x \quad \text{and} \quad 0 + x = x.$$

4. **Additive and Multiplicative Inverses**
   We state two properties here, one for addition and one for multiplication.
   The properties stated here make subtraction and division possible in the real
   number system.

   (a) Given any number a, there is at least one number b such that $a + b = 0$.
   (b) Given any number $a \neq 0$, there is at least one number b such that $ab = 1$.

5. **The Distributive Law**
   This property of arithmetic provides a relationship between the operations $+$
   and $\times$. It says that if a, b, and c are any numbers, then

   $$a\left(b + c\right) = ab + ac.$$

6. **Properties of the Order $<$**

   (a) Given any real numbers a and b, either $a < b$ or $b < a$ or $a = b$, and not
       more than one of these three conditions can hold.
   (b) Given any real numbers a, b and c, if $a < b$ and $b < c$, then $a < c$.
   (c) If a, b, and x are any real numbers and $a < b$, then we have
       $a + x < b + x$.
   (d) If a, b, and x are any real numbers and $a < b$ and $x > 0$, then we have
       $ax < bx$.

   As usual, if a and b are real numbers, then the assertion $a > b$ means that
   $b < a$, the assertion $a \leq b$ means that either $a < b$ or $a = b$, and the assertion
   $a \geq b$ means that either $a > b$ or $a = b$.

7. **The Axiom of Completeness**
   The axiom of completeness is a precise way of saying that if the real numbers
   are laid down on a "number line" in the traditional way, then *every* point
   on that number line is occupied by a number. You will see this axiom in
   Subsection 5.7.1.

5.3   Arithmetical Properties of R

All of the standard rules of algebra can be deduced from the axioms that appear in Section 5.2. For example, it is possible to deduce that if a and b are any real numbers, then

$$(-a)(-b) = ab.$$

If you would like to see how these properties are deduced, click on the icon .

5.4   Order Properties of R

The operations $+$ and $\times$, whose properties are described in the first five axioms in Subsection 5.2, give us the ability to add, subtract, multiply, and divide, but it is the order relation $<$ that makes the theory of limits possible. The order $<$ gives the system of real numbers a geometric flavor and allows us to think of the real numbers as being strung out on a "number line". When this line is horizontal, our usual convention is to make a number x lie to the left of a number y when $x < y$.

In this section we shall explore some of the relationships between the order $<$ in **R** and the arithmetical operations $+$ and $\times$. As usual, we call a real number x **positive** if $x > 0$ and **negative** if $x < 0$.

5.4.1   Additive Inverses and the Order in R

Suppose that x is any real number.

1. *If $x < 0$, then $-x > 0$.*
2. *If $x > 0$, then $-x < 0$.*

Proof. If $x < 0$, then $x + (-x) < 0 + (-x)$ and so $0 < -x$. Part 2 can be deduced in the same way.■

5.4.2   Multiplying an Inequality by a Negative Number

Suppose that a, b, and x are real numbers; that $a < b$; and that $x < 0$. Then $ax > bx$.

Proof. Since $-x > 0$, we have $a(-x) < b(-x)$ and so $-ax < -bx$. Therefore

$$-ax + ax + bx < -bx + ax + bx,$$

and we obtain $bx < ax$. ■

5.4.3   Squares Are Nonnegative

For every real number x we have $x^2 \geq 0$.

Proof. If $x > 0$, then $xx > 0x$ and so $x^2 > 0$. If $x < 0$, then, since $-x > 0$ and $x^2 = (-x)^2$, we again have $x^2 > 0$. Finally, if $x = 0$, then $x^2 = 0$. ∎

5.4.4   Absolute Value

Given any real number x, the **absolute value** of x is the number $|x|$ defined by

$$|x| = \begin{cases} x & \text{if} \quad x \geq 0 \\ -x & \text{if} \quad x < 0. \end{cases}$$

Note that if x is any number, then $|x| \geq 0$ and $|-x| = |x|$. Note also that if x is any real number, then $|x|^2 = x^2$.

5.4.5   Distance Between Numbers

The inequalities that are most useful to us in mathematical analysis often involve absolute value because the absolute value gives us a notion of *distance*. We can think of the distance between two numbers x and y as being the number $|x - y|$. This interpretation of absolute value allows us to view many inequalities geometrically. For example, the inequality $|x - a| < \delta$ says that the distance from x to a is less than δ. By thinking about the inequality $|x - a| < \delta$ in this way we can easily see that it is equivalent to the condition $a - \delta < x < a + \delta$.

$$\begin{array}{ccc} \underset{a-\delta}{+} & \underset{a}{+} & \underset{a+\delta}{+} \end{array}$$

In the special case $a = 0$, this statement tells us that the inequality $|x| < \delta$ is equivalent to the inequality $-\delta < x < \delta$.

5.4.6   The Triangle Inequality

The simplest form of the **triangle inequality** says that if x and y are any given real numbers, then

$$|x + y| \leq |x| + |y| .$$

To see why this inequality holds we observe that

$$-|x| \leq x \leq |x| \quad \text{and} \quad -|y| \leq y \leq |y| .$$

Adding, we obtain

$$-|x| - |y| \leq x + y \leq |x| + |y| ,$$

which we can write as

$$-(|x| + |y|) \leq x + y \leq |x| + |y| ,$$

and we deduce from Subsection 5.4.5 that

$$|x + y| \leq |x| + |y| \,.$$

A more interesting form of the triangle inequality says that if a, b, and c are any given real numbers, then

$$|a - c| \leq |a - b| + |b - c| \,.$$

This inequality follows from the first one:

$$|a - c| = |(a - b) + (b - c)| \leq |a - b| + |b - c| \,.$$

This inequality says that the distance from a to c cannot be more than the distance from a to b plus the distance from b to c. Our reason for calling this statement the *triangle inequality* is that its two-dimensional analog can be interpreted as saying that the sum of the lengths of any two sides of a triangle is not less than the length of the third side.

5.4.7   Exercises on Inequalities

1. Prove that if x and y are positive real numbers, then their product xy is positive.
2. Prove that if x and y are negative real numbers, then their product xy is positive.
3. Given real numbers a and b, prove that

    $$|a| - |b| \leq |a - b| \,.$$

4. Given real numbers a and b, prove that

    $$||a| - |b|| \leq |a - b| \,.$$

5. Ⓝ In each of the following cases, find the numbers x for which the given inequality is true. Compare your answers with the answers given by *Scientific Notebook*.

    (a) $|2x - 3| \leq |6 - x|$.
    (b) $||x| - 5| < |x - 6|$.
    (c) ▨ $|2\,|x| - 5| \leq |4 - |x - 1||$.

6. Prove that if a, b, c, x, y, and z are any real numbers, then

    $$(ax + by + cz)^2 \leq \left(a^2 + b^2 + c^2\right)\left(x^2 + y^2 + z^2\right) \,.$$

7. Given that a, b, and c are positive numbers and that $c < a + b$, prove that

$$\frac{c}{1+c} < \frac{a}{1+a} + \frac{b}{1+b}.$$

5.5   Integers and Rationals

Natural numbers, integers, and rational numbers are not mentioned in the axioms that appear in Section 5.2 and so, in a careful development of the real number system from these axioms, the sets $\mathbf{Z}^+, \mathbf{Z}$, and $\mathbf{Q}$ of positive integers, integers and rational numbers need to be defined. However, the process of defining integers and rational numbers depends on some concepts from the theory of sets that are beyond the scope of this book. So, instead of giving precise definitions, we shall assume familiarity with the sets $\mathbf{Z}^+, \mathbf{Z}$, and $\mathbf{Q}$, and we shall assume the following basic facts about them:

1. The least positive integer is 1.
2. The sum, difference, and product of two integers is always an integer.
3. As usual, a real number x is said to be rational if it is possible to find integers m and n such that $n \neq 0$ and $x = m/n$. The set $\mathbf{Q}$ of rational numbers is closed under the operations of addition, subtraction, multiplication, and division, as long as we exclude division by zero.

5.5.1   Exercises on Integers and Rational Numbers

1. (a) Explain why the numbers $\frac{3}{2} + \frac{2}{7}$ and $\frac{3}{2} \times \frac{2}{7}$ are both rational.
    (b) Explain why the number 2.345 is rational.
    (c) Explain why the sum, difference, product, and quotient of rational numbers must always be rational.
2. For each of the following statements, say whether the statement is true or false and justify your assertion:
    (a) If x is rational and y is irrational, then $x + y$ is irrational.
    (b) If x is rational and y is irrational, then xy is irrational.
    (c) If x is irrational and y is irrational, then $x + y$ is irrational.

5.6   Upper and Lower Bounds

The concepts of upper and lowers bounds that we shall study in this section will be used when we introduce the axiom of completeness for the real number system in Section 5.7.

5.6.1   Definition of Upper and Lower Bounds

Suppose that A is a set of real numbers and that $\alpha \in \mathbf{R}$. We say that the number α is an **upper bound** of the set A if no member of A can be greater than α. In other words, α is an upper bound of A if we have $x \leq \alpha$ for every $x \in A$. The idea of a lower bound is defined similarly. We say that α is a **lower bound** of A if no member of A can be less than α. In other words, α is a lower bound of A if we have $x \geq \alpha$ for every $x \in A$.

Notice that if α is an upper bound of a set A, then so is every number that is greater than α. Similarly, if α is a lower bound of A, then so is every number that is less than α.

5.6.2   Some Examples and Remarks

1. 3 is an upper bound of $[0, 3)$ but 2.99 is not.
2. A number α is an upper bound of the set $[0, 1]$ if and only if $\alpha \geq 1$. Therefore the set of upper bounds of $[0, 1]$ is $[1, \infty)$.
3. In this example we describe the upper bounds of the interval $[0, 1)$. We begin by observing that any number $\alpha \geq 1$ is an upper bound of the interval $[0, 1)$. Next we observe that if $\alpha < 0$, then α is not an upper bound of $[0, 1)$. Finally we consider the case $0 \leq \alpha < 1$. In this case, α fails to be an upper bound of the interval $[0, 1)$ because the numbers between α and 1 are members of $[0, 1)$ that are greater than α. For example, the number $(\alpha + 1)/2$ is a member of $[0, 1)$ greater than α.

We conclude that the set of upper bounds of the interval $[0, 1)$ is $[1, \infty)$. By comparing this fact with Example 2, we make the interesting observation that the two intervals $[0, 1)$ and $[0, 1]$ have precisely the same upper bounds.
4. The set of lower bounds of the set $\{2\} \cup [3, 4]$ is the interval $(-\infty, 2]$.
5. Since no member of the empty set $\emptyset$ can be greater than 3, the number 3 is an upper bound of $\emptyset$. As a matter of fact, every real number is both an upper bound and a lower bound of $\emptyset$.
6. Suppose that A is a nonempty set of real numbers, that α is a lower bound of A, and that β is an upper bound of A. We must have $\alpha \leq \beta$.

   Proof. Using the fact that A is nonempty we choose[16] a member x of A. Since α is a lower bound of A we have $\alpha \leq x$, and since β is an upper bound of A we have $x \leq \beta$. Therefore $\alpha \leq \beta$. ∎

---

[16]   Note this use of the word *choose* in a mathematical argument. The word choose is used when we already know that a given set is nonempty and we want to refer to one of its members.

Notice that we needed to know that $A \neq \emptyset$ in order to write this proof. As we saw in Example 5, a lower bound of $\emptyset$ can be greater than an upper bound of $\emptyset$.

7. Suppose that A is a set of real numbers that has more than one member, that α is a lower bound of A, and that β is an upper bound of A. We must have $\alpha < \beta$. We leave the proof of this fact as an exercise. 🖳

5.6.3   Bounded Sets

If A is a set of real numbers, and if there exists a number α that is an upper bound of A, then we say that the set A is **bounded above**. If there exists a number α that is a lower bound of a given set A, then we say that A is **bounded below**. If a given set is bounded both above and below, then we say that the set is **bounded**.

5.6.4   Suprema and Infima

As we have seen, if α is an upper bound of a set A, then so is every number greater than α. On the other hand, if α is an upper bound of a set A, then a number that is less than α may or may not be an upper bound of A. If α is an upper bound of a given set A and no number less than α is an upper bound of A, then the set of upper bounds of A is the interval $[\alpha, \infty)$ and we see that α is the least possible upper bound of the set A. The least upper bound of a set A, if it exists, is called the **supremum**[17] of the set A and is written as $\sup A$.

In the same way, if α is a lower bound of a set A, then so is every number less than α. If no lower bound of A can be greater than α, then the set of lower bounds of A is the interval $(-\infty, \alpha]$ and α is the greatest lower bound of A. The greatest lower bound of a set A, if it exists, is called the **infimum** of A and is written $\inf A$.

Thus, when we say that a number α is the supremum of a set A, what we are really saying is that the number α is an upper bound of A but, given any number $\beta < \alpha$, the number β fails to be an upper bound of A.

Another way of expressing the condition $\alpha = \sup A$ is to say that for every member x of A we have $x \leq \alpha$, but, given any number $\beta < \alpha$, there exists at least one member x of A such that $x > \beta$.

As an exercise you should now write down the corresponding conditions for a number to be the infimum of a set. The exercises that follow will also help you to understand the concepts of upper bound and lower bound.

---

[17]   The plural of *supremum* is *suprema*, just as the plural of maximum is maxima.

5.6.5   Some Examples of Suprema and Infima

1. The supremum of the interval $[0, 1]$ is the number 1. The supremum of the interval $[0, 1)$ is also the number 1.
2. A set may or may not have a largest member. For example, the interval $[0, 1]$ does have a largest member, the number 1, while the interval $[0, 1)$ does not have a largest member. In the event that a given set A has a largest member, that largest member is also $\sup A$.
3. Look again at the interval $[0, 1)$. Even though this interval does not have a largest member, it still has a supremum. In fact, as we have seen, $\sup [0, 1) = 1$.

5.6.6   Exercises on Upper and Lower Bounds

1. ▦ Suppose that A is a nonempty bounded set of real numbers that has no largest member and that $a \in A$. Explain why the sets A and $A \setminus \{a\}$ have exactly the same upper bounds.
2. (a) Give an example of a set A that has a largest member a such that the sets A and $A \setminus \{a\}$ have exactly the same upper bounds.
   (b) Give an example of a set A that has a largest member a such that the sets A and $A \setminus \{a\}$ do not have exactly the same upper bounds.
3. (a) ▦ Given that S is a subset of a given interval $[a, b]$, explain why, for every member x of the set S, we have
$$|x| \leq |a| + |b - a|.$$
   (b) Given that a set S of numbers is bounded and that
$$T = \{|x| \mid x \in S\},$$
   prove that the set T must also be bounded.
4. ▦ Given that A is a set of real numbers and that $\sup A \in A$, explain why $\sup A$ must be the largest member of A.
5. Given that A is a set of real numbers and that $\inf A \in A$, explain why $\inf A$ must be the smallest member of A.
6. ▦ Is it possible for a set of numbers to have a supremum even though it has no largest member?
7. Given that α is an upper bound of a set A and that $\alpha \in A$, explain why $\alpha = \sup A$.
8. Explain why the empty set does not have a supremum.
9. Explain why the set $[1, \infty)$ does not have a supremum.
10. Given that two sets A and B are bounded above, explain why their union

$A \cup B$ is bounded above.

11. (a) ▉ If a man says truthfully that he sells more BMWs than anyone in the Southeast, what can you deduce about him?

(b) ▉ Given that $\alpha = \sup A$ and that $x < \alpha$, what conclusions can you draw about the number x?

(c) Given that $\alpha = \inf A$ and that $x > \alpha$, what conclusions can you draw about the number x?

12. If A and B are sets of real numbers, then the sets $A + B$ and $A - B$ are defined[18] by

$$A + B = \{x \mid \exists a \in A \text{ and } \exists b \in B \text{ such that } x = a + b\}$$

and

$$A - B = \{x \mid \exists a \in A \text{ and } \exists b \in B \text{ such that } x = a - b\}.$$

(a) Work out $A + B$ and $A - B$ in each of the following cases:

   I. $A = [0,1]$ and $B = [-1,0]$.

   II. $A = [0,1]$ and $B = \{1,2,3\}$.

   III. $A = (0,1)$ and $B = \{1,2,3\}$.

(b) Prove that if two sets A and B are bounded, then so are $A + B$ and $A - B$.

5.7   The Axiom of Completeness

The precise statement of the axiom of completeness is as follows:

5.7.1   Statement of the Axiom of Completeness

Whenever a set of real numbers is nonempty and bounded above, it has a supremum.

5.7.2   Discussion of the Axiom of Completeness

To understand what the axiom is saying we return to the intervals $[0,1]$ and $[0,1)$ that we studied in Examples 2 and 3 of Subsection 5.6.2. We saw that each of these intervals has a supremum of 1. The closed interval $[0,1]$ has a largest member because 1 is its largest member. However, even though the interval $[0,1)$ has no largest member, the number 1 is still its supremum.

We can think of the supremum of a set as being at the "top" of the set. If a set of numbers has a largest member, then that largest member is the supremum.

---

[18]   Do not confuse the set $A - B$ that is being defined here with the set difference $A \setminus B$.

However, as long as a set of numbers is nonempty and bounded above, it will have a supremum even if it does not have a largest member. We can think of the axiom of completeness as saying that whenever a set A of real numbers is nonempty and bounded above, the point on the number line at the "top" of the set A is occupied by a number: the supremum of the set A.

5.7.3   Infimum Version of the Axiom of Completeness

Suppose that a set A of real numbers is nonempty and bounded below. Then A has an infimum.

Proof.   We define B to be the set of all lower bounds of A. Since A is bounded below, we know that the set B is nonempty. In order to show that A has a greatest lower bound, we need to show that the set B has a largest member. The idea of the proof is to show that B is bounded above and then to show that $\sup B$ is the largest member of B. To show that $\sup B$ is the largest member of B, all we have to show is that $\sup B \in B$.

We begin by showing that every member of A must be an upper bound of B. Suppose that $y \in A$.[19] For every member x of B, since x is a lower bound of A and $y \in A$, we have $x \leq y$. This shows that every member y of A is an upper bound of B, and, since A is nonempty, we conclude that B is bounded above.

We now define $\alpha = \sup B$. In order to show that $\alpha \in B$, we need to show that α is a lower bound of A. But given any member y of A, it follows from the fact that y is *an* upper bound of B and α is the *least* upper bound of B that $\alpha \leq y$. Thus α is a lower bound of A, as required. ∎

5.7.4   Exercises on Supremum and Infimum

1.   Suppose that A is a nonempty bounded set of real numbers that has no largest member and that $a \in A$. Prove that $\sup A = \sup (A \setminus \{a\})$.
2. Given that A and B are sets of numbers, that A is nonempty, that B is bounded above, and that $A \subseteq B$, explain why $\sup A$ and $\sup B$ exist and why $\sup A \leq \sup B$.
3. Given that A is a nonempty bounded set of numbers, explain why $\inf A \leq \sup A$.
4.   It is given that A and B are nonempty bounded sets of real numbers, that for every $x \in A$ there exists $y \in B$ such that $x < y$, and for every $y \in B$ there exists $x \in A$ such that $y < x$. Prove that $\sup A = \sup B$.
5. Suppose that A and B are nonempty sets of real numbers and that for every

---

[19] Note the use of the words "suppose that" here. In order to show that every member y of a set A has a certain property we write: Suppose that $y \in A$. We *never* use the word "choose" for this purpose.

$x \in A$ and every $y \in B$ we have $x < y$. Prove that $\sup A \le \inf B$. Give an example of sets A and B satisfying these conditions for which $\sup A = \inf B$.

6. ▣ Suppose that A and B are nonempty sets of real numbers and that $\sup A = \inf B$. Prove that for every number $\delta > 0$ it is possible to find a member x of A and a member y of B such that $x + \delta > y$.

7. ▣ Suppose that A and B are nonempty sets of real numbers, that $\sup A \le \inf B$, and that for every number $\delta > 0$ it is possible to find a member x of A and a member y of B such that $x + \delta > y$. Prove that $\sup A = \inf B$.

8. Suppose that A is a nonempty bounded set of real numbers, that A has no largest member, and that $x < \sup A$. Prove that there are at least two different members of A lying between x and $\sup A$.

9. Suppose that A is a nonempty bounded set of real numbers, that $\delta > 0$, and that for any two different members x and y of A we have $|x - y| \ge \delta$. Prove that A has a largest member. You can find a hint to the solution of this exercise in Theorem 5.9.1.

10. Suppose that S is a nonempty bounded set of real numbers, that $\alpha = \inf S$ and $\beta = \sup S$, and that every number that lies between two members of S must also belong to S. Prove that S must be one of the four intervals $[\alpha, \beta]$, $[\alpha, \beta)$, $(\alpha, \beta]$, (α, β). See Theorem 5.8.1 for a solution of this exercise.

11. Suppose that A is a nonempty bounded set of real numbers, that $\alpha = \inf A$, and that $\beta = \sup A$. Suppose that

$$S = \{x - y \mid x \in A \text{ and } y \in A\}.$$

Prove that $\sup S = \beta - \alpha$. You will find a solution to this exercise in Subsection 5.8.3.

12. Suppose that A is a set of numbers and that A is nonempty and bounded above. Suppose that q is a given number and that the set C is defined as follows:

$$C = \{q + x \mid x \in A\}.$$

Prove that the set C is nonempty and bounded above and that

$$\sup C = q + \sup A.$$

13. ▣ Suppose that A and B are nonempty bounded sets of numbers and that the sets $A + B$ and $A - B$ are defined as in Exercise 12 of Subsection 5.6.6.

Prove that

$$\sup\left(A+B\right)=\sup A+\sup B$$

and

$$\sup\left(A-B\right)=\sup A-\inf B.$$

5.8   Some Consequences of the Completeness Axiom

5.8.1   A Condition for a Set to Be an Interval

The concept of an interval was introduced in Example 5 of Subsection 4.2.9. In this section we provide a way of determining whether or not a given set is an interval, even when we don't have any information about endpoints.

Suppose that S is a set of real numbers. The following two conditions are equivalent:

1. *The set S is an interval.*
2. *Whenever s and t belong to S and x is a number satisfying $s < x < t$, we have $x \in S$.*

Proof. It is obvious that if S is an interval, then S satisfies condition 2. The main thrust of this theorem is that condition 2 implies condition 1. We now suppose that S satisfies condition 2, and, to show that S must be an interval, we shall consider several cases:

Case 1: Suppose that S is bounded below but not above. We define $a = \inf S$ and what we want to show is that either $S = [a, \infty)$ or $S = (a, \infty)$.

Certainly, no member of the set S can be less than a because a is a lower bound of S. Therefore $S \subseteq [a, \infty)$. Therefore, in order to show that S is one of the two intervals $[a, \infty)$ and (a, ∞), all we have to do is show that every number $x > a$ must belong to S. For this purpose, suppose[20] that x is a number satisfying $x > a$. Since a is the greatest lower bound of S, we know that x is not a lower bound of S, and, using this fact, we choose[21] a member s of S such that $s < x$. Using the fact that S is not bounded above, we now choose a member t of S such that $x < t$. From condition 2 we deduce that $x \in S$.

Case 2: Suppose that S is bounded above but not below. In this case, if $b = \sup S$, then either $S = (-\infty, b)$ or $S = (-\infty, b]$. The proof of this assertion will be left as an exercise. ■

---

[20]   See Subsection 3.6.2 for a discussion of the use of the word *suppose* in a mathematical proof.
[21]   See Subsection 3.5.2 for a description of the use of the word *choose* in a mathematical proof.

Case 3: Suppose that S is unbounded both above and below. In this case we can show that $S = \mathbf{R}$. The proof of this assertion will be left as an exercise.

Case 4: Suppose that the set S is bounded. In the event that S is empty we have nothing to prove, because the empty set is an interval, and so we assume, from now on, that $S \neq \emptyset$. We define $a = \inf S$ and $b = \sup S$. Since $S \subseteq [a, b]$, in order to show that S is one of the four intervals $[a, b]$, $[a, b)$, $(a, b]$ and (a, b) we need only show that $(a, b) \subseteq S$.

For this purpose, suppose that $x \in (a, b)$. Using the facts that $a < x$ and a is the greatest lower bound of S we choose a member s of S such that $s < x$. Using the fact that $x < b$ and that b is the least upper bound of S we choose a member t of S such that $x < t$. Since $s < x < t$, it follows from condition 2 that $x \in S$, thus completing the proof that $(a, b) \subseteq S$. ∎

5.8.2   The Diameter of a Set

If A is a nonempty bounded set of real numbers, then we define the **diameter** of A to be the number

$$\sup A - \inf A.$$

5.8.3   Alternative Expression for the Diameter of a Set

Suppose that A is a nonempty bounded set of real numbers. Then

$$\sup A - \inf A = \sup \left\{ |x - t| \mid t \in A \text{ and } x \in A \right\}.$$

Proof. We define

$$C = \left\{ |x - t| \mid t \in A \text{ and } x \in A \right\}.$$

Given any members t and x of the set A, if $t \leq x$, we deduce from the fact that $x \leq \sup A$ and $t \geq \inf A$ that

$$|x - t| = x - t \leq \sup A - \inf A;$$

and if $t > x$, we have

$$|x - t| = t - x \leq \sup A - \inf A.$$

We deduce that the number $\sup A - \inf A$ is an upper bound of the set C. Now, to prove that the number $\sup A - \inf A$ is the least upper bound of C, suppose that

$$\alpha < \sup A - \inf A.$$

Since

$$\alpha + \inf A < \sup A$$

we know that the number $\alpha + \inf A$ cannot be an upper bound of A and, using this fact, we choose a number $x \in A$ such that

$$\alpha + \inf A < x.$$

Since

$$\inf A < x - \alpha,$$

we know that the number $x - \alpha$ cannot be a lower bound of A and, using this fact, we choose a number $t \in A$ such that $t < x - \alpha$. From the latter inequality we obtain

$$\alpha < x - t \le |x - t|.$$

Thus no number less than $\sup A - \inf A$ can be an upper bound of C, and we have shown that $\sup A - \inf A$ is the least upper bound of C. ∎

5.9   The Archimedean Property of the System **R**

In its classical form, the Archimedean property of the number system **R** says that the set $\mathbf{Z}^+$ of positive integers is not bounded above. Although this statement may appear to be obvious, it is not. Certainly, it is obvious that no positive integer n can be an upper bound of $\mathbf{Z}^+$; for if n is any positive integer, then $n + 1$ is a larger one. But it is less obvious that if x is *any* real number, then there must exist positive integers that are larger than x. As you will see in this section, the Archimedean property of **R** depends on the axiom of completeness. Without this axiom, the Archimedean property may actually be false. There are algebraic systems that resemble the real number system and that satisfy all of the axioms that we have stated for **R**, with the exception of the completeness axiom, and which do not have the Archimedean property. In such systems, the set of positive integers *is* bounded above.

5.9.1   Sets of Integers that Are Bounded Above

Every nonempty set of integers that is bounded above must have a largest member.

Proof. Suppose that S is a nonempty set of integers and that S is bounded above. We define $w = \sup S$, and, to prove that S has a largest member, we shall show that $w \in S$. To obtain a contradiction we assume that w does not belong to S.

Using the fact that $w - 1$, being less than w, cannot be an upper bound of S, we choose a member m of S such that $w - 1 < m$. From the fact that $w \notin S$ we see that

$$w - 1 < m < w.$$

Now we use the fact that m is not an upper bound of S to choose a member n of S such that $m < n$. Thus

$$w - 1 < m < n < w.$$

From this inequality we see that

$$0 < n - m < 1,$$

which is impossible because $n - m$ is a positive integer. ■

5.9.2   Sets of Integers that Are Bounded Below

Every nonempty set of integers that is bounded below must have a smallest member.

Proof. We leave the proof of this property as an exercise.

5.9.3   Unboundedness of the Set Z

The set Z of integers is unbounded both above and below.

Proof. This property of $\mathbf{Z}$ follows at once from the fact that $\mathbf{Z}$ has neither a smallest nor a largest member. ■

5.9.4   A Fact About Reciprocals of Natural Numbers

Given any positive number x, it is possible to find a positive integer n such that $\frac{1}{n} < x$.

Proof. Suppose that $x > 0$ and, using the fact that the number $\frac{1}{x}$ is not an upper bound of the set $\mathbf{Z}^+$, choose a positive integer n such that $n > \frac{1}{x}$. We see that $\frac{1}{n} < x$. ■

5.9.5   Denseness of the Set Q of Rational Numbers

By turning back to the statement of Cantor that appears in Subsection 5.1.2, we can see that in a certain sense, which we shall not make precise here, the overwhelming majority of real numbers are irrational. However, the set of rational numbers is still large enough to ensure that between any two different real numbers there must be some rational numbers. This property of the set $\mathbf{Q}$ is known as the **denseness** of the set of rational numbers. The precise statement of this denseness is as follows:

Suppose that a and b are any real numbers and that $a < b$. Then the set

Q $\cap (a, b)$ *is nonempty; in other words, there exists a rational* r *such that* $a < r < b$.

Proof. ⬚   Using Theorem 5.9.4, we choose a positive integer n such that

$$0 < \frac{1}{n} < b - a.$$

Note that

$$a + \frac{1}{n} < b.$$

We shall prove the theorem by finding an integer m such that

$$a < \frac{m}{n} < b.$$

We begin by defining

$$S = \left\{ x \in \mathbf{Z} \mid \frac{x}{n} > a \right\} = \left\{ x \in \mathbf{Z} \mid x > na \right\}.$$

Since the set **Z** is unbounded above, we know that $S \neq \emptyset$. Therefore, since S is bounded below, it follows from Theorem 5.9.2 that S has a least member that we shall call m.

Thus

$$\frac{m}{n} > a \quad \text{but} \quad \frac{m - 1}{n} \leq a.$$

From the fact that

$$\frac{m - 1}{n} \leq a$$

we see that

$$\frac{m}{n} = \frac{m - 1}{n} + \frac{1}{n} \leq a + \frac{1}{n} < b,$$

and so

$$a < \frac{m}{n} < b,$$

as promised. ∎

5.9.6   Exercises on the Archimedean Property of the System R

1. Prove that if A is the set of all rational numbers in the interval $[0, 1]$, then $\sup A = 1$.

2. Suppose that

$$A = \left\{ \frac{1}{1}, \frac{1}{2}, \frac{1}{3}, \cdots \right\} = \left\{ \frac{1}{n} \mid n \in \mathbf{Z}^+ \right\}.$$

   Prove that $\inf A = 0$.

3. A nonempty set G of real numbers is said to be a **subgroup** of $\mathbf{R}$ if whenever x and y belong to G, then so do the numbers $x + y$ and $x - y$.

   (a) Determine which of the following sets are subgroups of $\mathbf{R}$.

$\emptyset$	$\{1, 0, -1\}$	$\mathbf{Q}$	$\mathbf{Z}$
$\mathbf{Z}^+$	$\mathbf{Q}^+$	$\{2n \mid n \in \mathbf{Z}\}$	$\{2n \mid n \in \mathbf{Z}^+\}$
$\{0\}$	$\mathbf{R}$	$\mathbf{R} \setminus \mathbf{Q}$	$\{m + n\sqrt{2} \mid m \in \mathbf{Z} \text{ and } n \in \mathbf{Z}\}$

   (b) Explain why every subgroup of $\mathbf{R}$ must contain the number 0. Show that if G is any subgroup of $\mathbf{R}$ other than $\{0\}$, then G must contain infinitely many positive numbers.

   (c) ■ Suppose that G is a subgroup of $\mathbf{R}$ other than $\{0\}$, that

$$p = \inf \{x \in G \mid x > 0\},$$

   and that the number p is positive. Prove that either $p \in G$ or the set G must contain at least two different members between p and $3p/2$.

   (d) ■ Suppose that G is a subgroup of $\mathbf{R}$ other than $\{0\}$, that

$$p = \inf \{x \in G \mid x > 0\},$$

   and that the number p is positive. Prove that p is the smallest positive member of G and that

$$G = \{np \mid n \in \mathbf{Z}\}.$$

   (e) ■ Suppose that G is a subgroup of $\mathbf{R}$ other than $\{0\}$, and that

$$0 = \inf \{x \in G \mid x > 0\}.$$

   Prove that if a and b are any real numbers satisfying $a < b$, then it is possible to find a member x of the set G such that $a < x < b$.

4. This exercise invites you to explore the so-called **division algorithm**, which describes the process by which an integer b can be divided into an integer a

to yield a quotient q and a remainder r.[22]

Suppose that a and b are positive integers and that

$$S = \{n \in \mathbf{Z} \mid nb \le a\}.$$

(a) Prove that $S \ne \emptyset$ and that S is bounded above.
(b) Prove that if the largest member of S is called q and we define
$r = a - qb$, then $a = qb + r$ and $0 \le r < b$.

5.10   Boundedness of Functions

5.10.1   Supremum, Infimum, Maximum, and Minimum of a Function

A real-valued function f defined on a set S is said to be **bounded** if the range of f is bounded. In other words, the function f is bounded if and only if the set

$$\{f(x) \mid x \in S\}$$

is a bounded set of real numbers, and we give similar definitions of **bounded above** and **bounded below.** If a function f is bounded above on a nonempty set S, then we define its **supremum** sup f by the equation

$$\sup f = \sup \{f(x) \mid x \in S\}$$

and we define the **infimum** inf f similarly. In the event that the range of f has a maximum value, then this maximum value is called the **maximum** value of the function f and the **minimum** value of f is defined similarly.

5.10.2   Some Exercises on Suprema and Infima of Functions

1. Given that $f(x) = 1/x$ whenever $x > 0$, prove that the function f is unbounded above. Prove that for every number $\delta > 0$ the restriction of f to the interval $[\delta, \infty)$ is bounded.
2. Give an example of a bounded function on the interval $[0, 1]$ that has a minimum value but does not have a maximum value.
3. A function f is said to be **increasing** on a set S if the inequality $f(t) \le f(x)$ holds whenever t and x belong to S and $t \le x$. Prove that every increasing function on the interval $[0, 1]$ must have both a maximum and a minimum value.
4. ▉ Prove that if f and g are bounded above on a nonempty set S, then

$$\sup (f + g) \le \sup f + \sup g.$$

---

[22]   For a more detailed but elementary account of this material, see Lewin [22].

5. Give an example of two bounded functions f and g on the interval $[0, 1]$ such that

$$\sup (f + g) < \sup f + \sup g.$$

6. ▇ Given that f is a bounded function on a nonempty set S and that c is a real number, prove that

$$\sup (cf) = \begin{cases} c \sup f & \text{if } c > 0 \\ c \inf f & \text{if } c < 0. \end{cases}$$

7. ▇ Prove that if f is a bounded function on a nonempty set S, then

$$|\sup f| \le \sup |f|.$$

5.11   Sequences, Finite Sets, and Infinite Sets

5.11.1   Definition of a Sequence

The usual definition of a **sequence in a set** S is that it is a function from $\mathbf{Z}^+$ into S. A slightly more general and more useful notion of a sequence in a set S is that it is a function f whose domain is the set of all integers greater than or equal to a given integer k such that $f(n) \in S$ for each n.

5.11.2   Examples of Sequences

1. We could define $f(n) = 2n - 1$ for every integer $n \ge 4$.
2. We could define $f(n) = \sqrt{n - 6}$ for every integer $n \ge 6$.
3. We could define $f(n) = (n + 6)!$ for every integer $n \ge -6$.

5.11.3   Traditional Notation for Sequences

If f is a sequence defined on the set of all integers $n \ge k$, where k is a given integer, and if, for each integer n in the domain of f we have written $f(n)$ in the form x_n, then a traditional notation for the sequence f is (x_n).

In other words, the notation (x_n) stands for the function whose value at each integer n in its domain is x_n. So, for example, if we have defined

$$x_n = (n - 6)!$$

for every integer $n \ge 6$, then we have defined a sequence (x_n).

There is, of course, nothing special about the letter n of the alphabet. The sequence that we have called (x_n) could just as well have been called (x_m), or

(x_i), or (x_j) and so on. In following this tradition we need to be careful not to confuse the symbol x_n, which stands for the value of the sequence at a given integer n, and the symbol (x_n) that stands for the sequence itself. We often call x_n the n**th term** of the sequence (x_n)

5.11.4   Some Further Examples of Sequences

1. If for each integer $n > 4$ we define

$$x_n = \frac{(-1)^n}{n - 4},$$

then (x_n) is a sequence in $\mathbf{R}$. As a matter of fact, we could also say that this sequence (x_n) is in the set $\mathbf{Q}$ of rational numbers.
2. If for each integer $n \geq 1$ we define E_n to be the interval $[0, n]$, then (E_n) is a sequence of subsets of $\mathbf{R}$.
3. If we define a function $f_n : \mathbf{R} \to \mathbf{R}$ for each integer $n \geq 1$ by

$$f_n(x) = \frac{(-1)^n \sin nx}{(2n + 1)!}$$

for every real number x, then (f_n) is a sequence of functions from $\mathbf{R}$ to $\mathbf{R}$.

5.11.5   Bounded Sequences.

In keeping with our notation for functions we say that a given sequence (x_n) of real numbers is **bounded** if the range of the sequence (x_n) is bounded. In other words, if the domain of the sequence (x_n) is the set of all integers $n \geq k$, where k is some integer, then we say that the sequence (x_n) is bounded if and only if the set

$$\{x_n \mid n \geq k\}$$

is bounded.

Similar definitions can be given to the concept of **bounded above** and **bounded below**.

5.11.6   Supremum and Infimum of a Sequence

If a given sequence (x_n) is bounded above, then the **supremum** of the sequence (x_n) is defined to be the supremum of its range. The infimum of a sequence is defined similarly.

5.11.7   One-One Sequences

A sequence (x_n) in a set S is **one-one** if for any two different integers m and n in the domain of the sequence we have $x_m \neq x_n$.

5.11.8   Monotone Sequences

A sequence (x_n) of real numbers is said to be **increasing** if we have $x_n \le x_{n+1}$ for every n in the domain of the sequence. If for each such n we have $x_n <$ x_{n+1}, then we say that the sequence (x_n) is **strictly increasing**. **Decreasing** and **strictly decreasing** sequences are defined similarly.

A sequence that is either increasing or decreasing is said to be **monotone**, and a sequence that is either strictly increasing or strictly decreasing is said to be **strictly monotone**. Note that a strictly monotone sequence is always one-one.

5.11.9   Finite Sets and Infinite Sets

Suppose that S is a given set. If S is nonempty, we can choose a member of S and call it x_1. There are now two possibilities: Either $S = \{x_1\}$ or it is possible to choose a number x_2 in the set $S \setminus \{x_1\}$. If x_2 has been chosen, then there are again two possibilities: Either $S = \{x_1, x_2\}$ or it is possible to choose a number x_3 in the set $S \setminus \{x_1, x_2\}$.

If we attempt to continue this process, there are two possibilities: Either the process will terminate because, for some positive integer n, we have

$$S = \{x_1, x_2, x_3, \cdots, x_n\}$$

or otherwise the process continues indefinitely and the number x_n is defined for every positive integer n. Thus either it is possible to express the set S in the form

$$S = \{x_1, x_2, x_3, \cdots, x_n\}$$

for some positive integer n, in which case we say that the set S is **finite**, or we can find a one-one sequence (x_n) in the set S, in which case we say that the set S is **infinite**. You can find a more detailed discussion of finite sets in the optional, more detailed approach to set theory. To reach this discussion, click on the icon

.

5.12   Sequences of Sets

Not every sequence has to be a sequence of numbers. For example, we saw in Subsection 5.11.4 that if

$$A_n = [0, n]$$

for every positive integer n, then (A_n) is a sequence of sets. In this section we give a brief discussion of sequences of this type.

5.12.1   Union and Intersection of a Sequence of Sets

Suppose that A_n is a given set for each positive integer n. If n is any positive integer, then the symbol

$$\bigcup_{j=1}^{n} A_j$$

stands for the set of all those objects x that belong to the set A_j for at least one integer j satisfying $1 \leq j \leq n$. The **union** of the entire sequence (A_n) is defined to be the set of all those objects x that belong to A_n for at least one positive integer n. The union of the sequence (A_n) is written as

$$\bigcup_{n=1}^{\infty} A_n.$$

If n is any positive integer, then the symbol

$$\bigcap_{j=1}^{n} A_j$$

stands for the set of all those objects x that belong to the set A_j for every integer j satisfying $1 \leq j \leq n$. The **intersection** of the entire sequence (A_n) is defined to be the set of all those objects x that belong to A_n for every positive integer n. The intersection of the sequence (A_n) is written as

$$\bigcap_{n=1}^{\infty} A_n.$$

5.12.2   Contracting and Expanding Sequences of Sets

If (A_n) is a sequence of sets, then we say that the sequence (A_n) is **contracting** if we have $A_{n+1} \subseteq A_n$ for every n in the domain of the sequence. If we have $A_n \subseteq A_{n+1}$ for every n in the domain of the sequence, then we say that the sequence is **expanding**

5.12.3   Exercises on Sequences of Sets

1. Evaluate

$$\bigcup_{n=1}^{\infty} \left[\frac{1}{n}, 1 \right].$$

2. Evaluate

$$\bigcup_{n=1}^{\infty} \left[1 + \frac{1}{n}, 5 - \frac{2}{n} \right].$$

3. Evaluate

$$\bigcap_{n=1}^{\infty} \left(1 - \frac{1}{n}, 2 + \frac{1}{n} \right).$$

4. Explain why if (A_n) is an expanding sequence of subsets of $\mathbf{R}$, then the sequence $(\mathbf{R} \setminus A_n)$ is a contracting sequence.

5. Suppose that (A_n) is a sequence of sets.

   (a) Prove that if we define

   $$B_n = \bigcup_{j=n}^{\infty} A_j$$

   for each n, then the sequence (B_n) is a contracting sequence of sets.

   (b) Prove that if we define

   $$B_n = \bigcap_{j=n}^{\infty} A_j$$

   for each n, then the sequence (B_n) is an expanding sequence of sets.

   (c) Prove that if we define

   $$B_n = \bigcap_{j=1}^{n} A_j$$

   for each n, then the sequence (B_n) is a contracting sequence of sets.

   (d) Prove that if we define

   $$B_n = \bigcup_{j=1}^{n} A_j$$

   for each n, then the sequence (B_n) is an expanding sequence of sets.

6. ▣ Prove that if (A_n) is a sequence of subsets of $\mathbf{R}$, then

$$\mathbf{R} \setminus \bigcup_{n=1}^{\infty} A_n = \bigcap_{n=1}^{\infty} (\mathbf{R} \setminus A_n).$$

5.13   Mathematical Induction

The principle of mathematical induction is an important technique for proving the truth of each of a sequence (p_n) of mathematical statements and for justifying the idea of a recursively defined sequence. If you would like to read a section on this topic, click on the icon ▣ .

5.14   The Extended Real Number System

5.14.1   Introduction to the Extended Real Number System

In elementary calculus we often use the infinity symbols ∞ and $-\infty$. For example, we may say that

$$\lim_{x \to 3} \frac{1}{|x-3|} = \infty$$

and that

$$\sum_{n=1}^{\infty} \frac{1}{n} = \infty.$$

In order to use the symbol ∞, we didn't actually have to have an object called ∞ in our number system. As you may recall, the statement

$$\lim_{x \to 3} \frac{1}{|x-3|} = \infty$$

simply means that we can make the number $1/|x-3|$ as large as we like by making x close enough to the number 3.

Sometimes, however, we find it convenient to work with the two symbols $-\infty$ and ∞ as if they were numbers. Using these extra symbols, we can sometimes simplify the statements of certain theorems and we can sometimes reduce the number of separate cases that have to be considered in their proofs. In introducing these extra symbols we are extending our number system slightly, giving us what we call the **extended real number system.**

$$[-\infty, \infty] = \mathbf{R} \cup \{-\infty, \infty\}.$$

5.14.2   Arithmetic in the Extended Real Number System

We define

$$\infty + \infty \;=\; \infty \times \infty = (-\infty)(-\infty) = \infty$$
$$-\infty - \infty \;=\; (-\infty)\infty = \infty(-\infty) = -\infty.$$

If x is a real number, then we define

$$\begin{aligned}
\infty + x &= x + \infty = \infty \\
-\infty + x &= x - \infty = -\infty \\
\frac{x}{\infty} &= \frac{x}{-\infty} = 0
\end{aligned}$$

and

$$\infty \times x = \begin{cases} \infty & \text{if } x > 0 \\ -\infty & \text{if } x < 0 \end{cases}$$

and

$$(-\infty) \times x = \begin{cases} -\infty & \text{if } x > 0 \\ \infty & \text{if } x < 0. \end{cases}$$

One important advantage of this arithmetic that includes the infinity symbols is that it simplifies the statements of the arithmetical rules for limits that we shall study in Section 7.5. For example, one of these rules tells us that if

$$\lim_{x \to a} f(x) = \lambda \quad \text{and} \quad \lim_{x \to a} g(x) = \mu,$$

then, under the right conditions, we should have[23]

$$\lim_{x \to a} (f(x) + g(x)) = \lambda + \mu.$$

In the event that $\lambda = \mu = \infty$, this rule tells us that if

$$\lim_{x \to a} f(x) = \infty \quad \text{and} \quad \lim_{x \to a} g(x) = \infty,$$

then

$$\lim_{x \to a} (f(x) + g(x)) = \infty + \infty = \infty,$$

which is a true statement. The arithmetical rules for limits do not work under all conditions, however. For example, if

$$\lim_{x \to a} f(x) = \infty \quad \text{and} \quad \lim_{x \to a} g(x) = -\infty,$$

then the limit

$$\lim_{x \to a} (f(x) + g(x))$$

---

[23]   The Greek letter λ is called lambda and is the Greek equivalent of the letter L. The next letter in the Greek alphabet is μ, which is pronounced "mu".

is quite unpredictable, and this is why we have left the expression $\infty - \infty$ undefined. For similar reasons, we have also left the expressions

$$-\infty + \infty, \quad \text{and} \quad 0 \times \infty, \quad \text{and} \quad \infty \times 0, \quad \text{and} \quad \frac{\infty}{\infty}$$

all undefined. Combinations of limits that lead to one of these undefined expressions are known as **indeterminate forms**, and, as you may know, these indeterminate forms require special techniques, such as L'Hôpital's rule. We shall study L'Hôpital's rule in Section 9.6.

5.14.3   The Order $<$ in the Extended Real Number System

We extend the usual order $<$ to the extended real number system $[-\infty, \infty]$ by agreeing that $-\infty < \infty$ and that for every real number x we have $-\infty < x$ and $x < \infty$.

5.14.4   Suprema and Infima in $[-\infty, \infty]$

Suppose that S is a set of real numbers and that S is unbounded above. Even though S has no upper bound in **R**, we can still think of ∞ as being an upper bound of S in the extended real number system; and, in this way, we can say that $\sup S = \infty$. Similarly, we can express the fact that a given set S of real numbers is unbounded below by saying that $\inf S = -\infty$.

5.14.5   Some Exercises on the Extended Real Number System

1. **H** Thinking of the rule for sums of limits

$$\lim_{x \to a} (f(x) + g(x)) = \lim_{x \to a} f(x) + \lim_{x \to a} g(x)$$

   that you saw in elementary calculus, give some examples to show why the expression $\infty + (-\infty)$ should not be defined.

2. Thinking of the rule for products of limits

$$\lim_{x \to a} (f(x)g(x)) = \left(\lim_{x \to a} f(x)\right)\left(\lim_{x \to a} g(x)\right)$$

   that you saw in elementary calculus, give some examples to show why the expression $\infty \times 0$ should not be defined.

3. Thinking of the rule for quotients of limits

$$\lim_{x \to a} \left(\frac{f(x)}{g(x)}\right) = \frac{\lim_{x \to a} f(x)}{\lim_{x \to a} g(x)}$$

   that you saw in elementary calculus, give some examples to show why the expression $\frac{\infty}{\infty}$ should not be defined.

4. Given that A and B are intervals and that $A \cap B \neq \emptyset$, prove that the set $A \cup B$ is an interval.

5. Given that A, B, and C are intervals and that the sets $A \cap B$ and $B \cap C$ are nonempty, prove that $A \cup B \cup C$ is an interval.

5.15   The Complex Number System (Optional)

An introduction to the complex number system can be found in the on-screen version of the book by clicking on the icon .

Chapter 6
Elementary Topology of the Real Line

This chapter presents the topological background that is needed for a study of analysis of functions of a single real variable. If you prefer to replace this chapter by a more extensive discussion that presents the topology of metric spaces, you can do so from the on-screen version of this book by clicking on the icon 📖 .

6.1   The Role of Topology

At the heart of mathematical analysis lies the notion of a limit and, in a sense, everything that we shall be doing from now on will be concerned with limits of one form or another. In Chapter 7 we shall discuss limits of sequences and then in Chapter 8 we shall apply what we have learned to the study of limits and continuity of functions. This, in turn, will prepare us for the study of derivatives and integrals, which is our main objective.

The key word in the title of this chapter is **topology**. When we speak of the *topology* of a mathematical system we mean those features of the system that make it possible to define limits and continuity. Since the main focus of our attention in this book is the system $\mathbf{R}$ of real numbers, we need to ask ourselves just what it is about the system $\mathbf{R}$ that makes it possible to define a limit of a sequence (x_n) or of a function $f : \mathbf{R} \to \mathbf{R}$. We hinted at the answer to this question in Section 5.4, where we introduced the properties of the order relation $<$ in $\mathbf{R}$. As we said there, it is the order relation $<$ that makes the theory of limits possible. As we said in Section 5.4, the order relation $<$ allows us to picture our number system as a line. With this picture in mind, we can define absolute value and we can think of the number $|x - y|$ as being the *distance* between two given numbers x and y. The notion of distance plays a major role in the theory of limits.

The topological structure of the number line $\mathbf{R}$ that we shall discuss in this chapter is a special case of the topological structure of more general mathematical systems in which the notion of a distance exists. For example, there is a notion of distance in the Euclidean plane $\mathbf{R}^2$ where we define the distance between two points (x_1, y_1) and (x_2, y_2) to be

$$\sqrt{(x_2 - x_1)^2 + (y_2 - y_1)^2},$$

and this definition is easy to extend to higher dimensional spaces. You can find a discussion of such spaces in the optional material on metric spaces that was mentioned at the beginning of this chapter. Finally, we mention that there are abstract mathematical systems that have a topological structure even though no notion of distance is defined in them. Such systems are known as *general topological spaces*. As we have said, we are concerned with the number line **R** in this book. We leave the study of more general systems for a more advanced course.

6.2   Interior Points and Neighborhoods

The definition of distance in the number line provides us with a way of saying that certain numbers that belong to a given set U are located "deep inside" U and away from its "boundary". A number of this type will be called an **interior point** of U, and when a number x is an interior point of a set U we shall say that the set U is a **neighborhood** of x. The concepts of interior point and neighborhood will be very useful when we develop the theory of limits of sequences and functions.

6.2.1   Definitions of Interior Point and Neighborhood

Suppose that x is a real number and that U is a set of real numbers. The number x is said to be an **interior point** of the set U if it is possible to find a number $\delta > 0$ such that

$$(x - \delta, x + \delta) \subseteq U.$$

$$x - \delta \qquad\qquad x \qquad\qquad x + \delta$$

In the event that x is an interior point of the set U, we say that the set U is a **neighborhood** of the number x.

6.2.2   Some Examples of Interior Points

1. Suppose that a and b are real numbers and that $a < b$. Then every number x in the interval (a, b) is an interior point of (a, b). To see this, suppose that $x \in (a, b)$. We need to find a number $\delta > 0$ such that $(x - \delta, x + \delta) \subseteq (a, b)$, and, for this purpose, we define δ to be the smaller of the two numbers $x - a$ and $b - x$. The following figure illustrates the case in which $\delta = x - a$.

$$a \qquad x \qquad x + \delta \qquad\qquad\qquad b$$
$$x - \delta$$

Certainly $\delta > 0$. Now, since $\delta \leq b - x$, we see that

$$x + \delta \leq x + (b - x) = b$$

and since $\delta \leq x - a$, we see that

$$x - \delta \geq x - (x - a) = a.$$

Therefore $(x - \delta, x + \delta) \subseteq (a, b)$, as promised.

2. If a and b are real numbers and $a \leq b$, then neither of the numbers a and b is an interior point of the interval $[a, b]$.
3. Since every interval of positive length contains some rational numbers, no number can be an interior point of the set $\mathbf{R} \setminus \mathbf{Q}$ of irrational numbers. Since every interval of positive length contains some irrational numbers, no number can be an interior point of the set $\mathbf{Q}$ of rational numbers.

Before stating the next theorem we observe that if x is any number, then larger sets are more likely to be neighborhoods of x than smaller sets. More precisely, if U is a neighborhood of x, then any set V that includes U as a subset must also be a neighborhood of x. The next theorem goes in the other direction by discussing the intersection of neighborhoods of a given number.

6.2.3   Intersection of Finitely Many Neighborhoods

The intersection of finitely many neighborhoods of a given number is a neighborhood of that number.

Proof. Suppose that the sets

$$U_1, U_2, \ldots, U_n$$

are all neighborhoods of a given number x. For each $j = 1, 2, \ldots, n$ we use the fact that U_j is a neighborhood of x to choose a number $\delta_j > 0$ such that

$$(x - \delta_j, x + \delta_j) \subseteq U_j.$$

Now we define δ to be the smallest of all of these numbers δ_j. Then $\delta > 0$, and since

$$(x - \delta, x + \delta) \subseteq (x - \delta_j, x + \delta_j) \subseteq U_j$$

for each j we have

$$(x - \delta, x + \delta) \subseteq \bigcap_{j=1}^{n} U_j. \quad \blacksquare$$

6.2.4   Intersection of Infinitely Many Neighborhoods

Although the intersection of finitely many neighborhoods of a number x must always be a neighborhood of x, the intersection of infinitely many neighborhoods of a number x can fail to be a neighborhood of x. For an example of this behavior, observe that if n is any positive integer, then the interval $\left(-\frac{1}{n}, \frac{1}{n}\right)$ is a neighborhood of the number 0. It is easy to see that

$$\bigcap_{n=1}^{\infty} \left(-\frac{1}{n}, \frac{1}{n}\right) = \{0\},$$

which is not a neighborhood of 0.

6.2.5   Exercises on Neighborhoods

1. Complete the following sentence: *"A set U fails to be a neighborhood of a number x when for every number $\delta > 0$, ..."*
2. Explain carefully why the assertion

$$\bigcap_{n=1}^{\infty} \left(-\frac{1}{n}, \frac{1}{n}\right) = \{0\}$$

    that was made in Subsection 6.2.4 is true. You will need to make use of Theorem 5.9.4.
3. Given that x is an interior point of U and that $U \subseteq V$, explain why x must be an interior point of V.
4. ▦ Suppose that x is a real number and that $U \subseteq \mathbf{R}$. Prove that the following two conditions are equivalent:
    (a) The set U is a neighborhood of the number x.
    (b) It is possible to find two numbers a and b such that

$$x \in (a, b) \subseteq U.$$

5. ▦ Suppose that x and y are two different real numbers. Prove that it is possible to find a neighborhood U of x and a neighborhood V of y such that $U \cap V = \emptyset$.
6. ▦ Given that S is a set of real numbers and that x is an upper bound of S, explain why S cannot be a neighborhood of x.
7. ▦ Given that a set S of real numbers is nonempty and bounded above, explain why neither S nor $\mathbf{R} \setminus S$ can be a neighborhood of $\sup S$.
8. ▦ Suppose that A and B are sets of real numbers and that x is an interior point of the set $A \cup B$. Is it true that x must either be an interior point of A or an interior point of B?
9. Suppose that A and B are sets of real numbers and that x is an interior point both of A and of B. Is it true that x must be an interior point of the set

$A \cap B$?

10. **H** Suppose that x and y are real numbers and that U is a neighborhood of y. Prove that the set V defined by

$$V = \{x + u \mid u \in U\}$$

is a neighborhood of the number $x + y$.

6.3 Open Sets and Closed Sets

6.3.1 Definition of an Open Set

A set U of real numbers is said to be **open** if every member of U is an interior point of U. In other words, a set U is said to be open if U is a neighborhood of every one of its members.

A set U of real numbers fails to be open when there is at least one number x in U such that x is not an interior point of U.

6.3.2 Definition of a Closed Set

A set H of real numbers is said to be **closed** if the set $\mathbf{R} \setminus H$ is open.

Thus if H is a set of real numbers and $U = \mathbf{R} \setminus H$, then, since each of the sets H and U is the complement of the other, the set H is closed if and only if U is open.

6.3.3 A Word of Warning

The words *open* and *closed* may be a little misleading because they may lead you to believe that a set of real numbers must either be open or it must be closed and that, in order to show that a given set is closed, all we have to do is to show that the set isn't open.

But nothing could be further from the truth. In fact, most sets of real numbers are neither open nor closed and there are some sets that are both open and closed. You will see examples of such sets soon. Thus, if we want to show that a given set H is closed, we are not trying to show that H isn't open. To show that H is closed, we need to show that its complement $\mathbf{R} \setminus H$ is open.

6.3.4 Some Examples of Open Sets and Closed Sets

As you will see from the examples in this subsection, many sets are neither open nor closed, some sets are open but not closed, some sets are closed but not open and some sets are both open and closed.

1. The set $\mathbf{R}$ of all real numbers is open. To see why, suppose that x is any real number. In order to prove that x must be an interior point of the set $\mathbf{R}$, we

need to prove that there exists a number $\delta > 0$ such that

$$(x - \delta, x + \delta) \subseteq \mathbf{R}.$$

Following the technique discussed in Subsection 3.5.1, we prove the existence of such a number δ by giving an example of one. As a matter of fact, since

$$(x - 1, x + 1) \subseteq \mathbf{R},$$

we can give an example by taking $\delta = 1$.

2. In order for a given set S to fail to be open there has to be at least one member of S that fails to be an interior point of S. Such a set S must therefore be nonempty, and so we conclude that the empty set $\emptyset$ is open and that its complement $\mathbf{R}$ is closed.

3. In Example 1 of Subsection 6.2.2 we saw that every member of an open interval of the form (a, b) is an interior point of that interval. Therefore every open interval is an open subset of $\mathbf{R}$. In the same way we can see that open intervals of the form $(-\infty, a)$ and (a, ∞) must also be open sets.

4. Suppose that a and b are real numbers and that $a \leq b$. Since the set $[a, b]$ is not a neighborhood of either of the numbers a and b, we deduce that $[a, b]$ is not an open set. On the other hand, the set

$$\mathbf{R} \setminus [a, b] = (-\infty, a) \cup (b, \infty)$$

is clearly open, and so the closed interval $[a, b]$ is a closed subset of $\mathbf{R}$.

5. Suppose that a and b are real numbers and that $a < b$. Since the set $[a, b)$ fails to be a neighborhood of its member a, the set $[a, b)$ is not open. The complement of $[a, b)$ is the set

$$\mathbf{R} \setminus [a, b) = (-\infty, a) \cup [b, \infty),$$

which fails to be a neighborhood of its member b. We conclude that the set $[a, b)$ is neither open nor closed.

6. In Example 3 of Subsection 6.2.2 we saw that neither the set $\mathbf{Q}$ of rational numbers nor the set $\mathbf{R} \setminus \mathbf{Q}$ of irrational numbers can be a neighborhood of any number. Therefore the set $\mathbf{Q}$ is neither open nor closed.

7. In this example we shall observe that the set $\mathbf{Z}$ of integers is closed. We need to show that $\mathbf{R} \setminus \mathbf{Z}$ is open, and, in order to demonstrate this fact, we suppose that $x \in \mathbf{R} \setminus \mathbf{Z}$. If n is the largest integer that does not exceed x, then we have

$$n < x < n + 1.$$

Since the set $(n, n+1)$ is a neighborhood of x, it follows at once that the larger set $\mathbf{R} \setminus \mathbf{Z}$ is also a neighborhood of x. Thus $\mathbf{R} \setminus \mathbf{Z}$ is open and we conclude that $\mathbf{Z}$ is closed.

8. Suppose that

$$S = \left\{ \frac{1}{n} \mid n \in \mathbf{Z}^+ \right\}.$$

Since no real number can be an interior point of S, we know that S cannot be open. Furthermore, $0 \in \mathbf{R} \setminus S$, but for every number $\delta > 0$ there are positive integers n such that

$$\frac{1}{n} \in (0 - \delta, 0 + \delta).$$

Therefore, although 0 is a member of $\mathbf{R} \setminus S$, the number 0 is not an interior point of $\mathbf{R} \setminus S$, and so the set $\mathbf{R} \setminus S$ also fails to be open. We conclude that the set S is neither open nor closed.

9. If

$$S = \left\{ \frac{1}{n} \mid n \in \mathbf{Z}^+ \right\} \cup \{0\},$$

then S is closed but not open. We leave the proof of this assertion as an exercise.

6.4   Some Properties of Open Sets and Closed Sets

In this section we explore some simple properties of open sets and closed sets that follow from the definitions.

6.4.1   Unions and Intersections of Open Sets and Closed Sets

1. *The intersection of finitely many open sets is open.*
2. *The union of any sequence of open sets is open.*
3. *The union of finitely many closed sets is closed.*
4. *The intersection of any sequence of closed sets is closed.*

Proof.

1. To prove part 1 of the theorem, suppose that

$$\{U_1, U_2, \cdots, U_n\}$$

is a finite family of open sets and suppose that x is a number that lies in the intersection of this family. In other words,

$$x \in \bigcap_{j=1}^{n} U_j.$$

We need to explain why the set $\bigcap_{j=1}^{n} U_j$ is a neighborhood of x. But this fact follows from Theorem 6.2.3 and the fact that each set U_j, being an open set that contains the number x, must be a neighborhood of x.

2. To prove part 2 of the theorem, suppose that (U_n) is a sequence of open sets, and suppose that the union of all these open sets has been called U. To show that this set U is open, suppose that $x \in U$. This means that x must belong to the set U_n for at least one value of n, and, using this fact, we choose[24] an integer n such that $x \in U_n$. Since U_n is a neighborhood of x and since $U_n \subseteq U$, we deduce that U is a neighborhood of x.

3. To prove part 3 of the theorem, suppose that

    $$\{H_1, H_2, \cdots, H_n\}$$

    is a finite family of closed sets. Then for each $j = 1, 2, \cdots, n$ the set $\mathbf{R} \setminus H_j$ is open, and since

    $$\mathbf{R} \setminus \bigcup_{j=1}^{n} H_j = \bigcap_{j=1}^{n} (\mathbf{R} \setminus H_j),$$

    which is open by part 1 of the theorem, we deduce that the set $\bigcup_{j=1}^{n} H_j$ is closed.

4. To prove part 4 of the theorem, suppose that (H_n) is a family of closed sets. Then for every positive integer n, the set $\mathbf{R} \setminus H_n$ is open, and since

    $$\mathbf{R} \setminus \bigcap_{n=1}^{\infty} H_n = \bigcup_{n=1}^{\infty} (\mathbf{R} \setminus H_n),$$

    which is open by part 2 of the theorem, we deduce that the set $\bigcap_{n=1}^{\infty} H_n$ is closed. ∎

6.4.2   Existence of a Largest Member

Every nonempty closed set that is bounded above must have a largest member.

Proof. Suppose that H is a nonempty closed set and that H is bounded above. In order to prove that H has a largest member, we shall show that $\sup H$ belongs to

---

[24]   See Subsection 3.5.2 to see how the word *choose* is used in a mathematical proof.

H. We write $\alpha = \sup H$, and, to obtain a contradiction, we assume that α does not belong to H. Then, of course, α belongs to the open set $\mathbf{R} \setminus H$, and, using this fact, we choose $\delta > 0$ such that

$$(\alpha - \delta, \alpha + \delta) \subseteq \mathbf{R} \setminus H.$$

Since no member of H can be greater than α and since no member of H can belong to the interval $(\alpha - \delta, \alpha + \delta)$, we see that the number $\alpha - \delta$ is an upper bound of H, contradicting the fact that α is the *least* upper bound of H. ∎

In the same way we can show that every nonempty closed set that is bounded below must have a least member. We leave the proof of this fact as an exercise.

6.4.3   The Sets that Are both Open and Closed

One of the consequences of the completeness of the number system $\mathbf{R}$ is that sets that are both open and closed are quite rare. As the following theorem shows, the sets $\mathbf{R}$ and $\emptyset$ are the only sets of this type.

The sets $\mathbf{R}$ and $\emptyset$ are the only subsets of $\mathbf{R}$ that are both open and closed.

Proof. Suppose that S is an open closed subset of $\mathbf{R}$ and, to obtain a contradiction, suppose that the set S is nonempty and that the set $\mathbf{R} \setminus S$ is also nonempty. Using the fact that these two sets are nonempty, we choose a member a of the set S and a member b of the set $\mathbf{R} \setminus S$. We may assume, without loss of generality, that $a < b$. We now define

$$E = S \cap (-\infty, b).$$

From the fact that both of the sets S and $(-\infty, b)$ are open, we deduce that the set E is open. But since $b \notin S$, we also have

$$E = S \cap (-\infty, b],$$

and we deduce from the fact that both S and $(-\infty, b]$ are closed that the set E is closed.

We observe that E is nonempty because $a \in E$ and that the number b is an upper bound of E. Thus by Theorem 6.4.2 we know that E has a largest member, which we shall call α. Using the fact that E is open we now choose a number $\delta > 0$ such that

$$(\alpha - \delta, \alpha + \delta) \subseteq E,$$

which is impossible since α is the largest member of E. ∎

6.4.4　Exercises on Open Sets and Closed Sets

1. ▨ Explain why, if U is open and H is closed, then the set $U \setminus H$ must be open.
2. Explain why, if U is open and H is closed, then the set $H \setminus U$ must be closed.
3. Give an example of an infinite family of open sets whose intersection fails to be open.
4. Give an example of an infinite family of closed sets whose union fails to be closed.
5. ▨ Give an example of two sets A and B, neither of which is open but for which the set $A \cup B$ is open.
6. Given a set H of real numbers, prove that the following conditions are equivalent:
   (a) The set H is closed.
   (b) For every number $x \in \mathbf{R} \setminus H$ it is possible to find a number $\delta > 0$ such that
   $$(x - \delta, x + \delta) \cap H = \emptyset.$$
7. (a) Given any number x, prove that the singleton $\{x\}$ is closed.
   (b) Use part a and the fact that every finite set is a finite union of singletons to deduce that every finite set is closed.
8. ▨ Given that $\mathbf{Q} \subseteq H$ and that H is closed, prove that $H = \mathbf{R}$.
9. Given that H is closed, nonempty, and bounded below, prove that H must have a least member.
10. Prove that no open set can have a largest member.
11. Given a real number x and a set S of real numbers, prove that the following two conditions are equivalent:
    (a) The number x is an interior point of S.
    (b) It is possible to find an open set U such that $x \in U \subseteq S$.
12. Prove that if S is any set of real numbers, then the set of all interior points of S must be open.
13. ▨ This exercise refers to the sum of two sets as it was defined in Exercise 12 in Subsection 5.6.6. Prove that if A is any set of real numbers and U is an open set, then the set $A + U$ must be open.

6.5   The Closure of a Set

The closure of a given subset S of $\mathbf{R}$ is the set that contains all of the numbers that are, in a certain sense, "close" to the set S. Any set S is a subset of its closure and, as you will see soon, a set S is equal to its closure if and only if it is closed.

6.5.1   Definition of Closure

Suppose that S is a set of real numbers and that x is any real number. We say that the number x is **close** to the set S if for every number $\delta > 0$ we have

$$(x - \delta, x + \delta) \cap S \neq \emptyset.$$

If S is a set of real numbers, then the set of all those numbers that are close to S is called the **closure** of the set S and is written as $\overline{S}$.

6.5.2   Some Examples Illustrating Closure

1. For every number $\delta > 0$ it is clear that

$$(1 - \delta, 1 + \delta) \cap [0, 1) \neq \emptyset,$$

and we deduce that the number 1 lies in the closure of the set $[0, 1)$.

If $x > 1$ and if we choose any positive number δ that does not exceed the positive number $x - 1$, then it is clear that

$$(x - \delta, x + \delta) \cap [0, 1) = \emptyset.$$

Thus if $x > 1$, then x cannot lie in the closure of the interval $[0, 1)$ and we see similarly that no negative number lies in the closure of this interval. We have therefore shown that the closure of the interval $[0, 1)$ is the closed interval $[0, 1]$.

2. Suppose that x is any real number and that $\delta > 0$. Since the interval $(x - \delta, x + \delta)$ must contain some rationals, we deduce that $x \in \overline{\mathbf{Q}}$. Thus $\overline{\mathbf{Q}} = \mathbf{R}$. In a similar manner we can see that $\overline{\mathbf{R} \setminus \mathbf{Q}} = \mathbf{R}$.

3. Given any number x and given $\delta > 0$ we have

$$(x - \delta, x + \delta) \cap \emptyset = \emptyset,$$

and so $\overline{\emptyset} = \emptyset$.

6.5.3   Describing Closure in Terms of Neighborhoods

Suppose that S is a set of real numbers and that x is a given number. Then the following two conditions are equivalent:

1. *The number x lies in the set $\overline{S}$.*
2. *For every neighborhood U of the number x we have $U \cap S \neq \emptyset$.*

Proof. To prove that condition 2 implies condition 1 we assume that condition 2 holds. To prove that $x \in \overline{S}$, we suppose[25] that $\delta > 0$. Since the interval $(x - \delta, x + \delta)$ is a neighborhood of x, we know that

$$(x - \delta, x + \delta) \cap S \neq \emptyset,$$

and we conclude that $x \in \overline{S}$.

Now to prove that condition 1 implies condition 2 we assume that condition 1 holds. To prove that condition 2 holds we suppose that U is a neighborhood of the number x. From the definition of a neighborhood we know that there exists a number $\delta > 0$ such that

$$(x - \delta, x + \delta) \subseteq U,$$

and we choose such a number δ. Since

$$(x - \delta, x + \delta) \cap S \neq \emptyset,$$

we see at once that $U \cap S \neq \emptyset$. ∎

6.5.4   Two Simple Facts About Closure

1. Suppose that S is a set of real numbers. If x is any member of S and $\delta > 0$, then the set

   $$(x - \delta, x + \delta) \cap S$$

   must be nonempty because it contains the number x itself. We deduce that $S \subseteq \overline{S}$.
2. Suppose that A and B are sets of real numbers and that $A \subseteq B$. If x is any number that lies in $\overline{A}$ and $\delta > 0$, then, since

   $$\emptyset \neq (x - \delta, x + \delta) \cap A \subseteq (x - \delta, x + \delta) \cap B,$$

   it follows that $x \in \overline{B}$. Therefore $\overline{A} \subseteq \overline{B}$.

---

[25]   See Subsection 3.6.2 for the use of the word *suppose* in a mathematical proof.

6.5.5   The Closure of Any Set is Closed

Suppose that S is a set of real numbers. Then the set $\overline{S}$ is closed.

Proof. To prove that the set $\overline{S}$ is closed we need to explain why the set $\mathbf{R} \setminus \overline{S}$ is open. In other words, we need to explain why every number that belongs to the set $\mathbf{R} \setminus \overline{S}$ must be an interior point of $\mathbf{R} \setminus \overline{S}$. Suppose that $x \in \mathbf{R} \setminus \overline{S}$. Using the fact that x is not close to the set S we choose a number $\delta > 0$ such that

$$(x - \delta, x + \delta) \cap S = \emptyset.$$

Now if y is any number that belongs to the interval $(x - \delta, x + \delta)$, then, since $(x - \delta, x + \delta)$ is a neighborhood of y and since $(x - \delta, x + \delta)$ does not intersect with S, we see that y cannot be close to the set S. We have therefore shown that

$$(x - \delta, x + \delta) \subseteq \mathbf{R} \setminus \overline{S},$$

and we conclude that x is indeed an interior point of $\mathbf{R} \setminus \overline{S}$. ∎

6.5.6   Theorem on Closed Sets and Closures

Suppose that S is a set of real numbers. Then the following two conditions are equivalent:

1. *The set S is closed.*
2. *We have $S = \overline{S}$.*

Proof. In the event that $S = \overline{S}$, it follows from the fact that $\overline{S}$ is closed that the set S must be closed.

Suppose now that the set S is closed. We already know that $S \subseteq \overline{S}$. To show that $\overline{S} \subseteq S$ we shall show that $\mathbf{R} \setminus S \subseteq \mathbf{R} \setminus \overline{S}$. Suppose that $x \in \mathbf{R} \setminus S$. Since S is closed, the set $\mathbf{R} \setminus S$ is a neighborhood of x, and since this neighborhood of x does not intersect with the set S, we see at once that x cannot belong to $\overline{S}$. ∎

6.5.7   Exercises on Closure

1. Suppose that

    $$S = [0, 1) \cup (1, 2).$$

    (a) What is the set of interior points of S?
    (b) Given that U is the set of interior points of S, evaluate $\overline{U}$.
    (c) Give an example of a set S of real numbers such that, if U is the set of interior points of S, then $\overline{U} \neq \overline{S}$.
    (d) Give an example of a subset S of the interval $[0, 1]$ such that $\overline{S} = [0, 1]$

but, if U is the set of interior points of S, then $\overline{U} \neq [0, 1]$.

2. ▉ Given that

$$S = \left\{ \frac{1}{n} \mid n \in \mathbf{Z}^+ \right\},$$

evaluate $\overline{S}$.

3. Given that S is a set of real numbers, that H is a closed set, and that $S \subseteq H$, prove that $\overline{S} \subseteq H$.

4. ▉ Given two sets A and B of real numbers, prove that

$$\overline{A \cup B} = \overline{A} \cup \overline{B}.$$

5. Given two sets A and B of real numbers, prove that

$$\overline{A \cap B} \subseteq \overline{A} \cap \overline{B}.$$

   Do the two sides of this inclusion have to be equal? What if A and B are open? What if they are closed?

6. Prove that if S is any set of real numbers, then the set $\mathbf{R} \setminus \overline{S}$ is the set of interior points of the set $\mathbf{R} \setminus S$.

7. Given that α is an upper bound of a given set S of real numbers, prove that the following two conditions are equivalent:

   (a) We have $\alpha = \sup S$.
   (b) We have $\alpha \in \overline{S}$.

8. Is it true that if A and B are sets of real numbers and

$$\overline{A} = \overline{B} = \mathbf{R},$$

   then $\overline{A \cap B} = \mathbf{R}$?

9. ▉ Prove that if A and B are open sets and

$$\overline{A} = \overline{B} = \mathbf{R},$$

   then $\overline{A \cap B} = \mathbf{R}$. What if only one of the two sets A and B is open?

10. Two sets A and B are said to be **separated** from each other if

$$\overline{A} \cap B = A \cap \overline{B} = \emptyset.$$

   Which of the following pairs of sets are separated from each other?

   (a) $[0, 1]$ and $[2, 3]$.
   (b) $(0, 1)$ and $(1, 2)$.
   (c) $(0, 1]$ and $(1, 2)$.

(d) $\mathbf{Q}$ and $\mathbf{R} \setminus \mathbf{Q}$.

11. Prove that if A and B are closed and disjoint from one another, then A and B are separated from each other.

12. Prove that if A and B are open and disjoint from one another, then A and B are separated from each other.

13. Suppose that S is a set of real numbers. Prove that the two sets S and $\mathbf{R} \setminus S$ will be separated from each other if and only if the set S is both open and closed. What then do we know about the sets S for which S and $\mathbf{R} \setminus S$ are separated from each other?

14. ▉ This exercise refers to the notion of a subgroup of $\mathbf{R}$ that was introduced in Exercise 3 of Subsection 5.9.6. That exercise should be completed before you start this one.

    (a) Given that H and K are subgroups of $\mathbf{R}$, prove that the set $H + K$ defined in the sense of Exercise 12 of Subsection 5.6.6 is also a subgroup of $\mathbf{R}$.

    (b) Prove that if a, b, and c are integers and if

$$a\sqrt{2} = b\sqrt{3} + c,$$

    then $a = b = c = 0$.

    (c) Prove that if m, n, p, and q are integers, then it is impossible to have

$$\frac{\sqrt{2} - m}{n} = \frac{\sqrt{3} - p}{q};$$

    and deduce that if α is any real number and if $H = \{n\alpha \mid n \in \mathbf{Z}\}$, then the subgroup $H + \mathbf{Z}$ cannot contain both of the numbers $\sqrt{2}$ and $\sqrt{3}$.

    (d) Suppose that G is a subgroup of $\mathbf{R}$ other than $\{0\}$, that

$$p = \inf \{x \in G \mid x > 0\},$$

    and that the number p is positive. Prove that the set G is closed.

    (e) Prove that if G is a subgroup of $\mathbf{R}$ other than $\{0\}$ and that G has no least positive member, then $\overline{G} = \mathbf{R}$.

    (f) Suppose that α is an irrational number, that

$$H = \{n\alpha \mid n \in \mathbf{Z}\},$$

    and that $G = H + \mathbf{Z}$. Prove that although the sets H and $\mathbf{Z}$ are closed subgroups of $\mathbf{R}$ and although the set G is also a subgroup of $\mathbf{R}$, the set G is not closed.

6.6   Limit Points

6.6.1   Introduction to Limit Points

A limit point of a given set S of real numbers is a number that lies close to the set S in a sense that is a little stronger than the kind of closeness that we defined in Section 6.5. As we saw, the condition that must be satisfied in order for a number to be close to a set S is that for every number $\delta > 0$ the interval $(x - \delta, x + \delta)$ contains at least one member of the set S. The stronger condition that must be satisfied in order for a number x to be a limit point of a given set S is that for every number $\delta > 0$ the interval $(x - \delta, x + \delta)$ contains at least one member of the set S other than the number x itself.

It turns out that a number that satisfies the latter condition is a number where the set S is very "crowded". What we mean by this is that every neighborhood of a limit point of a set must contain infinitely many different members of that set.

6.6.2   Definition of a Limit Point

Suppose that S is a set of real numbers and that x is any real number. We say that the number x is a **limit point** of the set S if for every number $\delta > 0$ we have

$$(x - \delta, x + \delta) \cap S \setminus \{x\} \neq \emptyset.$$

Equivalently, x is a limit point of S if and only if for every neighborhood U of x we have

$$U \cap S \setminus \{x\} \neq \emptyset.$$

The set of all limit points of a given set S is written as $\mathbf{L}\,(S)$. Limit points of a set are sometimes called **accumulation points** of the set.

6.6.3   Comparing the Set of Limit Points and the Closure of a Set

If S is a set of real numbers and x is a real number, then the condition that x be a limit point of S requires the set

$$(x - \delta, x + \delta) \cap S \setminus \{x\}$$

to be nonempty for every $\delta > 0$. This condition is more demanding than the condition that x merely be close to the set S, which requires the slightly larger set

$$(x - \delta, x + \delta) \cap S$$

to be nonempty for every $\delta > 0$. We see, therefore, that $\mathbf{L}\,(S) \subseteq \overline{S}$. Furthermore, if a number x does not belong to a set S, then, since

$$(x - \delta, x + \delta) \cap S \setminus \{x\} = (x - \delta, x + \delta) \cap S$$

for every $\delta > 0$, the conditions $x \in \overline{S}$ and $x \in \mathbf{L}(S)$ are also exactly the same. Thus only numbers that belong to the set S can be in $\overline{S}$ without belonging to $\mathbf{L}(S)$.

As you will see in the examples that follow, a member of a set may or may not be a limit point of the set and a limit point of a set may or may not be a member of the set. However, since every number in a set S must belong to $\overline{S}$, and since a number that does not belong to S will belong to $\overline{S}$ if and only if it belongs to $\mathbf{L}(S)$, we have

$$\overline{S} = S \cup \mathbf{L}(S).$$

We conclude that since a set S of real numbers is closed if and only if $S = \overline{S}$, a set S is closed if and only if every limit point of S belongs to S.

6.6.4   Some Examples of Limit Points

1. If
$$S = (0, 1) \cup \{2\},$$

   then
$$\overline{S} = [0, 1] \cup \{2\} \quad \text{and} \quad \mathbf{L}(S) = [0, 1].$$

   Note that 2 is a member of S but is not a limit point of S. Note that the numbers 0 and 1 are limit points of S even though they do not belong to S. The numbers that lie in the open interval $(0, 1)$ are both members of S and limit points of S.

2. If
$$S = \left\{ \frac{1}{n} \mid n \in \mathbf{Z}^+ \right\},$$

   then 0 is the only limit point of S. In other words, $\mathbf{L}(S) = \{0\}$.

3. $\mathbf{L}(\mathbf{Z}) = \emptyset$.

4. $\mathbf{L}(\mathbf{Q}) = \mathbf{R}$.

6.6.5   An Important Fact About Limit Points

Suppose that S is a set of real numbers and that x is a real number.

1. *If the number x has a neighborhood that contains only finitely many numbers in the set S, then x cannot be a limit point of S.*

2. *If x is a limit point of S, then every neighborhood of x must contain infinitely many different numbers that belong to the set S.*

Proof.   Part 2 of this theorem follows immediately from part 1, so our task is to prove part 1. Suppose that x has a neighborhood U for which the set $U \cap S$ is finite. Using the fact that the set $U \cap S \setminus \{x\}$ is also finite, we write it in the form

$$U \cap S \setminus \{x\} = \{y_1, y_2, \cdots, y_n\}.$$

The way in which we shall show that x is not a limit point of the set S is to give an example of a positive number δ such that the interval $(x - \delta, x + \delta)$ doesn't contain any of the numbers $y_1, y_2, \cdots, y_n$. This requirement will ensure that the set

$$U \cap (x - \delta, x + \delta)$$

does not contain any member of the set $S \setminus \{x\}$. With this idea in mind we define δ to be the smallest of all the numbers

$$|x - y_1|, \ |x - y_2|, \ |x - y_3|, \ \cdots, \ |x - y_n|.$$

Since none of the numbers $y_1, y_2, \cdots, y_n$ can be equal to x, we know that $\delta > 0$. Since the set

$$U \cap (x - \delta, x + \delta)$$

is a neighborhood of x that contains no numbers at all in the set $S \setminus \{x\}$, the number x can't be a limit point of S. ∎

6.6.6   Corollary: Finite Sets Never Have Limit Points

No finite set can have a limit point.

6.6.7   Exercises on Limit Points

1. Prove that $\mathbf{L}(\mathbf{Z}) = \emptyset$.
2. Prove that $\mathbf{L}(\mathbf{Q}) = \mathbf{R}$.
3. Prove that $\mathbf{L}\left(\left\{\frac{1}{n} \mid n \in \mathbf{Z}^+\right\}\right) = \{0\}$.
4. (a) Give an example of an infinite set that has no limit point.
   (b) Give an example of a bounded set that has no limit point.
   (c) Give an example of an unbounded set that has no limit point.

(d) Give an example of an unbounded set that has exactly one limit point.

(e) Give an example of an unbounded set that has exactly two limit points.

5. Prove that if A and B are sets of real numbers and if $A \subseteq B$, then $\mathbf{L}(A) \subseteq \mathbf{L}(B)$.

6. ▣ Prove that if A and B are sets of real numbers, then

$$\mathbf{L}(A \cup B) = \mathbf{L}(A) \cup \mathbf{L}(B).$$

7. Is it true that if A and B are sets of real numbers, then

$$\mathbf{L}(A \cap B) = \mathbf{L}(A) \cap \mathbf{L}(B)?$$

What if A and B are closed? What if A and B are open? What if A and B are intervals?

8. ▣ Is it true that if $\overline{D} = \mathbf{R}$, then $\mathbf{L}(D) = \mathbf{R}$?

9. Given that a set S of real numbers is nonempty and bounded above but that S does not have a largest member, prove that $\sup S$ must be a limit point of S. State and prove a similar result about $\inf S$.

10. ▣ Given any set S of real numbers, prove that the set $\mathbf{L}(S)$ must be closed.

11. Prove that if a set U is open, then $\mathbf{L}(U) = \overline{U}$.

12. ▣ Suppose that S is a set of real numbers, that $\mathbf{L}(S) \neq \emptyset$, and that $\delta > 0$. Prove that there exist two different numbers x and y in S such that $|x - y| < \delta$.

6.7   Neighborhoods of Infinity

When we added the two infinity symbols ∞ and $-\infty$ in Section 5.14 to the number system $\mathbf{R}$ to make the extended real number system $[-\infty, \infty]$ we were motivated by our desire to make statements of the form

$$\lim_{x \to a} f(x) = \lambda$$

include the possibilities that a and λ may be infinite. For example, if $\alpha = \infty$, then we are saying that

$$\lim_{x \to a} f(x) = \infty.$$

The concept of a neighborhood will be useful when we give a precise definition of limits of this type, and it will help us to draw an analogy between the definition of a finite limit and the definition of an infinite limit.

6.7.1   Defining Neighborhoods of ∞ and $-\infty$

A set U of real numbers is said to be a **neighborhood of** ∞ if there exists a real number w such that
$$(w, \infty) \subseteq U.$$

Similarly, a set U of real numbers is said to be a neighborhood of $-\infty$ if there exists a real number w such that
$$(-\infty, w) \subseteq U.$$

6.7.2   Some Examples of Neighborhoods of ∞

1. The set $\mathbf{R}$ of all real numbers is a neighborhood both of ∞ and of $-\infty$.
2. The interval $[2, \infty)$ is a neighborhood of ∞.
3. The interval $[2, \infty]$ is not a set of real numbers and so it does not fit the definition of a neighborhood of ∞.
4. The set $\mathbf{Q}$ of rational numbers is not a neighborhood of ∞.
5. The set $\mathbf{Q} \cup [2, \infty)$ is a neighborhood of ∞.

6.7.3   Some Simple Facts About Neighborhoods of ∞

1. Any set of real numbers that includes a neighborhood of ∞ is also a neighborhood of ∞. The same applies to neighborhoods of $-\infty$.
2. If U and V are neighborhoods of ∞, then so is the set $U \cap V$.
3. If U is a neighborhood of ∞ and S is unbounded above, then $U \cap S \neq \emptyset$.

Chapter 7
Limits of Sequences

This chapter presents the theory of limits of sequences of real numbers. If you chose to read the more general presentation of topology of metric spaces in place of the topology of the real line, and if you prefer to read the theory of limits of sequences in this more general situation, you can do so in the on-screen version of this book by clicking on the icon [icon].

If you would like to review the introduction to sequences, you can find it in Section 5.11 on page 98. As you can see there, if (x_n) is a given sequence, then for some integer k, the domain of (x_n) is the set of all integers $n \geq k$. In order to simplify our notation we shall adopt the convention that if the domain of a given sequence has not been mentioned explicitly, then $k = 1$.

Now that we have laid the topological groundwork of the real number system, we can begin the study of limits. In this chapter we shall be looking at limits of sequences. Then, in Chapter 8, we shall study limits and continuity of functions.

7.1 The Concepts "Eventually" and "Frequently"

According to the definitions in Section 5.11, a sequence (x_n) is said to be in a given set S if $x_n \in S$ for every integer n in its domain. We shall now broaden this concept a little and say what it means for a sequence to be *eventually* in a given set S and what it means for a sequence to be *frequently* in S. These notions will play a fundamental role when we discuss limits of sequences.

7.1.1 Definitions of "Eventually" and "Frequently"

A sequence (x_n) is said to be **eventually** in a given set S if there exists an integer N such that $x_n \in S$ for every integer $n \geq N$.

A sequence (x_n) is said to be **frequently** in a given set S if the condition $x_n \in S$ holds for infinitely many integers n in the domain of the sequence.

7.1.2 Some Examples Illustrating "Frequently" and "Eventually"

1. If we define

$$x_n = \begin{cases} 1 & \text{if} \quad 1 \leq n \leq 20 \\ 0 & \text{if} \quad n > 20, \end{cases}$$

then the sequence (x_n) is eventually in any set that contains the number 0.

2. If we define

$$x_n = 1 + (-1)^n$$

for every positive integer n, then the sequence (x_n) is frequently in the set $\{0\}$ and frequently in the set $\{2\}$ and is eventually in any set that includes the set $\{0, 2\}$.

3. Suppose that

$$x_n = \frac{1}{n}$$

for every positive integer n and that $\varepsilon > 0$. Then the sequence (x_n) is eventually in the interval $(0, \varepsilon)$ because $x_n \in (0, \varepsilon)$ for every integer $n > \frac{1}{\varepsilon}$.

7.1.3   Some Simple Observations About "Eventually" and "Frequently"

1. If a sequence is in a set S, then it is eventually in the set S.
2. If a sequence is eventually in a set S, then it is frequently in the set S.
3. A sequence (x_n) is frequently in a set S if and only if the set of those integers n for which $x_n \in S$ is unbounded above.
4. If (x_n) is a sequence of real numbers and S is a set of real numbers, then the sequence (x_n) is eventually in S if and only if it is not frequently in $\mathbf{R} \setminus S$.
5. If (x_n) is a sequence of real numbers and S is a set of real numbers, then the sequence (x_n) is frequently in S if and only if (x_n) is not eventually in $\mathbf{R} \setminus S$.

7.1.4   A Word of Warning

Given a sequence (x_n) of real numbers and a set $S \subseteq \mathbf{R}$, the condition that (x_n) be frequently in the set S requires that there should be infinitely many integers n for which $x_n \in S$. Do not make the common mistake of phrasing this condition by saying that "infinitely many x_n's belong to the set S". This phrasing is wrong because, among other things, it suggests that there should be infinitely many *different* numbers of the form x_n that belong to S. But this need not be so. For example, a constant sequence will be eventually (and therefore frequently) in any set that contains its constant value.

7.2   Subsequences

If you wish to include the topic of subsequences in your reading, click on the icon ▨ .

7.3   Limits and Partial Limits of Sequences

7.3.1   Definition of a Limit

Given a sequence (x_n) of real numbers and given $x \in [-\infty, \infty]$, we say that x is a **limit** of the sequence (x_n) if for every neighborhood U of x the sequence (x_n) is eventually in U. We write this condition as $x_n \to x$ as $n \to \infty$.

Note that the letter n in this definition is unimportant. We could, for example, say that $x_m \to x$ as $m \to \infty$ or that $x_j \to x$ as $j \to \infty$. Sometimes, when the symbol n is understood, we write the condition $x_n \to x$ as $n \to \infty$ briefly as $x_n \to x$.

In the event that $x_n \to x$ as $n \to \infty$ and x is a real number, then we say that x is a **finite limit** of the sequence (x_n). If $x_n \to x$ as $n \to \infty$ and x is either ∞ or $-\infty$, then we call x an **infinite limit** of the sequence (x_n).

7.3.2   Definition of a Partial Limit

Given a sequence (x_n) of real numbers and given $x \in [-\infty, \infty]$, we say that x is a **partial limit**[26] of the sequence (x_n) if for every neighborhood U of x, the sequence (x_n) is frequently in U.

As with limits, we call a partial limit a **finite partial limit** if it is a real number and an **infinite partial limit** if it is either ∞ or $-\infty$.

7.3.3   A Closer Look at Finite Limits

Suppose that (x_n) is a sequence of real numbers and that x is a real number. The following conditions are equivalent:

1. *$x_n \to x$ as $n \to \infty$.*
2. *For every number $\varepsilon > 0$ the sequence (x_n) is eventually in the interval $(x - \varepsilon, x + \varepsilon)$.*
3. *For every number $\varepsilon > 0$ there exists an integer N such that the condition*

$$x_n \in (x - \varepsilon, x + \varepsilon)$$

   holds for every integer $n \geq N$.
4. *For every number $\varepsilon > 0$ there exists an integer N such that the inequality*

$$|x_n - x| < \varepsilon$$

---

[26] Partial limits are somtimes known as *cluster points* (Kelley [15]), sometimes as *limit points* (Gelbaum and Olmsted, [10]) and sometimes as *subsequential limits* (Rudin [26]). The name *partial limit* has been adopted in this book because the terms *cluster point* and *limit point* might be confused with limit points of a set and the term *subsequential limit* is not appropriate when the theory of limits is generalized to more general spaces.

holds for every integer $n \geq N$.

Proof. From the definition of the concept "eventually" we see at once that the conditions 2 and 3 are equivalent to each other. Furthermore, given any n we have

$$x_n \in (x - \varepsilon, x + \varepsilon) \text{ if and only if } |x_n - x| < \varepsilon \ ,$$

$$x - \varepsilon \qquad\qquad x_n \qquad\quad x \qquad\qquad\qquad\qquad x + \varepsilon$$

and therefore the conditions 3 and 4 are equivalent to each other.

Therefore, to complete the proof, all we have to show is that the conditions 1 and 2 are equivalent to each other.

First we shall show that condition 1 implies condition 2 and, for this purpose, assume that condition 1 holds. To prove that condition 2 holds, suppose that $\varepsilon > 0$. Since the interval $(x - \varepsilon, x + \varepsilon)$ is a neighborhood of the number x, we see at once from condition 1 that (x_n) is eventually in $(x - \varepsilon, x + \varepsilon)$.

Now, to prove that condition 2 implies condition 1, we assume that condition 2 holds. To prove that condition 1 must hold, we suppose that U is a neighborhood of the number x. Using the definition of a neighborhood we choose a number $\varepsilon > 0$ such that

$$(x - \varepsilon, x + \varepsilon) \subseteq U.$$

From condition 2 we know that the sequence is eventually in $(x - \varepsilon, x + \varepsilon)$ and therefore (x_n) is eventually in the larger set U. ∎

7.3.4   A Closer Look at Finite Partial Limits

Suppose that (x_n) is a sequence of real numbers and that x is a real number. The following conditions are equivalent:

1. *The number x is a partial limit of the sequence (x_n).*
2. *For every number $\varepsilon > 0$ the sequence (x_n) is frequently in the interval $(x - \varepsilon, x + \varepsilon)$.*
3. *For every number $\varepsilon > 0$ there are infinitely many integers n for which*

$$x_n \in (x - \varepsilon, x + \varepsilon) .$$

4. *For every number $\varepsilon > 0$ there are infinitely many integers n for which $|x_n - x| < \varepsilon$.*

The proof of this theorem is almost the same as the proof of the corresponding theorem about limits, and we leave it as an exercise.

7.3.5   A Closer Look at Infinite Limits

We shall state this theorem for limits with the value ∞ and leave the statement and proof of the corresponding theorem for limits with value $-\infty$ as an exercise.

Suppose that (x_n) is a sequence of real numbers. The following conditions are equivalent:

1. *$x_n \to \infty$ as $n \to \infty$.*
2. *For every number w the sequence (x_n) is eventually in the interval (w, ∞).*
3. *For every number w there exists an integer N such that for every integer $n \geq N$ we have $x_n \in (w, \infty)$.*
4. *For every number w there exists an integer N such that for every integer $n \geq N$ we have $x_n > w$.*

Proof. It is easy to see that the conditions 2, 3, and 4 are equivalent to one another, and so, as in the case of Theorem 7.3.3, we shall complete the proof by showing that condition 1 implies condition 2 and that condition 2 implies condition 1.

Suppose that condition 1 holds and, to prove that condition 2 holds, suppose that w is a real number. Since the interval (w, ∞) is a neighborhood of ∞, it follows at once that the sequence (x_n) is eventually in (w, ∞). Thus condition 1 implies condition 2.

Now suppose that condition 2 holds and, to prove that condition 1 holds, suppose that U is a neighborhood of ∞. Choose a number w such that $(w, \infty) \subseteq U$. From condition 2 we know that (x_n) is eventually in (w, ∞), and so (x_n) must be eventually in the larger set U. Thus condition 1 implies condition 2 and our proof is complete. ∎

7.3.6   A Closer Look at Infinite Partial Limits

Suppose that (x_n) is a sequence of real numbers. The following conditions are equivalent:

1. *The extended real number ∞ is a partial limit of the sequence (x_n).*
2. *For every number w the sequence (x_n) is frequently in the interval (w, ∞).*
3. *For every number w there are infinitely many integers n for which $x_n \in (w, \infty)$.*
4. *For every number w there are infinitely many integers n for which $x_n > w$.*
5. *The sequence (x_n) is unbounded above.*

Proof. The proof that the first four conditions are equivalent to one another is almost identical to the proof of the corresponding theorem about infinite limits and will be left as an exercise. Furthermore, it is clear that any of these four conditions guarantees that the sequence (x_n) is unbounded above.

We shall complete the proof by showing that if the sequence (x_n) is unbounded above, then condition 4 must hold. Suppose that (x_n) is unbounded above and suppose that w is any real number. To obtain a contradiction, assume that there are at most finitely many integers n for which $x_n > w$. Using the fact that this finite set of integers is bounded above, choose an integer N such that every integer n for which $x_n > w$ must satisfy $n \leq N$. We see at once that if α is the largest member of the finite set

$$\{w, x_1, x_2, x_3, \cdots, x_N\},$$

then α is an upper bound of the sequence (x_n), contradicting our assumption that the sequence (x_n) is unbounded above. ∎

7.3.7   Some Examples of Limits and Partial Limits

1. Every constant sequence has a limit. Suppose that x is a given real number and that $x_n = x$ for every positive integer n. It is clear that $x_n \to x$ as $n \to \infty$.

2. Every eventually constant sequence has a limit. Suppose that x is a given real number, that (x_n) is a given sequence, and that for some integer N we have $x_n = x$ for every integer $n \geq N$. Again it is clear that $x_n \to x$ as $n \to \infty$.

3. Suppose that

$$x_n = \begin{cases} 1 & \text{if} \quad n \text{ is even} \\ 3 & \text{if} \quad n \text{ is odd.} \end{cases}$$

We shall show that the sequence (x_n) does not have a limit. Since (x_n) is not eventually (in fact, not ever) in the neighborhood $(3, \infty)$ of ∞ we know that ∞ is not a limit of (x_n) and in the same way we can see that $-\infty$ is not a limit of (x_n). Now, to show that the sequence (x_n) does not have a finite limit, suppose that x is a real number. To observe that x is not a limit of the sequence (x_n) we observe that the interval $\left(x - \frac{1}{2}, x + \frac{1}{2}\right)$, is a neighborhood of x. Furthermore, since it is impossible for both of the numbers 1 and 3 to belong to the interval $\left(x - \frac{1}{2}, x + \frac{1}{2}\right)$ it is clear that the sequence (x_n) is not eventually in $\left(x - \frac{1}{2}, x + \frac{1}{2}\right)$, and we conclude that x is not a limit of the sequence (x_n).

4. Suppose that

$$x_n = \begin{cases} 1 & \text{if } n \text{ is even} \\ 3 & \text{if } n \text{ is odd.} \end{cases}$$

If U is any neighborhood of 1, then, since there are infinitely many even integers in the domain of the sequence, the sequence is frequently in U. Therefore the number 1 is a partial limit of (x_n). In a similar way we can see that 3 is also a partial limit of (x_n). On the other hand, if x is any number unequal to 1 or 3, then we can choose a neighborhood U of x that does not contain either of the numbers 1 and 3, and, since the sequence is not frequently in U, we conclude that x is not a partial limit of (x_n). Thus the numbers 1 and 3 are the only partial limits of (x_n).

5. Suppose that

$$x_n = \frac{1}{n}$$

for every positive integer n. We shall show that $x_n \to 0$ as $n \to \infty$. For this purpose we shall show that (x_n) satisfies condition 4 of Theorem 7.3.3. Suppose that $\varepsilon > 0$. Choose an integer $N > 1/\varepsilon$. Then whenever $n \geq N$ we have

$$|x_n - 0| = \frac{1}{n} \leq \frac{1}{N} < \varepsilon.$$

Thus the sequence (x_n) satisfies condition 4 of Theorem 7.3.3 and we conclude that $x_n \to 0$ as $n \to \infty$.

6. Suppose that $x_n = n$ for every positive integer n. We shall use Theorem 7.3.5 to show that $x_n \to \infty$ as $n \to \infty$. Suppose that w is any real number and choose a positive integer N such that $N > w$. For every integer $n \geq N$ we have

$$x_n = n \geq N > w.$$

7. Suppose that

$$x_n = \begin{cases} 2 & \text{if } n \text{ is a multiple of 3} \\ \frac{1}{n} & \text{if } n \text{ is one more than a multiple of 3} \\ n & \text{if } n \text{ is two more than a multiple of 3.} \end{cases}$$

We can make the following observations about the sequence (x_n):

(a) If U is any neighborhood of 2, then, since $x_{3n} \in U$ for each n, we
conclude that (x_n) is frequently in U. Therefore 2 is a partial limit of
(x_n).

(b) Suppose that $\varepsilon > 0$ choose an integer N such that

$$3N + 1 > \frac{1}{\varepsilon}.$$

For every integer $n \geq N$ we have

$$|x_{3n+1} - 0| = \frac{1}{3n+1} \leq \frac{1}{3N+1} < \varepsilon,$$

and it follows from Theorem 7.3.4 that 0 is a partial limit of (x_n).

(c) Since the sequence (x_n) is unbounded above, it follows from Theorem
7.3.6 that ∞ is a partial limit of (x_n).

(d) Finally, suppose that x is any extended real number other than $0, 2$, and
∞. If $x < 0$, we define $U = (-\infty, 0)$.

If $x > 2$, we define $U = (2, x + 2)$.

If $0 < x < 2$, then we choose a positive integer N such that $1/N < x$
and we define $U = (1/N, 2)$.

In each of these cases we have found a neighborhood U of x such that
(x_n) is not frequently in U and so x cannot be a partial limit of (x_n).
We have therefore shown that the partial limits of (x_n) are $0, 2$, and ∞.

8. Suppose that

$$x_n = \frac{n}{n-3}$$

for every integer $n > 3$. We shall use Theorem 7.3.3 to show that $x_n \to 1$ as
$n \to \infty$. Suppose that $\varepsilon > 0$. Now for each n we have

$$|x_n - 1| = \left| \frac{n}{n-3} - 1 \right| = \frac{3}{n-3},$$

and therefore the inequality $|x_n - 1| < \varepsilon$ requires that

$$\frac{3}{n-3} < \varepsilon,$$

which is equivalent to the condition

$$n > 3 + \frac{3}{\varepsilon}.$$

This tells us how to finish the proof. We choose an integer N such that

$$N > 3 + \frac{3}{\varepsilon},$$

and we observe that whenever $n \geq N$ we have $|x_n - 1| < \varepsilon$.

9. Suppose that

$$x_n = \frac{2n^2 + n - 3}{n^2 + 3n + 2}$$

for every positive integer n. We shall use Theorem 7.3.3 to show that $x_n \to 2$ as $n \to \infty$. Suppose that $\varepsilon > 0$. Now for each n we have

$$|x_n - 2| = \left| \frac{2n^2 + n - 3}{n^2 + 3n + 2} - 2 \right| = \frac{5n + 7}{n^2 + 3n + 2}.$$

Now, as long as $n \geq 2$, we have $7 \leq 5n$ and so

$$\frac{5n + 7}{n^2 + 3n + 2} \leq \frac{5n + 5n}{n^2 + 3n + 2} = \frac{10n}{n^2 + 3n + 2} < \frac{10n}{n^2} = \frac{10}{n}.$$

Therefore, as long as $n \geq 2$, the inequality $|x_n - 2| < \varepsilon$ will hold as long as

$$\frac{10}{n} < \varepsilon,$$

which is equivalent to the condition

$$n > \frac{10}{\varepsilon}.$$

This tells us how to finish the proof. We choose an integer N such that $N \geq 2$ and $N > 10/\varepsilon$, and we observe that whenever $n \geq N$ we have $|x_n - 2| < \varepsilon$.

10. For each positive integer n, if n can be written in the form

$$n = 2^m 3^k$$

for some positive integers m and k, we define

$$x_n = \frac{m}{k}.$$

In the event that n cannot be written in the form $2^m 3^k$, we define $x_n = 0$. Observe that the range of the sequence (x_n) is the set of all nonnegative rational numbers. If U is a neighborhood of any nonnegative real number or a neighborhood of ∞, then, since U must contain infinitely many positive rational numbers, it is clear that $x_n \in U$ for infinitely many values of n. We conclude that the set of partial limits of the sequence (x_n) is $[0, \infty]$.

7.3.8   Some Exercises on Limits and Partial Limits

1. Given that

    $$x_n = 3 + \frac{1}{n}$$

    for each positive integer n, prove that 3 is a limit of (x_n).
2. Given that

    $$x_n = 3 + \frac{2}{n}$$

    for each positive integer n, prove that 3 is a limit of (x_n).
3. ▣ Given that $x_n = 1/n$ for each positive integer n and that $x \neq 0$, prove that x is not a partial limit of (x_n).
4. Given that

    $$x_n = \begin{cases} (-1)^n n^3 & \text{if } n \text{ is a multiple of 3} \\ 0 & \text{if } n \text{ is one more than a multiple of 3} \\ 4 & \text{if } n \text{ is two more than a multiple of 3,} \end{cases}$$

    prove that the partial limits of (x_n) are $-\infty, \infty, 0$, and 4.
5. ▣ Give an example of a sequence of real numbers whose set of partial limits is the set $\{1\} \cup [4, 5]$.
6. Given that

    $$x_n = \frac{3 + 2n}{5 + n}$$

    for every positive integer n, prove that $x_n \to 2$ as $n \to \infty$.
7. Given that

    $$x_n = \begin{cases} \frac{1}{2^n} & \text{if } n \text{ is even} \\ \frac{1}{n^2+1} & \text{if } n \text{ is odd} \end{cases}$$

prove that $x_n \to 0$ as $n \to \infty$.

8. Suppose that (x_n) is a sequence of real numbers and that $x \in \mathbf{R}$. Prove that the following conditions are equivalent:

   (a) $x_n \to x$ as $n \to \infty$.
   (b) For every number $\varepsilon > 0$ the sequence (x_n) is eventually in the interval $(x - 5\varepsilon, x + 5\varepsilon)$.

9. Prove that

$$\frac{n^2 + 3n + 1}{2n^2 + n + 4} \to \frac{1}{2}$$

   as $n \to \infty$.

10. For each positive integer n, if n can be written in the form

$$n = 2^m 3^k,$$

   where m and k are positive integers and $m \leq k$, then we define $x_n = \frac{m}{k}$. Otherwise we define $x_n = 0$. Prove that the set of partial limits of the sequence (x_n) is $[0, 1]$.

7.4   Some Elementary Facts About Limits and Partial Limits

7.4.1   Uniqueness of Limits

We shall now make the observation that a sequence can never have more than one limit. In fact, as we shall see, if a sequence has a limit α, then no number unequal to α can be even a partial limit of the sequence.

Suppose that (x_n) is a sequence of real numbers, that $x \in [-\infty, \infty]$, and that $x_n \to x$ as $n \to \infty$. Suppose that $y \in [-\infty, \infty]$ and that $y \neq x$. Then y cannot be a partial limit of the sequence (x_n).

Proof. We shall assume, without loss of generality, that $y < x$. (The other case is analogous.) Choose a real number w such that $y < w < x$.

Since $x_n \to x$ and since the interval (w, ∞) is a neighborhood of x, we know that the sequence (x_n) is eventually in (w, ∞). Therefore (x_n) cannot be frequently in the interval $(-\infty, w)$, and, since the latter interval is a neighborhood of y, we deduce that y is not a partial limit of (x_n). ∎

7.4.2   Limit Notation

As we have just seen, if (x_n) is a sequence of real numbers and $x \in [-\infty, \infty]$ and if $x_n \to x$ as $n \to \infty$, then x can be the only limit of (x_n). From now on, when a sequence (x_n) has a limit, we shall not merely refer to *a* limit of (x_n). We shall refer to *the* limit of the sequence and we shall write the limit as

$$\lim_{n \to \infty} x_n.$$

7.4.3   Convergent and Divergent Sequences

A sequence (x_n) of real numbers is said to be **convergent** if (x_n) has a limit and this limit is a real number. If a sequence (x_n) is convergent and if

$$\lim_{n \to \infty} x_n = x,$$

then we say that the sequence (x_n) converges to the number x.

A sequence that fails to be convergent is said to be **divergent**. Note that there are two distinct ways in which a given sequence (x_n) can be divergent:

1. The sequence may have its limit at ∞ or at $-\infty$.
2. The sequence may have no limit at all.

Note, for example, that any sequence that has more than one partial limit must be divergent.

7.4.4   Boundedness of Convergent Sequences

Every convergent sequence is bounded.

Proof. One way of proving this theorem is to observe that if (x_n) is a convergent sequence of real numbers, then it follows from Theorem 7.4.1 that neither $-\infty$ nor ∞ can be a partial limit of (x_n). It therefore follows from Theorem 7.3.6 that the sequence (x_n) must be bounded.

But it is also worth proving this theorem directly. Suppose that (x_n) is convergent and that its limit is x. Since the interval $(x - 1, x + 1)$ is a neighborhood of x, we know that (x_n) is eventually in $(x - 1, x + 1)$, and, using this fact, we choose an integer N such that $x_n \in (x - 1, x + 1)$ whenever $n \geq N$. Since the set

$$\{x_1, x_2, x_3, \cdots, x_N\} \cup (x - 1, x + 1),$$

being the union of two bounded sets, is bounded, we conclude that the sequence (x_n) is bounded. ∎

7.4.5   Partial Limits and Limits of Subsequences

If your reading of the on-screen version of this book has included the material on subsequences that can be found at ![icon], then you may wish to look at the

connection between partial limits and limits of subsequences that is provided at
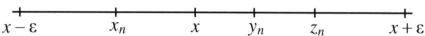 .

7.4.6   The Sandwich Theorem

Suppose that (x_n), (y_n), and (z_n) are sequences of real numbers and that the inequality

$$x_n \leq y_n \leq z_n$$

holds for all sufficiently large integers n. Suppose that x is a real number and that both of the sequences (x_n) and (z_n) converge to x. Then the sequence (y_n) must also converge to x.

Proof.   In order to show that $y_n \to x$ as $n \to \infty$ we shall show that, for every number $\varepsilon > 0$, the sequence (y_n) is eventually in the interval

$$(x - \varepsilon, x + \varepsilon) .$$

Suppose that $\varepsilon > 0$.

Using the fact that the inequality

$$x_n \leq y_n \leq z_n$$

holds for all sufficiently large integers n, choose an integer N_1 such that this inequality holds for all integers $n \geq N_1$. Now, using the fact that $x_n \to x$ as $n \to \infty$ and the fact that the interval $(x - \varepsilon, x + \varepsilon)$ is a neighborhood of x, choose an integer N_2 such that $x_n \in (x - \varepsilon, x + \varepsilon)$ for all integers $n \geq N_2$. Finally, using the fact that $z_n \to x$ as $n \to \infty$, choose an integer N_3 such that $z_n \in (x - \varepsilon, x + \varepsilon)$ for all integers $n \geq N_3$.

We now define N to be the largest of the three integers N_1, N_2, and N_3. Then for every integer $n \geq N$ we know that all three of the conditions

$$x_n \quad \leq \quad y_n \leq z_n$$
$$x_n \quad \in \quad (x - \varepsilon, x + \varepsilon)$$
$$z_n \quad \in \quad (x - \varepsilon, x + \varepsilon)$$

must hold and we obtain

$$x - \varepsilon < x_n \leq y_n \leq z_n < x + \varepsilon,$$

which tells us that

$$y_n \in (x - \varepsilon, x + \varepsilon).$$

Thus the sequence (y_n) is eventually in $(x - \varepsilon, x + \varepsilon)$ and we have shown that $y_n \to x$ as $n \to \infty$. ∎

7.4.7   Another Kind of Sandwich Theorem

In this subsection we describe a way in which sequences can be used to measure how far one set of numbers lies below another. This technique will be useful to us in the chapters that follow.

Suppose that A and B are nonempty sets of real numbers and that for all numbers $t \in A$ and $x \in B$ we have $t \leq x$. Then the following conditions are equivalent:

1. $\sup A = \inf B$.
2. *It is possible to find a sequence (t_n) in the set A and a sequence (x_n) in the set B such that $x_n - t_n \to 0$ as $n \to \infty$.*

Furthermore, if $\sup A = \inf B$ and if sequences (t_n) and (x_n) have been chosen as in condition 2, then we have

$$\lim_{n \to \infty} t_n = \lim_{n \to \infty} x_n = \sup A.$$

Proof. We mention first that since every member of A is a lower bound of B and every member of B is an upper bound of A and since the sets A and B are nonempty, it is clear that A must be bounded above and B must be bounded below. Furthermore, $\sup A \leq \inf B$.

To prove that condition 1 implies condition 2, assume that $\sup A = \inf B$. For every positive integer n we use the fact that the number $\sup A - \frac{1}{n}$ is not an upper bound of A and the fact that $\inf B + \frac{1}{n}$ is not a lower bound of B to choose a member t_n of the set A and a member x_n of the set B such that

$$\sup A - \frac{1}{n} < t_n \quad \text{and} \quad x_n < \inf B + \frac{1}{n}.$$

Since

$$0 \leq x_n - t_n < \inf B + \frac{1}{n} - \left(\sup A - \frac{1}{n} \right) = \frac{2}{n}$$

for each n, it follows from the sandwich theorem that $x_n - t_n \to 0$ as $n \to \infty$.

Now assume that condition 2 holds and choose sequences (t_n) and (x_n) in the sets A and B, respectively, such that $x_n - t_n \to 0$ as $n \to \infty$. Since

$$0 \leq \sup A - t_n \leq \inf B - t_n \leq x_n - t_n$$

for each positive integer n, it follows from the sandwich theorem (Theorem 7.4.6) that $t_n \to \inf B$ and $t_n \to \sup A$ as $n \to \infty$. Therefore $\inf B = \sup A$, and we see finally that

$$\lim_{n \to \infty} x_n = \lim_{n \to \infty} (x_n - t_n + t_n) = \lim_{n \to \infty} (x_n - t_n) + \lim_{n \to \infty} t_n = 0 + \sup A = \sup A.$$

7.4.8   Exercises on the Elementary Properties of Limits

1. The purpose of this exercise is to use *Scientific Notebook* to gain an intuitive feel for the limit behavior of a rather difficult sequence.

    (a) Point at the equation

    $$x_n = \frac{n^n \sqrt{n}}{(n!)\, e^n}$$

    and then click on the button $\boxed{f(\infty)}$ to supply the definition to *Scientific Notebook*. When you see the screen

    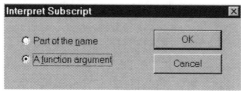

    make the selection "A function argument" so that *Scientific Notebook* knows that you are defining a sequence.

    (b) Point at the expression x_n and click on the button $\boxed{+}$ to display the sequence graphically. Revise your graph and set the domain interval as $[1, 500]$. Double click into your graph to make the buttons

    appear in the top right corner and click on the bottom button to select it. Trace your graph with the mouse and show graphically that

    $$\lim_{n \to \infty} \frac{n^n \sqrt{n}}{(n!)\, e^n} \approx 0.3989.$$

(c) Ⓝ Point at the expression

$$\lim_{n\to\infty} \frac{n^n \sqrt{n}}{(n!)\, e^n}$$

and ask *Scientific Notebook* to evaluate it numerically. Compare the result with the limit value that you found graphically.

(d) Ⓝ Point at the expression and ask *Scientific Notebook* to evaluate it exactly to show that the limit is $1/\sqrt{2\pi}$.

2. Ⓗ Prove that $5^n/n! \to 0$ as $n \to \infty$.

3. Ⓗ Prove that $n!/n^n \to 0$ as $n \to \infty$.

4. Ⓗ Given that (x_n) is a sequence of real numbers, that $x > 0$, and that $x_n \to x$ as $n \to \infty$, prove that there exists an integer N such that the inequality $x_n > 0$ holds for all integers $n \geq N$.

5. Given that $x_n \geq 0$ for every positive integer n and that x is a partial limit of the sequence (x_n), prove that $x \geq 0$.

6. Suppose that (x_n) is a sequence of real numbers and that $x \in \mathbf{R}$. Prove that the following conditions are equivalent:

   (a) $x_n \to x$ as $n \to \infty$.
   (b) $|x_n - x| \to 0$ as $n \to \infty$.

7. Ⓗ Suppose that (x_n) is a sequence of real numbers, that $x \in \mathbf{R}$, and that $x_n \to x$ as $n \to \infty$. Prove that $|x_n| \to |x|$ as $n \to \infty$.

8. Suppose that (x_n) is a sequence of real numbers, that $x \in \mathbf{R}$, and that $x_n \to x$ as $n \to \infty$. Suppose that p is an integer and that for every positive integer n we have

$$y_n = x_{n+p}.$$

Prove that $y_n \to x$ as $n \to \infty$.

9. Given that $a_n \leq b_n$ for every positive integer n and given that $a_n \to \infty$, prove that $b_n \to \infty$.

10. Suppose that (a_n) and (b_n) are sequences of real numbers, and that $|a_n - b_n| \leq 1$ for every positive integer n, and that ∞ is a partial limit of the sequence (a_n). Prove that ∞ is a partial limit of (b_n).

11. Two sequences (a_n) and (b_n) are said to be *eventually close* if for every number $\varepsilon > 0$ there exists an integer N such that the inequality $|a_n - b_n| < \varepsilon$ holds for all integers $n \geq N$.

    (a) Prove that if two sequences (a_n) and (b_n) are eventually close, and if a real number x is the limit of the sequence (a_n), then x is also the limit of the sequence (b_n).

(b) Prove that if two sequences (a_n) and (b_n) are eventually close, and if a real number x is a partial limit of the sequence (a_n), then x is also a partial limit of the sequence (b_n).

12. Suppose that (a_n) and (b_n) are sequences of real numbers, that $a_n \to a$ and $b_n \to b$ as $n \to \infty$, and that $a < b$. Prove that there exists an integer N such that the inequality $a_n < b_n$ holds for all integers $n \geq N$.

13. Give an example of two sequences (a_n) and (b_n) of real numbers satisfying $a_n > b_n$ for every positive integer n even though the sequence (a_n) has a partial limit a that is less than a partial limit b of the sequence (b_n).

14. (a) **H** Prove that if a real number x is a limit point of the range of a given sequence (x_n), then x must be a partial limit of (x_n).

(b) Give an example of a sequence (x_n) and a real number x that is a partial limit of (x_n) but is not a limit point of the range of (x_n).

(c) Prove that if a sequence (x_n) of real numbers is one-one and if a number x is a partial limit of the sequence (x_n), then x must be a limit point of the range of (x_n).

7.5   The Algebraic Rules for Limits

The algebraic rules for limits are the theorems that display a relationship between the behavior of limits and the algebraic operations (addition, subtraction, multiplication, and division) in the number system $\mathbf{R}$.

7.5.1   Product of a Bounded Sequence and a Sequence with Limit Zero

Suppose that (x_n) and (y_n) are sequences of real numbers, that the sequence (x_n) is bounded, and that the sequence (y_n) converges to 0. Then $x_n y_n \to 0$ as $n \to \infty$.

Proof. We begin by using the boundedness of the sequence (x_n) to choose a number b such that $|x_n| < b$ for every integer n in the domain of (x_n). Now, to prove that $x_n y_n \to 0$, suppose that $\varepsilon > 0$. Using the fact that $y_n \to 0$ and the fact that ε/b is a positive number, choose an integer N such that

$$|y_n| < \frac{\varepsilon}{b}$$

for all integers $n \geq N$. If necessary, replace N by a larger integer so that every integer $n \geq N$ will be in the domain of (x_n). Then for every integer $n \geq N$ we have

$$|x_n y_n - 0| = |x_n y_n| < b\left(\frac{\varepsilon}{b}\right) = \varepsilon,$$

and so we have shown that $x_n y_n \to 0$ as $n \to \infty$. ∎

7.5.2   The Algebraic Rules for Finite Limits

Suppose that (x_n) and (y_n) are sequences of real numbers that converge, respectively, to numbers x and y. Then we have:

1. $x_n + y_n \to x + y$.
2. $x_n - y_n \to x - y$.
3. $x_n y_n \to xy$.
4. $x_n/y_n \to x/y$ as long as $y \neq 0$.

Proof of Part 1. Suppose that $\varepsilon > 0$. We need to show that for n sufficiently large we have

$$|(x_n + y_n) - (x + y)| < \varepsilon.$$

The key to the proof is the inequality

$$|(x_n + y_n) - (x + y)| = |(x_n - x) + (y_n - y)| \leq |(x_n - x)| + |(y_n - y)|.$$

We begin by using the fact that $x_n \to x$ and the fact that $\varepsilon/2$ is a positive number to choose an integer N_1 such that $|x_n - x| < \varepsilon/2$ for all integers $n \geq N_1$. In a similar fashion we choose an integer N_2 such that $|y_n - y| < \varepsilon/2$ for all integers $n \geq N_2$. We now define N to be the larger of the two numbers N_1 and N_2, and for every integer $n \geq N$ we observe that

$$|(x_n + y_n) - (x + y)| \leq |(x_n - x)| + |(y_n - y)| < \frac{\varepsilon}{2} + \frac{\varepsilon}{2} = \varepsilon.$$

This completes the proof of part 1. ∎

The proof of part 2 will be left as an exercise.

Proof of Part 3. Since $y_n \to y$, it follows from part 2 that $y_n - y \to 0$. Since the convergent sequence (x_n) is bounded (by Theorem 7.4.4), we deduce from Theorem 7.5.1 that $x_n (y_n - y) \to 0$. Since $x_n - x \to 0$ and the sequence with the constant value y is bounded, it also follows from Theorem 7.5.1 that $y (x_n - x) \to 0$. Now for each n we have

$$x_n y_n - xy = x_n (y_n - y) + y (x_n - x),$$

and it therefore follows from Part 1 that $x_n y_n - xy \to 0$, in other words, that $x_n y_n \to xy$. ∎

Proof of Part 4. We assume that $y \neq 0$. Now since

$$\frac{x_n}{y_n} = x_n \left(\frac{1}{y_n}\right)$$

whenever these expressions are defined, the fact that $x_n/y_n \to x/y$ will follow from part 3 as soon as we have shown that $1/y_n \to 1/y$. Thus to complete the proof we need to show that $1/y_n \to 1/y$.

We begin by using the fact that $y \neq 0$ to choose a number δ such that

$$0 < \delta < |y| \, .$$

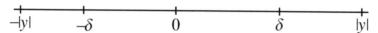

Using the fact that the set $\mathbf{R} \setminus [-\delta, \delta]$ is a neighborhood of the number y, we now choose an integer N such that the inequality $|y_n| > \delta$ holds for all integers $n \geq N$. We note that for all such n we have $y_n \neq 0$, and so the number $1/y_n$ is defined. Furthermore, if $n \geq N$, we have $|1/y_n| < 1/\delta$ and so, if we restrict the domain of the sequence $(1/y_n)$ to the set of integers $n \geq N$, we can assert that the sequence $(1/y_n)$ is bounded. Now for every integer $n \geq N$ we have

$$\frac{1}{y_n} - \frac{1}{y} = (y - y_n) \left(\frac{1}{y_n y} \right) .$$

Since $y - y_n \to 0$ and since the sequence $(1/y_n y)$ is bounded, the fact that $1/y_n - 1/y \to 0$ follows from Theorem 7.5.1. ∎

7.5.3   The Algebraic Rules for Limits, General Form

The difference between this theorem and the preceding one is that we now allow the possibility that some of the limits of the sequences may be infinite.

Suppose that (x_n) and (y_n) are sequences of real numbers, that x and y belong to $[-\infty, \infty]$, and that $x_n \to x$ and $y_n \to y$ as $n \to \infty$. Then each of the following statements is true as long as its right-hand side is defined:

1. $x_n + y_n \to x + y$.
2. $x_n - y_n \to x - y$.
3. $x_n y_n \to xy$.
4. $x_n / y_n \to x/y$.

This theorem is actually an efficient way of stating a whole host of different results. In part 1, for example, x and y might both be real numbers, or we might have $x = \infty$, in which case y can be anything except $-\infty$. In part 3, the symbols x and y might stand for real numbers, or we might have $x = \infty$, in which case y can be anything except 0. The benefit of the extended real number system is that it allows us to unify all of these results and to make them resemble the finite cases that were handled in Theorem 7.5.2. We shall discuss just three of the extra cases that are included in this more general version of the theorem and leave a discussion of the others as an exercise. You should write out the proofs of sufficiently many of these infinite limit cases to make sure that you understand them all.

Proof of Part 1 when $x = \infty$. Since $y \neq -\infty$, the sequence (y_n) must be bounded below. Choose a lower bound α of (y_n). Since $y + \infty = \infty$, we need to show that $x_n + y_n \to \infty$. For this purpose we shall use Theorem 7.3.5. Suppose that w is a real number and, using the fact that $x_n \to \infty$, choose an integer N such that for all integers $n \geq N$ we have

$$x_n > w - \alpha.$$

Then for each such n we have

$$x_n + y_n > w - \alpha + \alpha = w,$$

and we have shown that $x_n + y_n \to \infty$. ∎

Proof of Part 3 when $x > 0$ **and** $y = \infty$. Since $x \times \infty = \infty$, we need to show that $x_n y_n \to \infty$. We begin by choosing a number δ such that $0 < \delta < x$. (For example, one may define $\delta = x/2$.)

To show that $x_n y_n \to \infty$, suppose that w is any real number. In view of Theorem 7.3.5, we need to show that $x_n y_n > w$ for all sufficiently large integers n. Using the fact that the interval (δ, ∞) is a neighborhood of x and the fact that $x_n \to x$, we choose an integer N_1 such that $x_n > \delta$ for all integers $n \geq N_1$. Now we choose an integer N_2 such that $y_n > |w| / \delta$ for every integer $n \geq N_2$ and we define N to be the larger of the two numbers N_1 and N_2. Then for every integer $n \geq N$ we have

$$x_n y_n > \frac{\delta |w|}{\delta} = |w| \geq w,$$

and the proof is complete. ∎

Proof of Part 4 when $y = \infty$. Because of the identity

$$\frac{x_n}{y_n} = x_n \left(\frac{1}{y_n} \right)$$

and part 3, all we have to show is that $1/y_n \to 0$. Suppose that $\varepsilon > 0$ and, using the fact that the interval $(1/\varepsilon, \infty)$ is a neighborhood of ∞, choose an integer N such that for every integer $n \geq N$ we have $y_n > 1/\varepsilon$. For every integer $n \geq N$

we see that

$$-\varepsilon < 0 < \frac{1}{y_n} < \varepsilon.$$

The condition $1/y_n \to 0$ therefore follows from Theorem 7.3.3. ∎

7.5.4   Some Exercises on The Algebraic Rules for Limits

1. Write out proofs of those cases of Theorem 7.5.3 that were not proved above.
2. Given that (x_n) and (y_n) are sequences of real numbers, that (x_n) converges to a number x, and that a real number y is a partial limit of the sequence (y_n), prove that $x + y$ is a partial limit of the sequence $(x_n + y_n)$.
3. State and prove some analogs of Exercise 2 for subtraction, multiplication, and division.
4. Give an example of two sequences (x_n) and (y_n) and a partial limit x of (x_n) and a partial limit y of (y_n) such that $x + y$ fails to be a partial limit of the sequence $(x_n + y_n)$.
5. Give an example of two divergent sequences (x_n) and (y_n) such that the sequence $(x_n + y_n)$ is convergent.
6. Give an example of two sequences (x_n) and (y_n) such that $x_n \to 0$, and $y_n \to \infty$, and
    (a) $x_n y_n \to 0$.
    (b) $x_n y_n \to 6$.
    (c) $x_n y_n \to \infty$.
    (d) The sequence $(x_n y_n)$ is bounded but has no limit.
7. Given two sequences (x_n) and (y_n) of real numbers such that both of the sequences (x_n) and $(x_n + y_n)$ are convergent, is it true that the sequence (y_n) must be convergent?
8. ▨ Given that (x_n) is a sequence of real numbers and that $x_n \to 0$, prove that

    $$\frac{x_1 + x_2 + x_3 + \cdots + x_n}{n} \to 0.$$

9. ▨ Given that (x_n) is a sequence of real numbers, that x is a real number, and that $x_n \to x$, prove that

    $$\frac{x_1 + x_2 + x_3 + \cdots + x_n}{n} \to x.$$

10. Given that (x_n) and (y_n) are sequences of real numbers and that $x_n - y_n \to 0$, prove that (x_n) and (y_n) have the same set of partial limits.

11. ▣ Suppose that (x_n) and (y_n) are sequences of real numbers, that $x_n - y_n \to 0$, and that the number 0 fails to be a partial limit of at least one of the sequences (x_n) and (y_n). Prove that

$$\frac{x_n}{y_n} \to 1$$

as $n \to \infty$.

12. Give an example to show that the requirement in Exercise 11 that 0 not be a partial limit of at least one of the two sequences is really needed.

13. ▣ Suppose that (x_n) and (y_n) are sequences of real numbers, that $x_n/y_n \to 1$, and that at least one of the sequences (x_n) and (y_n) is bounded. Prove that $x_n - y_n \to 0$. Give an example to show that the conclusion $x_n - y_n \to 0$ can fail if both (x_n) and (y_n) are unbounded.

14. Suppose that (x_n) and (y_n) are sequences of real numbers, that $y_n \to 1$, and that for each n we have $z_n = x_n y_n$. Prove that the sequences (x_n) and (z_n) have the same set of partial limits.

7.6   The Relationship Between Sequences and the Topology of R

In this section we shall reveal a connection that exists between the theory of limits of sequences and the topology of the number line R. We shall observe that any person who had a complete list showing which numbers were limits of which sequences could use this information to determine which sets are open and which sets are closed.

7.6.1   The Relationship Between Limits and Closure

1. *If (x_n) is a sequence in a set S of real numbers, then every finite partial limit of (x_n) must belong to the set $\overline{S}$.*
2. *If (x_n) is a sequence that is frequently in a set S of real numbers, and if (x_n) converges to a number x, then $x \in \overline{S}$.*
3. *If S is a set of real numbers and $x \in \overline{S}$, then there exists a sequence (x_n) in S such that (x_n) converges to the number x.*

Proof of Part 1. Suppose that (x_n) is a sequence in a set S of real numbers and that x is a finite partial limit of the sequence (x_n). To show that $x \in \overline{S}$, suppose that U is a neighborhood of x. Since the sequence (x_n) is frequently in U, it is clear that $U \cap S \neq \emptyset$. ∎

Proof of Part 2. Suppose that (x_n) is a sequence that is frequently in a set S

of real numbers and that (x_n) converges to a number x. To show that $x \in \overline{S}$, suppose that U is a neighborhood of x and choose an integer N such that $x_n \in U$ for every integer $n \geq N$. Since (x_n) is frequently in the set S, there are integers $n \geq N$ for which $x_n \in S$; and since for all such integers n we have $x_n \in S \cap U$, we must have $U \cap S \neq \emptyset$. ∎

Proof of Part 3. 🖎   Suppose that S is a set of real numbers and that $x \in \overline{S}$. We know that if n is any positive integer, then the interval

$$\left(x - \frac{1}{n}, x + \frac{1}{n} \right),$$

being a neighborhood of x, must intersect with the set S, and, using this fact, we choose a number that we shall call x_n such that

$$x_n \in \left(x - \frac{1}{n}, x + \frac{1}{n} \right) \cap S.$$

In this way we have chosen a sequence (x_n) in the set S and since

$$|x_n - x| < \frac{1}{n}$$

for each n we see at once that $x_n \to x$. ∎

7.6.2   Corollary: Closed Sets and Limits of Sequences

Suppose that S is a set of real numbers. The following conditions are equivalent:

1. *The set S is closed.*
2. *No sequence in S can have a partial limit in* **R** $\setminus S$.
3. *No sequence that is frequently in S can have a limit in* **R** $\setminus S$.

Proof. Since S is closed if and only if $S = \overline{S}$, the result follows at once from Theorem 7.6.1. ∎

7.6.3   Unbounded Sets and Limits of Sequences

Suppose that S is a set of real numbers. Then the following conditions are equivalent:

1. *The set S is unbounded above.*
2. *There exists a sequence (x_n) in the set S such that $x_n \to \infty$ as $n \to \infty$.*

Proof. To show that condition 2 implies condition 1, assume that condition 2 holds and choose a sequence (x_n) in S such that $x_n \to \infty$ as $n \to \infty$. If w is any real number, then it follows at once from the fact that the sequence (x_n) is eventually in the interval (w, ∞) that w is not an upper bound of S. Therefore S is not bounded above.

Now to show that condition 1 implies condition 2, assume that condition 1 holds. For each positive integer n we use the fact that the number n is not an upper bound of S to choose a number that we shall call x_n such that $x_n > n$. The sequence (x_n) that we have made in this way is in the set S and it is clear that $x_n \to \infty$. ■

7.6.4   Exercises on Sequences and the Topology of R

1. Prove that a set S of real numbers is unbounded below if and only if there exists a sequence (x_n) in S such that $x_n \to -\infty$.
2. Suppose that S is a nonempty set of real numbers and that α is an upper bound of S. Prove that the following conditions are equivalent:
   (a) We have $\alpha = \sup S$.
   (b) There exists a sequence (x_n) in S such that $x_n \to \alpha$ as $n \to \infty$.
3. Given a sequence (x_n) that is frequently in a set S of real numbers, and given a partial limit x of the sequence (x_n), is it necessarily true that $x \in \overline{S}$?
4. Prove that a set U of real numbers is open if and only if every sequence that converges to a member of U must be eventually in U.
5. Given that S is a set of real numbers and that x is a real number, prove that the following conditions are equivalent:
   (a) The number x is a limit point of the set S.
   (b) There exists a sequence (x_n) in the set $S \setminus \{x\}$ such that $x_n \to x$.
6. Prove that if (x_n) is a sequence of real numbers, then the set of all partial limits of (x_n) is closed.
7. Suppose that A and B are nonempty sets of real numbers and that for every number $x \in A$ and every number $y \in B$ we have $x < y$. Prove that the following conditions are equivalent:
   (a) We have $\sup A = \inf B$.
   (b) There exists a sequence (x_n) in the set A and a sequence (y_n) in the set B such that $y_n - x_n \to 0$ as $n \to \infty$.
8. Suppose that S is a nonempty bounded set of real numbers. Prove that there exist two sequences (x_n) and (y_n) in the set S such that

$$y_n - x_n \to \sup S - \inf S$$

as $n \to \infty$.

7.7   Limits of Monotone Sequences

In this section we begin our study of those properties of limits that depend upon the completeness of the real number system. The thrust of this section is the fact that a monotone sequence, as defined in Subsection 5.11.8, always has a limit and will converge if and only if it is bounded.

7.7.1   The Monotone Sequence Theorem

1. *Every monotone sequence of real numbers has a limit in $[-\infty, \infty]$.*
2. *A monotone sequence of real numbers is convergent if and only if it is bounded.*
3. *The limit of an increasing sequence is its supremum and the limit of a decreasing sequence is its infimum.*

Proof. 🖐️   We shall prove part 3. The first two parts of the theorem will then be clear. In proving part 3 we shall consider an increasing sequence and we shall leave the analogous proof for decreasing sequences as an exercise. 📖 Suppose then that (x_n) is an increasing sequence of real numbers and define x to be the supremum of this sequence. Note that this definition of x means that, in the event that the sequence (x_n) is unbounded above, we have $x = \infty$. See Subsection 5.14.4. We need to show that $x_n \to x$. For this purpose, suppose that U is a neighborhood of x and choose a number $w < x$ such that the interval (w, x) is included in U.

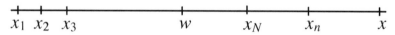

Using the fact that w, being less than the supremum of the sequence (x_n), cannot be an upper bound of (x_n), we now choose an integer N such that $x_N > w$. Then for each $n \geq N$ we have

$$x_n \geq x_N > w,$$

from which it follows that $x_n \in U$. Thus the sequence (x_n) is eventually in the neighborhood U and the proof is complete. ∎

7.7.2   Geometric Sequences

If c is a given number and if we define $x_n = c^n$ for each integer $n \geq 0$, then the sequence (x_n) is called a **geometric sequence**. In this subsection we shall discuss the limit behavior of sequences of this type.

The Case $c > 1$: Suppose that $c > 1$ and that $x_n = c^n$ for each integer $n \geq 0$. Since the sequence (x_n) is increasing, it must have a limit. We shall now show that this limit must be ∞. Write

$$x = \lim_{n \to \infty} x_n$$

and, to obtain a contradiction, assume that the number x is finite. Using the fact that $x/c < x$ and that x is the least upper bound of the sequence (x_n), choose an integer k such that $x_k > x/c$. We observe that

$$x_{k+1} = cx_k > c\left(\frac{x}{c}\right) = x,$$

which contradicts the fact that the number x is an upper bound of the sequence (x_n).

The Case $c < -1$: If $c < -1$, then, as we have just seen, $(-c)^n \to \infty$. Therefore, since the sign of c^n alternates, the sequence (c^n) has the two partial limits ∞ and $-\infty$.

The Case $|c| < 1$: Since $1/|c| > 1$, we know that $(1/|c|)^n \to \infty$; in other words, $1/|c|^n \to \infty$. We deduce from the algebraic rules for limits (Theorem 7.5.2) that $|c|^n \to 0$ and since $|c|^n \to 0$ we must also have $c^n \to 0$.

The Case $c = 1$ or $c = -1$: We leave consideration of these cases as an exercise.

7.7.3   A More Complicated Monotone Sequence

Suppose that (x_n) is a sequence of real numbers satisfying the condition $x_1 = 0$ and the condition

$$x_{n+1} = \sqrt[3]{\frac{\sqrt{21} + 9x_n}{9}}$$

for every positive integer n. We shall show that this sequence is convergent and find its limit.

1. **Exploring the Sequence**

   In order to motivate the method that we shall use to study the limit behavior of the sequence (x_n), we shall first explore a few of its terms with the aid of a computer: We begin by observing that

   $$x_2 = \sqrt[3]{\frac{\sqrt{21}}{9}} \approx .798\,53$$

   $$x_3 = \sqrt[3]{\frac{\sqrt{21} + 9\sqrt[3]{\frac{\sqrt{21}}{9}}}{9}} \approx 1.093\,5$$

$$x_4 = \sqrt[3]{\dfrac{\sqrt{21} + 9\sqrt[3]{\dfrac{\sqrt{21}+9\sqrt[3]{\frac{\sqrt{21}}{9}}}{9}}}{9}} \approx 1.170\,3.$$

Using the on-screen version of this book, you can obtain approximations to the first few terms of this sequence quite efficiently as follows: Point at the equation

$$f(x) = \sqrt[3]{\dfrac{\sqrt{21} + 9x}{9}},$$

open the Compute menu, click on **Define** and then click on **New Definition**. Then point at the Compute menu again, click on **Calculus**, and then on **Iterate**. You will see the iterate dialogue box shown in Figure 7.1. Fill in

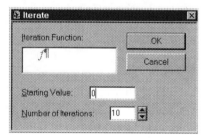

Figure 7.1

f as the iteration function, 0 as the starting value, and choose how many iterations you want. When you click on OK you will see the column of function values representing the number of terms of the sequence that you requested:

$$
\begin{array}{ll}
 & 0 \\
 & .798\,53 \\
 & 1.093\,5 \\
 & 1.170\,3 \\
 & 1.188\,7 \\
\text{Iterates:} & 1.193 \\
 & 1.194 \\
 & 1.1942 \\
 & 1.1943 \\
 & 1.1943 \\
 & 1.1943
\end{array}
$$

Now that we have explored a few terms of the sequence, we can guess that the sequence is probably increasing and that its limit is approximately 1.1943. It is certainly reasonable to guess that $x_n < 2$ for every n.

2. **Proving that the Sequence Is Increasing**

   We shall use mathematical induction to prove that the sequence (x_n) is increasing: For each positive integer n, we define $p(n)$ to be the assertion that $x_n < x_{n+1}$. Since the assertion $p(1)$ says that

   $$0 < \sqrt[3]{\frac{\sqrt{21}}{9}},$$

   it is clear that the assertion $p(1)$ is true. Now suppose that n is any integer for which the assertion $p(n)$ is true. To show that $p(n+1)$ is also true, we observe that

   $$x_{n+2} = \sqrt[3]{\frac{\sqrt{21} + 9x_{n+1}}{9}} > \sqrt[3]{\frac{\sqrt{21} + 9x_n}{9}} = x_{n+1}.$$

   It therefore follows by mathematical induction that $x_n < x_{n+1}$ for every positive integer n.

3. **Proving that the Sequence Is Bounded**

   We shall use mathematical induction to show that $x_n < 2$ for every positive integer n. For each positive integer n we define $p(n)$ to be the assertion that $x_n < 2$. The assertion $p(1)$ is clearly true. Now suppose that n is any integer for which the assertion $p(n)$ is true. To show that $p(n+1)$ is also true we observe that

   $$x_{n+1} = \sqrt[3]{\frac{\sqrt{21} + 9x_n}{9}} < \sqrt[3]{\frac{\sqrt{21} + 9(2)}{9}} < \sqrt[3]{\frac{72}{9}} = 2.$$

   It therefore follows by mathematical induction that $x_n < 2$ for every positive integer n.

4. **Conclusions About the Limit**

   We deduce that the sequence (x_n), being increasing and bounded, must be convergent and we shall call its limit x. Now from the equation

   $$x_{n+1} = \sqrt[3]{\frac{\sqrt{21} + 9x_n}{9}}$$

we obtain

$$\lim_{n\to\infty} x_{n+1} = \lim_{n\to\infty} \sqrt[3]{\frac{\sqrt{21} + 9x_n}{9}},$$

which yields

$$x = \sqrt[3]{\frac{\sqrt{21} + 9x}{9}};$$

in other words,

$$9x^3 - 9x - \sqrt{21} = 0.$$

By sketching the graph of this cubic polynomial (see Figure 7.2) we can see

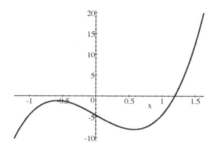

Figure 7.2

that this equation has just one real solution that is a little more than 1. As a matter of fact, the exact value of this solution can be shown to be

$$x = \frac{\sqrt[3]{18\left(\sqrt{21} + 3\right)^2} + 6}{3\sqrt[3]{12\sqrt{21} + 36}}.$$

You can obtain this value from *Scientific Notebook* by asking it for an exact solution to the equation $9x^3 - 9x - \sqrt{21} = 0$. However, if you would like to see how equations of this sort can be solved algebraically, then you can use the on-screen version of this book to take you to some reading material by clicking on the icon 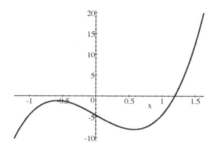 .

7.7.4   Exercises on Monotone Sequences

1. Given that $c > 1$, use the following method to prove that $c^n \to \infty$:

(a) Write $\delta = c - 1$, so that $c = 1 + \delta$, and then use mathematical induction to prove that, if n is any positive integer, then $c^n \geq 1 + n\delta$.

(b) Explain why $1 + n\delta \to \infty$ and then use Exercise 9 in Subsection 7.4.8 to show that $c^n \to \infty$.

2. Given that $c > 1$, use the following method to prove that $c^n \to \infty$:

(a) Explain why the sequence (c^n) is increasing and deduce that it has a limit.

(b) Call the limit x and show that if x is finite, then the equation

$$c^{n+1} = cc^n$$

leads to the equation $x = cx$, which implies that $x = 0$. But x cannot be equal to zero? Why not?

3. Suppose that $|c| < 1$ and that, for every positive integer n,

$$x_n = \sum_{i=1}^{n} c^{i-1}.$$

Explain why

$$x_n \to \frac{1}{1-c}.$$

4. Suppose that (x_n) is a given sequence of real numbers, that $x_1 = 0$, and that the equation

$$8x_{n+1}^3 = 6x_n + 1$$

holds for every positive integer n.

(a) Ⓝ Use *Scientific Notebook* to work out the first 20 terms in the sequence (x_n).

(b) Prove that $x_n < 1$ for every positive integer n.

(c) Prove that the sequence (x_n) is strictly increasing.

(d) Ⓗ Deduce that the sequence (x_n) is convergent and discuss its limit. Assuming an unofficial knowledge of the trigonometric functions, prove that the limit of the sequence (x_n) is $\cos \frac{\pi}{9}$.

5. In this exercise we study the sequence (x_n) defined by the equation

$$x_n = \left(1 + \frac{1}{n}\right)^{n}$$

for every integer $n \geq 1$. You will probably want to make use of the binomial theorem when you do this exercise.

(a) (N) Ask *Scientific Notebook* to make a 2D plot of the graph of the
function f defined by the equation

$$f(n) = \left(1 + \frac{1}{n}\right)^n$$

for $1 \leq n \leq 100$.

(b) (H) Prove that $x_n < 3$ for every n.

(c) (H) Prove that the sequence (x_n) is increasing.

(d) Deduce that the sequence (x_n) converges to a number between 2 and 3.
Have you seen this number before?

6. This exercise concerns the sequence (x_n) defined by the fact that $x_1 = 1$
and that, for each $n \geq 1$, we have

$$x_{n+1} = \sqrt[5]{4x_n - 2}.$$

(a) (N) Use *Scientific Notebook* to work out the first 20 terms in the
sequence (x_n).

(b) Prove that $1 \leq x_n < 2$ for every n.

(c) Prove that the sequence (x_n) is strictly increasing.

(d) (N) Prove that the sequence has a limit x that satisfies the equation
$x^5 - 4x + 2 = 0$. Ask *Scientific Notebook* to make a 2D plot of the
expression $x^5 - 4x + 2$ on the interval $[-2, 2]$ and to solve the equation

$$\begin{aligned} x^5 - 4x + 2 &= 0 \\ x &\in [1, 2] \end{aligned}$$

numerically. Compare the answer obtained here with the results that you
obtained in part a.

7. (a) (H) Given that

$$f(x) = \frac{x}{2} + \frac{9}{2x}$$

for every number $x > 0$, prove that $f(x) \geq 3$ for each n and that the
equation $f(x) = 3$ holds if and only if $x = 3$.

(b) (H) Given that $x_1 = 4$ and, for each $n \geq 1$, we have

$$x_{n+1} = \frac{x_n}{2} + \frac{9}{2x_n},$$

prove that the sequence (x_n) is decreasing and that the sequence converges to the number 3.

8. This exercise is a study of the sequence (x_n) for which $x_1 = 0$ and

$$x_{n+1} = \frac{1}{2 + x_n}$$

for every positive integer n. We note that this sequence is bounded below by 0 and above by $1/2$.

(a) **H** **N** Supply the definition

$$f(x) = \frac{1}{2 + x}$$

to *Scientific Notebook*. Then open your Compute menu, click on Calculus, and choose to iterate the function f ten times, starting at the number 0. Evaluate the column of numbers that you have obtained accurately to ten decimal places and, in this way, show the first ten members of the sequence (x_n).

(b) **H** Show that

$$x_{n+2} = \frac{2 + x_n}{5 + 2x_n}$$

for every integer $n \geq 1$, and then show that the sequence (x_{2n-1}) is increasing and that the sequence (x_{2n}) is decreasing and that these two sequences have the same limit $\sqrt{2} - 1$.

(c) Deduce that $x_n \to \sqrt{2} - 1$ as $n \to \infty$.

7.8   The Cantor Intersection Theorem

In this section we shall discuss an important topological property of the number system **R**, a property that depends strongly upon the completeness of the real number system and is known as the *Cantor intersection theorem*. This theorem will be used to prove some important facts about sequences in Section 7.9, facts that will be the key to some important facts about continuous functions in Chapter 8. These facts about continuous functions will, in turn, be the key to the mean value theorem, which is the foundation stone of differential calculus. In addition, the Cantor intersection theorem will be used to simplify the proof of an important theorem about integrals that appears as Theorem 14.3.3.

7.8.1   Introduction to the Cantor Intersection Theorem

The Cantor intersection theorem refers to a contracting sequence (see Subsection

5.12.2) of sets of real numbers, and it tells us that, under certain conditions, if (S_n) is a contracting sequence of nonempty sets, then the intersection

$$\bigcap_{n=1}^{\infty} S_n$$

is also nonempty. Before we state the Cantor intersection theorem, we shall look at some examples that show that the intersection of a contracting sequence of nonempty sets can sometimes be empty.

1.   We define

$$S_n = \left(0, \frac{1}{n}\right]$$

   for every positive integer n. Although the sequence (S_n) is contracting and all the sets S_n are nonempty, their intersection is empty.
2.   We define

$$S_n = [n, \infty)$$

   for every positive integer n. Once again, the sequence (S_n) is contracting and all of the sets S_n are nonempty, but their intersection is empty.

7.8.2   Statement of the Cantor Intersection Theorem

Suppose that (H_n) is a contracting sequence of nonempty, closed, bounded sets of real numbers. Then the intersection of the sets H_n is nonempty.

Proof. As we know from Theorem 6.4.2, every one of the sets H_n must have a least member which we shall call x_n. For each n, it follows from the fact that

$$x_{n+1} \in H_{n+1} \subseteq H_n$$

and the fact that x_n is the *least* member of H_n that

$$x_n \leq x_{n+1},$$

and we therefore know that the sequence (x_n) is increasing. From the fact that $H_n \subseteq H_1$ for each n we deduce that (x_n), being a sequence in the bounded set H_1, must be bounded, and we conclude from Theorem 7.7.1 that the sequence (x_n) is convergent. We now define

$$x = \lim_{n \to \infty} x_n.$$

We shall show that the intersection of all the sets H_n is nonempty by observing that the number x that we have just defined belongs to this intersection. In fact, if n is any positive integer, then, since $x_j \in H_n$ whenever $j \geq n$, the sequence (x_j) is eventually in H_n and, since the set H_n is closed, it follows from Theorem 7.6.2 that $x \in H_n$. ∎

7.8.3   Exercises on the Cantor Intersection Theorem

1. Suppose that (H_n) is a sequence (not necessarily contracting) of closed bounded sets and that for every positive integer n we have

$$\bigcap_{i=1}^{n} H_i \neq \emptyset.$$

Prove that

$$\bigcap_{i=1}^{\infty} H_i \neq \emptyset.$$

2. Suppose that H is a closed bounded set of real numbers and that (U_n) is an expanding sequence of open sets.

    (a) Explain why the sequence of sets $H \setminus U_n$ is a contracting sequence of closed bounded sets.

    (b) Use the Cantor intersection theorem to deduce that if $H \setminus U_n \neq \emptyset$ for every n, then

    $$\bigcap_{n=1}^{\infty} (H \setminus U_n) \neq \emptyset.$$

    (c) Prove that if

    $$H \subseteq \bigcup_{n=1}^{\infty} U_n,$$

    then there exists an integer n such that $H \subseteq U_n$.

3. Suppose that (U_n) is a sequence of open sets (not necessarily expanding) and that H is a closed bounded set and that

    $$H \subseteq \bigcup_{n=1}^{\infty} U_n.$$

    Prove that there exists a positive integer N such that

    $$H \subseteq \bigcup_{n=1}^{N} U_n.$$

4. The Cantor intersection theorem depends upon the completeness of the real number system. Where in the proof of the theorem is the completeness used?

7.9   The Existence of Partial Limits

In this section we shall study some important facts about limits of sequences, facts that depend upon the completeness of the number system $\mathbf{R}$. A key theorem is Theorem 7.9.2. Note how its proof makes use of the Cantor intersection theorem (Theorem 7.8.2). In the chapters that follow we shall encounter several important properties of continuous functions, derivatives and integrals, that also depend upon the completeness of the number system $\mathbf{R}$. When we study those properties, the results contained in this section will enable us to shorten and simplify many of our proofs.

7.9.1   Sequences in a Closed Bounded Set

Suppose that H is a set of real numbers. Then the following conditions are equivalent:

1. *The set H is closed and bounded.*
2. *Every sequence that is frequently in the set H has a partial limit that belongs to H.*
3. *Every sequence in the set H has a partial limit that belongs to H.*

Proof that Condition 1 Implies Condition 2. Suppose that the set H is closed and bounded and that (x_n) is a sequence that is frequently in H. For each positive integer n we shall write

$$E_n = \{x_n, x_{n+1}, x_{n+2}, \cdots\} = \{x_j \mid j \geq n\}.$$

Using the fact that the sequence $\left(H \cap \overline{E}_n\right)$ is a contracting sequence of nonempty closed bounded sets and the Cantor intersection theorem, we choose a number x such that

$$x \in \bigcap_{n=1}^{\infty} H \cap \overline{E}_n.$$

To complete the proof, we need to see why this number x is a partial limit of the sequence (x_n). Suppose that $\varepsilon > 0$. For every positive integer n it follows from the fact that $x \in \overline{E}_n$ that there are integers $j \geq n$ for which

$$x_j \in (x - \varepsilon, x + \varepsilon),$$

and from this observation we observe that the sequence (x_n) is frequently in the interval $(x - \varepsilon, x + \varepsilon)$. ∎

Proof that Condition 2 Implies Condition 3. This assertion is clear. ∎

Proof that Condition 3 Implies Condition 1. We assume that every sequence in the set H has a partial limit in H. We need to prove that the set H is both closed and bounded. There cannot be a sequence in H that has a limit at ∞ because (by the uniqueness theorem for limits) such a sequence would not have a partial limit in H. Therefore it follows from Theorem 7.6.3 that the set H is bounded above, and it follows analogously that H is bounded below. Finally, to prove that H is closed we suppose that a number x belongs to $\overline{H}$ and we shall show that $x \in H$. Using Theorem 7.6.1, we choose a sequence (x_n) in H that converges to x. Since x is the only partial limit of (x_n) and (x_n) must have a partial limit in H, we see that $x \in H$, as promised. ∎

7.9.2   Existence of Partial Limits of a Bounded Sequence

Every bounded sequence has a finite partial limit.

Proof. Suppose that (x_n) is a bounded sequence and choose a lower bound α and an upper bound β of (x_n). Since (x_n) is in the closed bounded interval $[\alpha, \beta]$, the sequence (x_n) must have a partial limit in this interval. ∎

7.9.3   Sequences that Have Limits

We already know from Theorem 7.4.1 that if a sequence has a limit, then it cannot have more than one partial limit in the extended real number system $[-\infty, \infty]$. We now show that the converse of this statement is also true:

Suppose that (x_n) is a sequence of real numbers, that $x \in [-\infty, \infty]$, and that x is the only extended real number that is a partial limit of the sequence (x_n). Then we have $x_n \to x$ as $n \to \infty$.

Proof. In order to prove that $x_n \to x$ as $n \to \infty$, we shall consider three cases:

Case 1: Suppose that $x = \infty$. Since $-\infty$ is not a partial limit of the sequence (x_n), the sequence (x_n) must be bounded below. Choose a lower bound α of the sequence (x_n). Now, to prove that $x_n \to \infty$, suppose that w is a real number. We need to show that the sequence (x_n) is eventually in the interval (w, ∞). To obtain a contradiction, assume that (x_n) is not eventually in the interval (w, ∞). Then (x_n) must be frequently in the interval $[\alpha, w]$,

_____

and it follows from Theorem 7.9.1 that the sequence (x_n) has a partial limit in $[\alpha, w]$, which contradicts our assumption that ∞ is the only partial limit of (x_n). ∎

Case 2: Suppose that $x = -\infty$. We leave the proof of this case as an exercise.

Case 3: Suppose that x **Is a Real Number.**   We begin with the observation that since neither $-\infty$ nor ∞ is a partial limit of the sequence (x_n), the sequence must be bounded. Choose a lower bound α and an upper bound β of the sequence (x_n). Now suppose that $\varepsilon > 0$. We need to show that the sequence (x_n) is eventually in the interval $(x - \varepsilon, x + \varepsilon)$. To obtain a contradiction, assume that the sequence (x_n) is frequently in the complement of this interval, in other words, that (x_n) is frequently in the set

$$H = [\alpha, \beta] \cap (\mathbf{R} \setminus (x - \varepsilon, x + \varepsilon)) = [\alpha, \beta] \setminus (x - \varepsilon, x + \varepsilon).$$

In the following figure, the set H is depicted by the heavy line.

$$\alpha \qquad\qquad x - \varepsilon \quad\; x \quad\; x + \varepsilon \qquad\qquad \beta$$

Since the set H is closed and bounded, we deduce from Theorem 7.9.1 that the sequence (x_n) has a partial limit in H, contradicting the fact that x is the only partial limit of (x_n). ∎

7.9.4   The Bolzano-Weierstrass Theorem

The Bolzano-Weierstrass theorem is similar in nature to the Cantor intersection theorem (Theorem 7.8.2) and, like the Cantor intersection theorem, it depends upon the completeness of the real number system. From an intuitive point of view, the statement of the Bolzano-Weierstrass theorem is quite simple. We can picture it as telling us that the only way in which infinitely many numbers can be crammed into a bounded piece of the number line is to give them very crowded living conditions somewhere. The precise statement of the Bolzano-Weierstrass theorem is as follows:

   Every bounded infinite set of real numbers must have at least one limit point.

Proof. Suppose that S is a bounded infinite set. Using the property of infinite sets that we discussed in Subsection 5.11.9, choose a one-one sequence (x_n) in the set S. Using Theorem 7.9.2, choose a finite partial limit x of the sequence (x_n). Since every neighborhood of x contains the numbers x_n for infinitely many values of n, and since the sequence (x_n) is one-one, we know that every neighborhood of the number x contains infinitely many different numbers that belong to the set S. Therefore x is a limit point of S. ∎

7.9.5   Exercises on the Bolzano-Weierstrass Theorem

1. The Bolzano-Weierstrass theorem does not tell us that if a set S is bounded and infinite, then at least one member of S must be a limit point of S. Give an example of a bounded infinite set S such that no member of S is a limit point of S.

2. Prove that if H is a closed bounded infinite set, then

$$H \cap \mathbf{L}(H) \neq \emptyset.$$

3. This exercise suggests a different proof of the Bolzano-Weierstrass theorem:

    (a) Prove that if E is a nonempty bounded set with no least member, then $\inf E$ is a limit point of E.

    (b) Prove that if a bounded set S has a nonempty subset that does not have a least member, then S has a limit point.

    (c) Given that S is a bounded infinite set and that every nonempty subset of S has a least member, find an example of a strictly increasing sequence in the set S. By considering the limit of this sequence, prove that S must have a limit point.

7.9.6   Some Further Exercises on Limit Points

Readers of the on-screen version of this text who have a familiarity with the concept of countability can find some exercises that relate to this concept by clicking on the icon ▨.

7.9.7   Cauchy Sequences

You can reach this optional topic from the on-screen version of the text by clicking on the icon ▨.

7.10   Upper and Lower Limits

You can find this optional topic in the on-screen version of the text by clicking on the icon ▨.

Chapter 8
Limits and Continuity of Functions

This chapter presents the theory of limits and continuity of functions of a single real variable. If you chose to read the more general presentation of limits of sequences in metric spaces, and if you prefer to read the theory of limits of functions in that more general situation, then you can do so in the on-screen version of the book by clicking on the icon ▦ .

8.1   Limits of Functions

8.1.1   Introduction to Limits of Functions

The theory of limits of functions is the cornerstone of calculus because these are the limits upon which the notion of a derivative depends. Almost every course in elementary calculus begins with this topic, but, as you have seen, this book did not. Instead, we laid the foundations for the study of limits of functions in our earlier study of the topology of the system $\mathbf{R}$ and limits of sequences. Now, as we study limits and continuity of functions, we shall draw upon the knowledge that we gained from those earlier chapters and use it to shorten and simplify many of our proofs. In this way, we shall be able to take a deeper look at the limits and continuity of functions than would have been possible in an elementary course.

According to the intuitive notion of a limit as it usually appears in elementary calculus, the condition $f(x) \to \lambda$ as $x \to a$ means that the number $f(x)$ can be made to lie as close as we like to the number λ as long we take the number x unequal to the number a but close enough to a.Notice how this idea makes no demand at all about the value $f(x)$ when $x = a$.This means that the number $f(a)$ itself might be very far away from λ and, in fact, there is no need for $f(a)$ even to be defined. To understand why we are so insistent that the number $f(a)$ should play no role at all in the concept of $f(x) \to \lambda$ as $x \to a$, we shall take a brief preview of the important limit upon which the definition of a derivative is based.

The derivative $f'(a)$ of a given function f at a number a can be thought of as

$$f'(a) = \lim_{x \to a} \frac{f(x) - f(a)}{x - a}.$$

What we are really doing here is defining a new function q by the equation

$$q(x) = \frac{f(x) - f(a)}{x - a}$$

for every number x in the domain of the function f except $x = a$ (which must, of course, be excluded). The definition of a derivative requires that

$$f'(a) = \lim_{x \to a} \frac{f(x) - f(a)}{x - a} = \lim_{x \to a} q(x).$$

Thus the type of limit that we need when we define derivatives is a limit at a of a function q that is definitely not defined at a.

So, to repeat what we have said: When we speak of a limit of a function f at a number a, we don't make any demands about the value (if any) of the number $f(a)$. If the domain of our function f is written as S, then S is a set of real numbers and we may or may not have $a \in S$. We shall now state the definition of a limit precisely.

8.1.2   Definition of a Limit of a Function

Suppose that S is a set of real numbers and that $f : S \to \mathbf{R}$. Given a number a that is a limit point of S and given a real number λ, we say that λ is a **limit** of the function f at the number a and we write

$$f(x) \to \lambda \quad \text{as} \quad x \to a$$

if, for every neighborhood V of the number λ, it is possible to find a neighborhood U of the number a such that the condition $f(x) \in V$ holds for every number x in the set $U \cap S \setminus \{a\}$.

Another way of stating this definition is to say that for every neighborhood V of the number λ it is possible to find a neighborhood U of the number a such that

$$f[U \cap S \setminus \{a\}] \subseteq V.$$

8.1.3   Limits Can Be Taken Only at Limit Points

In Definition 8.1.2 we stipulated that if $f : S \to \mathbf{R}$ and a and λ are real numbers, then the condition

$$f(x) \to \lambda \quad \text{as} \quad x \to a$$

can make sense only when a is a limit point of S. In order to understand the need for this restriction, we shall consider what happens when the number a is not a limit point of S:

Suppose that S is a set of real numbers and that a is a number that is not a limit point of S. Suppose that f is any function defined on the set S and that λ is any number. We shall now observe that, even though the function f and the number λ are completely arbitrary, the requirements of the definition of a limit are satisfied: For every neighborhood V of the number λ, there is a neighborhood U of a such that the condition $f(x) \in V$ holds for every number $x \in U \cap S \setminus \{a\}$. In fact, all we have to do is choose a neighborhood U of a such that

$$U \cap S \setminus \{a\} = \emptyset.$$

Since there aren't any numbers x in the set $U \cap S \setminus \{a\}$, it is certainly true that $f(x) \in V$ for every $x \in U \cap S \setminus \{a\}$.

Thus if V is a neighborhood of λ, the existence of a neighborhood U of a such that $f(x) \in V$ for every $x \in U \cap S \setminus \{a\}$ is interesting only if the number a is a limit point of the set S. For this reason, the definition of a limit of a function is restricted to numbers that are limit points of the domain of that function.

8.1.4   Epsilon-Delta Form of the Definition of a Limit

In this subsection we show how the definition of a limit can be phrased in terms of inequalities instead of neighborhoods. The inequality form of the condition

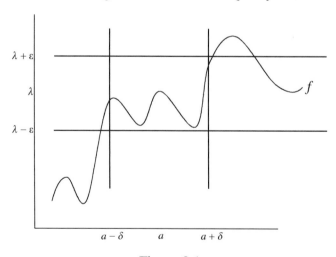

Figure 8.1

$f(x) \to \lambda$ as $x \to a$ is illustrated in Figure 8.1. It says that whenever $\varepsilon > 0$, it is possible to find a number $\delta > 0$ that enables us to force the graph of the f to lie between the two horizontal lines $y = \lambda - \varepsilon$ and $y = \lambda + \varepsilon$, by taking $x \neq a$ and taking our point on the graph between the two vertical lines $x = a - \delta$ and $x = a + \delta$.

Although the forms of the definition that refer to inequalities look a little more clumsy, they are frequently more useful when we are dealing with a known function. The inequality forms of the limit concept can be stated as follows:

Suppose that f is a function defined on a set S of real numbers, that a is a limit point of S, and that λ is a real number. Then the following conditions are equivalent:

1. $f(x) \to \lambda$ *as* $x \to a$.
2. *For every neighborhood V of the number λ there exists a number $\delta > 0$ such that the condition $f(x) \in V$ holds for every number $x \in S \setminus \{a\}$ that satisfies the inequality $|x - a| < \delta$.*
3. *For every number $\varepsilon > 0$ there exists a neighborhood U of the number a such that the inequality $|f(x) - \lambda| < \varepsilon$ holds for every number $x \in U \cap S \setminus \{a\}$.*
4. *For every number $\varepsilon > 0$ there exists a number $\delta > 0$ such that the inequality $|f(x) - \lambda| < \varepsilon$ holds for every number $x \in S \setminus \{a\}$ that satisfies the inequality $|x - a| < \delta$.*

We shall prove the equivalence of conditions 1 and 4 and leave the other parts of the proof of this theorem as exercises.

Proof that Condition 1 Implies Condition 4. We assume that $f(x) \to \lambda$ as $x \to a$ and, to prove that condition 4 holds, we suppose that ε is a given positive number. Using the fact that the interval $(\lambda - \varepsilon, \lambda + \varepsilon)$ is a neighborhood of the number λ, we choose a neighborhood U of the number a such that the condition

$$f(x) \in (\lambda - \varepsilon, \lambda + \varepsilon)$$

holds for every number $x \in U \cap S \setminus \{a\}$. Using the fact that U is a neighborhood of the number a, we choose a number $\delta > 0$ such that

$$(a - \delta, a + \delta) \subseteq U.$$

We now observe that the condition

$$f(x) \in (\lambda - \varepsilon, \lambda + \varepsilon)$$

holds for every number

$$x \in (a - \delta, a + \delta) \cap S \setminus \{a\}.$$

This is just another way of saying that the inequality $|f(x) - \lambda| < \varepsilon$ holds for every number $x \in S \setminus \{a\}$ that satisfies the inequality $|x - a| < \delta$. ∎

Proof that Condition 4 Implies Condition 1. We assume that condition 4 holds and, to prove that $f(x) \to \lambda$ as $x \to a$, we suppose that V is a neighborhood of the number λ. Choose a number $\varepsilon > 0$ such that

$$(\lambda - \varepsilon, \lambda + \varepsilon) \subseteq V.$$

Now we use condition 4 to choose a number $\delta > 0$ such that the inequality $|f(x) - \lambda| < \varepsilon$ holds for every number $x \in S \setminus \{a\}$ that satisfies the inequality $|x - a| < \delta$ and we define

$$U = (a - \delta, a + \delta).$$

We see at once that the condition $f(x) \in V$ holds for every number $x \in U \cap S \setminus \{a\}$. ∎

8.1.5   Some Examples of Limits of Functions

1. In this example we shall show that $x^2 + 1 \to 5$ as $x \to 2$. We define

$$f(x) = x^2 + 1$$

for every real number x, and, to prove that $f(x) \to 5$ as $x \to 2$, we suppose that $\varepsilon > 0$. We need to find a number $\delta > 0$ such that the inequality $|f(x) - 5| < \varepsilon$ will hold for every number x that is unequal to 2 and that satisfies the inequality $|x - 2| < \delta$. Now, if x is any real number, then we have

$$|f(x) - 5| = |x^2 + 1 - 5| = |x - 2|\,|x + 2|.$$

Of the two factors $|x - 2|$ and $|x + 2|$, the one that is small when x is close to 2 is the factor $|x - 2|$. The other factor, $|x + 2|$, is approximately 4. We shall now observe that the factor $|x + 2|$ will not exceed 5 if we keep x reasonably close to 2. In fact, if we stipulate that $|x - 2| < 1$, then it is easy to see that $|x + 2| < 5$.

1	2	3

As long as $|x - 2| < 1$ we therefore have

$$|f(x) - 5| \leq 5\,|x - 2|,$$

and we can therefore guarantee the inequality $|f(x) - 5| < \varepsilon$ by requiring that

$$|x - 2| < \frac{\varepsilon}{5}.$$

We can now continue our proof by defining δ to be the smaller of the two numbers 1 and $\varepsilon/5$. Whenever a number x satisfies the inequality

$|x - 2| < \delta$ we know that both of the inequalities $|x - 2| < 1$ and $|x - 2| < \varepsilon/5$ must hold, and therefore

$$|f(x) - 5| \leq 5\,|x - 2| < 5\left(\frac{\varepsilon}{5}\right) = \varepsilon.$$

We have now obtained more than we actually need. We have shown that the inequality $|f(x) - 5| < \varepsilon$ holds for every number x satisfying the inequality $|x - 2| < \delta$, regardless of whether or not $x = 2$. Throwing away what we know about the number $f(2)$, we can now observe that the inequality $|f(x) - 5| < \varepsilon$ holds for every number $x \neq 2$ that satisfies the inequality $|x - 2| < \delta$.

2. In this example we define

$$f(x) = \begin{cases} x^2 + 1 & \text{if } x \neq 2 \\ 17 & \text{if } x = 2, \end{cases}$$

and, once again, we shall show that $f(x) \to 5$ as $x \to 2$. Once again we suppose that $\varepsilon > 0$, and we need to find a number $\delta > 0$ such that the inequality $|f(x) - 5| < \varepsilon$ will hold for every number x that is unequal to 2 and that satisfies the inequality $|x - 2| < \delta$. Our proof will be an almost exact parallel to the one that appears in the preceding example, except that we must take care to avoid the possibility that $x = 2$.

Now if $x \neq 2$, then we have

$$|f(x) - 5| = \left|x^2 + 1 - 5\right| = |x - 2|\,|x + 2|.$$

As long as $|x - 2| < 1$ and $x \neq 2$, we therefore have

$$|f(x) - 5| \leq 5\,|x - 2|,$$

and, as before, we can guarantee the inequality $|f(x) - 5| < \varepsilon$ by requiring that $x \neq 2$ and $|x - 2| < \varepsilon/5$. We define δ to be the smaller of the two numbers 1 and $\varepsilon/5$. Given any number $x \neq 2$ that satisfies the inequality $|x - 2| < \delta$ we have

$$|f(x) - 5| \leq 5\,|x - 2| < 5\left(\frac{\varepsilon}{5}\right) = \varepsilon.$$

3. In this example we shall look at a more complicated rational function than the one that appeared in the preceding examples, and we shall see that the process of finding a limit of such a function is not as hard as one might

expect. We shall show that

$$\frac{x^3 - 2x^2 + x - 3}{x^4 + 3x^2 + 5} \to -\frac{1}{33}$$

as $x \to 2$. The graph of this rational function is illustrated in Figure 8.2.

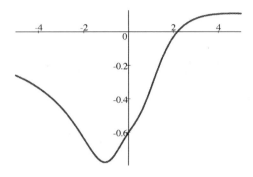

Figure 8.2

Before we begin, we need to do a little algebra: For any number x we have

$$\left| \frac{x^3 - 2x^2 + x - 3}{x^4 + 3x^2 + 5} - \left(-\frac{1}{33}\right) \right| = \left| \frac{1}{33} \frac{(x-2)\left(x^3 + 35x^2 + 7x + 47\right)}{x^4 + 3x^2 + 5} \right|$$

$$\leq \frac{1}{165} |x - 2| \left| x^3 + 35x^2 + 7x + 47 \right|.$$

Now whenever $|x - 2| < 1$ we have

$$\left| x^3 + 35x^2 + 7x + 47 \right| \leq 3^3 + (35)\left(3^2\right) + (7)(3) + 47 = 410,$$

and so, if $|x - 2| < 1$, then

$$\left| \frac{x^3 - 2x^2 + x - 3}{x^4 + 3x^2 + 5} - \left(-\frac{1}{33}\right) \right| \leq \frac{410}{165} |x - 2| = \frac{82}{33} |x - 2|.$$

Now we can begin: Suppose that $\varepsilon > 0$. We define δ to be the smaller of the two numbers 1 and $33\varepsilon/82$, and we observe that whenever $x \neq 2$ and

$|x - 2| < \delta$ we have

$$\left| \frac{x^3 - 2x^2 + x - 3}{x^4 + 3x^2 + 5} - \left(-\frac{1}{33} \right) \right| \leq \frac{82}{33} |x - 2| < \frac{82}{33} \left(\frac{33\varepsilon}{82} \right) = \varepsilon.$$

4. In this example we see an example of a function that does not have a limit at the number 2. We define

$$f(x) = \begin{cases} 1 & \text{if } x > 2 \\ -1 & \text{if } x < 2, \end{cases}$$

and, to obtain a contradiction, we assume that λ is a real number and that $f(x) \rightarrow \lambda$ as $x \rightarrow 2$. Using this assumption and the fact that 1 is a positive number we choose a number $\delta > 0$ such that the inequality

$$|f(x) - \lambda| < 1$$

holds whenever $x \neq 2$ and $|x - 2| < \delta$. Choose numbers t and x such that

$$2 - \delta < t < 2 < x < 2 + \delta.$$

$\overline{2 - \delta} \qquad\qquad\qquad t \qquad\qquad 2 \qquad\qquad x \qquad\qquad\qquad 2 + \delta$

Since $f(x) = 1$ and $f(t) = -1$, we see that $f(x) - f(t) = 2$. But we also know that

$$\begin{aligned} |f(x) - f(t)| &= |f(x) - \lambda + \lambda - f(t)| \\ &\leq |f(x) - \lambda| + |\lambda - f(t)| < 1 + 1 = 2, \end{aligned}$$

and we have reached the desired contradiction.

5. In this example we define

$$f(x) = \begin{cases} 1 & \text{if } x \in (0, 2) \text{ and } x \text{ is rational} \\ -1 & \text{if } x \in (0, 2) \text{ and } x \text{ is irrational.} \end{cases}$$

Using the technique of the preceding example, one may show that this function has no limit at the number 2. We leave the details as an exercise.

6. In this example we define

$$f(x) = \sin \frac{1}{x}$$

for every number $x > 0$. The graph of this function is illustrated in Figure 8.3. Using the method that was demonstrated in Example 4, one may show

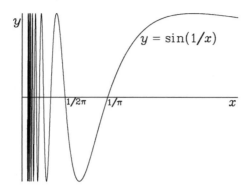

Figure 8.3

that this function has no limit at the number 0. We leave the details as an exercise.

7. In this example we define $f(x) = 1$ for every number x in the set

$$[0, 1] \cup \{2\}.$$

No matter what number λ we take, the condition $f(x) \to \lambda$ as $x \to 2$ is impossible because the number 2 is not a limit point of the domain of the function.

8. In this example we shall explore the properties of an interesting function $f : [0, 1] \to \mathbf{R}$ that is known as the **ruler function**. One of the interesting features of this function is that, even though it takes a positive value at every rational number, its limit at every number is zero. Given any number $x \in [0, 1]$ we define

$$f(x) = \begin{cases} 0 & \text{if} \quad x \text{ is irrational} \\ \frac{1}{n} & \text{if} \quad x = \frac{m}{n}, \text{ where } m \text{ and } n \text{ are integers with no common factor.} \end{cases}$$

The following table illustrates some values of the function f.

x	$\frac{1}{8}$	$\frac{2}{8}$	$\frac{3}{8}$	$\frac{4}{8}$	$\frac{5}{8}$	$\frac{6}{8}$	$\frac{7}{8}$	$\frac{8}{8}$
$f(x)$	$\frac{1}{8}$	$\frac{1}{4}$	$\frac{1}{8}$	$\frac{1}{2}$	$\frac{1}{8}$	$\frac{1}{4}$	$\frac{1}{8}$	1

The graph of this function is illustrated in Figure 8.4.
From the on-screen version of this text you can view an animation[27] of this

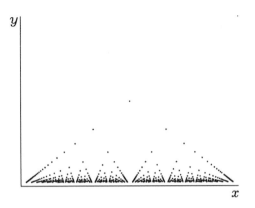

Figure 8.4

graph by clicking on the icon [icon] . Now suppose that a is any number in
the interval $[0, 1]$. We want to show that $f(x) \to 0$ as $x \to a$. Suppose that
$\varepsilon > 0$. We now define

$$S = \{x \in [0, 1] \mid f(x) \geq \varepsilon\}.$$

If x is any number in the set S, then x must be a rational number that can
be expressed in the form m/n, where m and n are positive integers with no
common factor and

$$\frac{1}{n} = f(x) \geq \varepsilon,$$

in other words,

$$n \leq \frac{1}{\varepsilon}.$$

Since there are only finitely many positive integers that do not exceed $1/\varepsilon$
and since there can be only finitely many rational numbers in $[0, 1]$ that have
a given denominator n, we deduce that the set S must be finite.

Now, because the finite set $S \setminus \{a\}$ is closed, the set

$$U = \mathbf{R} \setminus (S \setminus \{a\})$$

---

[27] I wish to express my appreciation to Eric Kehr for the valuable help he provided me in the
task of producing this animation.

is a neighborhood of the number a.Furthermore, the inequality

$$|f(x) - 0| < \varepsilon$$

holds for every number

$$x \in [0, 1] \cap U \setminus \{a\}.$$

8.1.6 Some Exercises on Limits of Functions

1. Write careful proofs of each of the following assertions:
   (a) $x^3 - 3x \to 2$ as $x \to -1$.

   (b) $\dfrac{1}{x} \to \dfrac{1}{3}$ as $x \to 3$.

   (c) $\dfrac{x^3 - 8}{x^2 + x - 6} \to \frac{12}{5}$ as $x \to 2$.

2. ■ Given that

$$f(x) = \begin{cases} x & \text{if } 0 < x < 2 \\ x^2 & \text{if } x > 2, \end{cases}$$

   prove that $f(x) \to 1$ as $x \to 1$ and that this function f has no limit at the number 2.

3. Given that

$$f(x) = \begin{cases} x & \text{if } x \text{ is rational} \\ x^2 & \text{if } x \text{ is irrational}, \end{cases}$$

   prove that $f(x) \to 1$ as $x \to 1$ and that this function f has no limit at the number 2.

4. Ⓝ Given that

$$f(x) = \frac{x^2 - 9}{|x - 3|}$$

   for every number $x \neq 3$, prove that f has no limit at the number 3. Ask *Scientific Notebook* to draw the graph of this function.

5. ■ Given that S is a set of real numbers, that $f : S \to \mathbf{R}$, that λ is a real number, and that a is a limit point of S, prove that the following conditions are equivalent:
   (a) $f(x) \to \lambda$ as $x \to a$.

(b) For every number $\varepsilon > 0$ there exists a number $\delta > 0$ such that the inequality $|f(x) - \lambda| < 3\varepsilon$ holds for every number x in the set $S \setminus \{a\}$ that satisfies the inequality $|x - a| < \delta$.

(c) For every number $\varepsilon > 0$ there exists a number $\delta > 0$ such that the inequality $|f(x) - \lambda| < 3\varepsilon$ holds for every number x in the set $S \setminus \{a\}$ that satisfies the inequality $|x - a| < 5\delta$.

6. Given that S is a set of real numbers, that $f : S \to \mathbf{R}$, that λ is a real number, and that a is a limit point of S, prove that the following conditions are equivalent:

(a) $f(x) \to \lambda$ as $x \to a$.

(b) For every number $\varepsilon > 0$ there exists a neighborhood U of the number a such that the inequality $|f(x) - \lambda| < \varepsilon$ holds for every number x in the set $U \cap S \setminus \{a\}$.

(c) For every neighborhood V of the number λ there exists a number $\delta > 0$ such that the condition $f(x) \in V$ holds for every number x in the set $S \setminus \{a\}$ that satisfies the inequality $|x - a| < \delta$.

7. Given that S is a set of real numbers, that $f : S \to \mathbf{R}$, that λ is a real number, and that a is an interior point of S, prove that the following conditions are equivalent:

(a) $f(x) \to \lambda$ as $x \to a$.

(b) For every number $\varepsilon > 0$ there exists a number $\delta > 0$ such that the inequality $|f(x) - \lambda| < \varepsilon$ will hold for every number $x \setminus \{a\}$ that satisfies the inequality $|x - a| < \delta$.

8. Suppose that S is a set of real numbers, that a is a limit point of S, that $f : S \to \mathbf{R}$, and that λ is a real number. Prove that if $f(x) \to \lambda$ as $x \to a$, then $|f(x)| \to |\lambda|$ as $x \to a$. Compare this exercise with Exercise 7 of Subsection 7.4.8.

9. Suppose that S is a set of real numbers, that a is a limit point of S, that $f : S \to \mathbf{R}$, and that λ is a real number. Complete the following sentence: *The function f fails to have a limit of λ at the number a when there exists a number $\varepsilon > 0$ such that for every number $\delta > 0$ ......*

8.2   Limits at Infinity and Infinite Limits

In this section we show how the notion $f(x) \to \lambda$ as $x \to a$ can be broadened to include the possibilities that a or λ (or both) could be infinite. As we shall see, the definition stated in terms of neighborhoods is very similar to the definition that we saw for finite limits.

8.2.1   Limits at ∞

Suppose that S is a set of real numbers and that S is unbounded above. Suppose that $f : S \rightarrow \mathbf{R}$. Given any real number λ, we say that λ is a **limit** of the function f **at infinity** if for every neighborhood V of the number λ there exists a neighborhood U of ∞ such that the condition $f(x) \in V$ holds for every number x in the set $U \cap S$.

If you prefer, you can express the latter set as $U \cap S \setminus \{\infty\}$. Since ∞ does not belong to S, the expressions $U \cap S$ and $U \cap S \setminus \{\infty\}$ are exactly the same.

8.2.2   Some Alternative Ways of Looking at Limits at ∞

Suppose that S is a set of real numbers and that S is unbounded above. Suppose that $f : S \rightarrow \mathbf{R}$ and that λ is a real number. Then the following conditions are equivalent:

1. *$f(x) \rightarrow \lambda$ as $x \rightarrow \infty$.*
2. *For every neighborhood V of the number λ there exists a real number w such that the condition $f(x) \in V$ will hold for every number $x \in S$ that satisfies the inequality $x > w$.*
3. *For every number $\varepsilon > 0$ there exists a real number w such that the inequality*

$$|f(x) - \lambda| < \varepsilon$$

will hold for every number $x \in S$ that satisfies the inequality $x > w$.

Proof that Condition 1 Implies Condition 3. We assume that $f(x) \rightarrow \lambda$ as $x \rightarrow \infty$, and, to prove that condition 3 must hold, we suppose that $\varepsilon > 0$. Using the fact that the interval $(\lambda - \varepsilon, \lambda + \varepsilon)$ is a neighborhood of the number λ, we choose a neighborhood U of ∞ such that the condition

$$f(x) \in (\lambda - \varepsilon, \lambda + \varepsilon)$$

holds for every number $x \in U \cap S$. Using the fact that U is a neighborhood of ∞, we now choose a number w such that $(w, \infty) \subseteq U$. We now observe that if x is any number in the set S that satisfies the inequality $x > w$, then the inequality

$$|f(x) - \lambda| < \varepsilon$$

must hold.

We leave the rest of proof of this theorem as an exercise.

8.2.3   Limits at $-\infty$

We leave as an exercise the task of writing the definition of a limit at $-\infty$ and exploring some alternative forms of this definition.

8.2.4   Infinite Limits

Suppose that S is a set of real numbers, that $f : S \to \mathbf{R}$, and that a is a limit point of S. We say that $f(x) \to \infty$ as $x \to a$ if, for every neighborhood V of ∞, there exists a neighborhood U of a such that the condition $f(x) \in V$ holds for every number x in the set $S \cap U \setminus \{a\}$.

8.2.5   Some Alternative Ways of Looking at Infinite Limits

Suppose that S is a set of real numbers, that $f : S \to \mathbf{R}$, and that a is a limit point of S. Then the following conditions are equivalent:

1. *$f(x) \to \infty$ as $x \to a$.*
2. *For every real number w there exists a neighborhood U of a such that the inequality $f(x) > w$ holds for every number x in the set $S \cap U \setminus \{a\}$.*
3. *For every real number w there exists a number $\delta > 0$ such that the inequality $f(x) > w$ holds for every number x in the set $S \setminus \{a\}$ that satisfies the inequality $|x - a| < \delta$.*

We leave the proof of this assertion as an exercise. 🖳

8.2.6   Infinite Limits at ∞

Suppose that S is a set of real numbers and that S is unbounded above. Suppose that $f : S \to \mathbf{R}$. We say that $f(x) \to \infty$ as $x \to \infty$ if for every neighborhood V of ∞ there exists a neighborhood U of ∞ such that the condition $f(x) \in V$ holds for every number x in the set $S \cap U$.

8.2.7   An Alternative Way of Looking at Infinite Limits at ∞

Suppose that S is a set of real numbers and that S is unbounded above. Suppose that $f : S \to \mathbf{R}$. Then the following conditions are equivalent:

1. *$f(x) \to \infty$ as $x \to \infty$.*
2. *For every real number w there exists a real number p such that the inequality $f(x) > w$ holds for every number x in the set S that satisfies the inequality $x > p$.*

We leave the proof of this assertion as an exercise. 🖳

8.2.8   The General Case

The time has come to end this orgy of separate cases and to write a single definition that unifies all of those that appeared previously.

Suppose that S is a set of real numbers and that $f : S \to \mathbf{R}$. Suppose that a and λ are extended real numbers. In the event that $a \in \mathbf{R}$ we suppose that a is a limit point of S, and in the event that $a = \infty$ we assume that S is unbounded above, and in the event that $a = -\infty$ we assume that S is unbounded below. The condition $f(x) \to \lambda$ as $x \to a$ means that for every neighborhood V of λ there exists a neighborhood U of a such that the condition $f(x) \in V$ holds for every x in the set $U \cap S$.

From now on we shall agree that if we say that f is a function defined on a set S of real numbers and $f(x) \to \lambda$ as $x \to a$, then a and λ are extended real numbers and, if $a \in \mathbf{R}$, then a is a limit point of S, and if $a = -\infty$ then the set S is unbounded below, and if $a = \infty$, then the set S is unbounded above.

8.2.9 Some Examples of Limits at Infinity and Infinite Limits

1. Suppose that

$$f(x) = \frac{x}{(x-3)^2}$$

for every number $x \neq 3$. We shall show that $f(x) \to \infty$ as $x \to 3$.

To motivate the proof that we are about to write we notice that, if $|x - 3| \leq 1$, we have

$$f(x) = \frac{x}{|x-3|\,|x-3|} \geq \frac{2}{|x-3|} \geq \frac{1}{|x-3|}.$$

Now we begin: Suppose that w is any real number. We need to find a number $\delta > 0$ such that the inequality $f(x) > w$ will hold for all numbers $x \neq 3$ that satisfy the inequality $|x - 3| < \delta$. Since the condition $f(x) > w$ requires that $1/\delta \geq w$, we define

$$\delta = \frac{1}{1 + |w|},$$

and we observe that the inequality $f(x) > w$ holds for every number $x \neq 3$ that satisfies the inequality $|x - 3| < \delta$.

2. (a) If

$$f(x) = \frac{x}{x-3}$$

for all numbers $x > 3$, then $f(x) \to \infty$ as $x \to 3$. We leave the proof of this assertion as an exercise.

(b) If

$$f(x) = \frac{x}{x-3}$$

for all numbers $x < 3$, then $f(x) \to -\infty$ as $x \to 3$.

(c) If

$$f(x) = \frac{x}{x - 3}$$

for all numbers $x \neq 3$, then the function f has no limit at the number 3.

3. Suppose that

$$f(x) = \frac{2x^3 - x^2 + 3x + 1}{x^3 + 2x^2 + 4}$$

for numbers $x > 0$. We shall observe that $f(x) \to 2$ as $x \to \infty$. Before we prove this fact we observe that if x is any positive number, we have

$$
\begin{aligned}
|f(x) - 2| &= \left| \frac{2x^3 - x^2 + 3x + 1}{x^3 + 2x^2 + 4} - 2 \right| \\
&= \left| \frac{5x^2 - 3x + 7}{x^3 + 2x^2 + 4} \right| \leq \frac{|5x^2 - 3x + 7|}{x^3}.
\end{aligned}
$$

Now as long as $x \geq 1$ we have

$$\left| 5x^2 - 3x + 7 \right| \leq 5x^2 + 3x + 7 \leq 5x^2 + 3x^2 + 7x^2 = 15x^2,$$

and so for $x \geq 1$ we have

$$|f(x) - 2| \leq \frac{15x^2}{x^3} = \frac{15}{x}.$$

Now we begin: Suppose that $\varepsilon > 0$. We define w to be the larger of the two numbers 1 and $15/\varepsilon$. We see at once that whenever $x > w$ we have

$$|f(x) - 2| \leq \frac{15}{x} < \varepsilon.$$

4. Suppose that

$$f(x) = \frac{x^3 - x^2 - 27x + 29}{x^2 + 3x - 10}$$

for all numbers $x > 2$. We shall show that $f(x) \to \infty$ as $x \to \infty$. We begin by expressing $f(x)$ in ⓝ 🖼 partial fractions. For each $x > 2$ we have

$$f(x) = \frac{x^3 - x^2 - 27x + 29}{x^2 + 3x - 10} = x - 4 - \frac{3}{x - 2} - \frac{2}{x + 5}.$$

Now as long as $x > 5$, neither of the fractions $\frac{3}{x-2}$ and $\frac{2}{x+5}$ can exceed 1, and so if $x > 5$, we have

$$f(x) > x - 4 - 1 - 1 = x - 6.$$

Now to prove that $f(x) \to \infty$ as $n \to \infty$, suppose that w is any real number. We define p to be the larger of the numbers 5 and $w + 6$ and we observe that the inequality

$$f(x) > x - 6 \geq w + 6 - 6 = w$$

holds whenever $x \geq p$.

Another way of doing this problem is to make the observation that if $x > 2$, then

$$\frac{x^3 - x^2 - 27x + 29}{x^2 + 3x - 10} = \frac{x - 1 - \frac{27}{x} + \frac{29}{x^2}}{1 + \frac{3}{x} - \frac{10}{x^2}}$$

and that, consequently, we have

$$f(x) > \frac{x - 3}{1 + \frac{1}{2}}$$

if x is sufficiently large. We leave the details as an exercises.

8.3   One-Sided Limits

8.3.1   Introduction to One-Sided Limits

Suppose that S is a set of real numbers, that $f : S \to \mathbf{R}$, that λ is an extended real number, and that a is a real number that is a limit point of the set $(-\infty, a) \cap S$. As we know, the condition $f(x) \to \lambda$ as $x \to a$ is equivalent to the condition that for every neighborhood V of λ there exists a number $\delta > 0$ such that the condition $f(x) \in V$ holds for every number $x \in S \setminus \{a\}$ that satisfies the inequality $|x - a| < \delta$.

We shall now restrict the function f to the part of the set S that lies to the left of the number a, and we call this restricted function g. More precisely, we define

$$g(x) = f(x)$$

for every number x in the set $S \cap (-\infty, a)$. The condition $g(x) \to \lambda$ as $x \to a$ is equivalent to the condition that for every neighborhood V of λ there exists a number $\delta > 0$ such that the condition $g(x) \in V$ holds for every number $x \in S \cap (-\infty, a)$ that satisfies the inequality $|x - a| < \delta$. Another way of stating this condition is to say that for every neighborhood V of λ there exists a number $\delta > 0$ such that the condition $f(x) \in V$ holds for every number $x \in S$ that satisfies the inequality

$$0 < a - x < \delta.$$

If $g(x) \to \lambda$ as $x \to a$, then we say that λ is a limit **from the left** at a of the function f and we write

$$f(x) \to \lambda \text{ as } x \to a - .$$

In the same way, if a is a limit point of the set $S \cap (a, \infty)$, then we say that λ is a limit **from the right** at a of the function f and we write

$$f(x) \to \lambda \text{ as } x \to a+$$

if for every neighborhood V of λ there exists a number $\delta > 0$ such that the condition $f(x) \in V$ holds for every number $x \in S$ that satisfies the inequality

$$0 < x - a < \delta.$$

8.3.2   Linking One-Sided and Two-Sided Limits

Suppose that S is a set of real numbers, that $f : S \to \mathbf{R}$, that λ is an extended real number, and that a is a limit point of both of the sets $(-\infty, a) \cap S$ and $S \cap (a, \infty)$. Then the following conditions are equivalent:

1. *$f(x) \to \lambda$ as $x \to a$.*
2. *The number λ is a limit of f from the left at a and is also a limit of f from the right at a.*

We leave the proof of this theorem as an exercise.

8.3.3   Some Further Exercises on Limits

1. Given that

$$f(x) = \begin{cases} 1 & \text{if } x < 2 \\ 0 & \text{if } x > 2, \end{cases}$$

    prove that f has a limit from the left at 2 and also has a limit from the right at 2 but does not have a limit at 2.

2. Given that

$$f(x) = \frac{1}{|x - 3|}$$

    for all numbers $x \neq 3$, explain why f has a limit (an infinite limit) at 3.

3. Given that

$$f(x) = \frac{1}{x - 3}$$

for all numbers $x \neq 3$, explain why f has an infinite limit from the left at 3 and also has an infinite limit from the right at 3 but does not have a limit at 3.

4. Prove that

$$\frac{x^3 - 8}{x^2 + x - 6} \rightarrow \infty$$

as $x \rightarrow \infty$.

5. Prove that

$$\frac{x^4 - 4x^3 - x^2 + x + 7}{x^3 - 2x^2 - 2x - 3} \rightarrow \infty$$

as $x \rightarrow \infty$.

6. Prove that

$$\frac{3x^2 + x - 1}{5x^2 + 4} \rightarrow \frac{3}{5}$$

as $x \rightarrow \infty$.

7. ▦ Given that

$$f(x) = \begin{cases} 1 & \text{if} \quad x \text{ is rational} \\ 0 & \text{if} \quad x \text{ is irrational,} \end{cases}$$

explain why f does not have a limit from the right at 2.

8. Suppose that a is an interior point of a set S of real numbers and that $f : S \rightarrow \mathbf{R}$. Suppose that $f(x) \rightarrow 0$ as $x \rightarrow a-$ and that $f(x) \rightarrow 1$ as $x \rightarrow a+$. Prove that the function f does not have a limit at the number a.

8.4   The Relationship Between Limits of Functions and Limits of Sequences

In this section we shall see that the concepts of limit of a function and limit of a sequence are closely related. Then in the next section we shall use this close relationship to obtain a free ride through many of the basic properties of limits of functions by tying them to the analogous statements about sequences.

The key to the relationship between limits of functions and limits of sequences is Section 7.6, and it would therefore be a good idea to review that section now. At the same time it might be a good idea to take a look at Exercise 5 of Subsection 7.6.4.

8.4.1   Connecting the Two Limit Notions

Suppose that S is a set of real numbers, that $f : S \to \mathbf{R}$, that a is a limit point of S, and that λ is a real number. Then the following conditions are equivalent:

1. $f(x) \to \lambda$ *as $x \to a$.*
2. *For every sequence (x_n) in the set $S \setminus \{a\}$, if $x_n \to a$ as $n \to \infty$, then $f(x_n) \to \lambda$ as $n \to \infty$.*

We shall prove this theorem by showing first that if condition 1 is true, then so is condition 2. Then we shall show that if condition 1 is false, then condition 2 must also be false.

Proof that Condition 1 Implies Condition 2. We assume that $f(x) \to \lambda$ as $x \to a$. To prove that condition 2 must hold, we suppose that (x_n) is a sequence in the set $S \setminus \{a\}$ and that $x_n \to a$ as $n \to \infty$. We must now show that $f(x_n) \to \lambda$ as $n \to \infty$. Suppose that $\varepsilon > 0$. Using the fact that $f(x) \to \lambda$ as $x \to a$, we now choose a number $\delta > 0$ such that the condition

$$|f(t) - \lambda| < \varepsilon$$

holds whenever $t \in S \setminus \{a\}$ and $|t - a| < \delta$. Using the fact that $x_n \to a$ as $n \to \infty$, we now choose an integer N such that the condition $|x_n - a| < \delta$ holds for every $n \geq N$ and we observe that the inequality

$$|f(x_n) - \lambda| < \varepsilon$$

holds whenever $n \geq N$. ∎

Proof that if Condition 1 Is False, then Condition 2 Is False: Using the fact that the condition $f(x) \to \lambda$ as $x \to a$ is false, we choose a number $\varepsilon > 0$ such that for every positive number δ there is at least one number $x \in (a - \delta, a + \delta) \cap S \setminus \{a\}$ for which $|f(x) - \lambda| \geq \varepsilon$.

For each positive integer n we use the fact that $1/n > 0$ to choose a number that we shall call x_n in the set

$$\left(a - \frac{1}{n}, a + \frac{1}{n} \right) \cap S \setminus \{a\}$$

such that $|f(x_n) - \lambda| \geq \varepsilon$. In this way we have found a sequence (x_n) in $S \setminus \{a\}$ such that $x_n \to a$ even though the corresponding sequence $(f(x_n))$ fails to have a limit value of λ. We deduce that condition 2 is false. ∎

8.4.2 An Analog for Limits at Infinity

Suppose that S is a set of real numbers, that S is unbounded above, that $f : S \to$ $\mathbf{R}$, and that λ is a real number. Then the following conditions are equivalent:

1. *$f(x) \to \lambda$ as $x \to \infty$.*
2. *For every sequence (x_n) in the set S, if $x_n \to \infty$ as $n \to \infty$, then $f(x_n) \to \lambda$ as $n \to \infty$.*

The proof that condition 1 implies condition 2 is analogous to the proof just given for limits at a number a and will be left as an exercise.

Proof that if Condition 1 Is False, then Condition 2 Is False. Using the fact that the condition $f(x) \to \lambda$ as $x \to \infty$ is false, we choose a number $\varepsilon > 0$ such that for every real number v there is at least one number $x \in (v, \infty) \cap S$ for which $|f(x) - \lambda| \geq \varepsilon$. For each positive integer n we choose a number that we shall call $x_n \in S$ such that $x_n > n$ and such that $|f(x_n) - \lambda| \geq \varepsilon$. In this way we have found a sequence (x_n) in S such that $x_n \to \infty$, even though the corresponding sequence $(f(x_n))$ fails to have a limit value of λ. We deduce that condition 2 is false. ∎

8.4.3 Analogs for Infinite Limits

The two preceding theorems have natural analogs for infinite limits. For example:

Suppose that S is a set of real numbers, that $f : S \to \mathbf{R}$, and that a is a limit point of S. Then the following conditions are equivalent:

1. *$f(x) \to \infty$ as $x \to a$.*
2. *For every sequence (x_n) in the set $S \setminus \{a\}$, if $x_n \to a$ as $n \to \infty$, then $f(x_n) \to \infty$ as $n \to \infty$.*

We leave the proof of this fact as an exercise.

8.4.4 A Stronger Connection Between Functions and Sequences

As before, we suppose that S is a set of real numbers, that $f : S \to \mathbf{R}$, and that a and λ are extended real numbers. If $a \in \mathbf{R}$, we assume that a is a limit point of S; if $a = \infty$, we assume that S is unbounded above; and if $a = -\infty$, we assume that S is unbounded below. Under these conditions, Theorem 8.4.1 tells us that the condition $f(x) \to \lambda$ as $x \to a$ will be assured if we know that whenever (x_n) is a sequence in the set $S \setminus \{a\}$ and $x_n \to a$ as $n \to \infty$, then $f(x_n) \to \lambda$ as $n \to \infty$.

There is, however, a stronger theorem that tells us that the condition $f(x) \to$ λ as $x \to a$ will be assured if we know that whenever (x_n) is a sequence in the set $S \setminus \{a\}$ and $x_n \to a$ as $n \to \infty$, if the sequence $(f(x_n))$ has any limit at all, then this limit must be λ. An alternative way of stating this theorem is as follows:

If the function f does not have a limit value of λ at a, then there must exist a sequence (x_n) in the set $S \setminus \{a\}$ and a number $\mu \neq \lambda$ such that $x_n \to a$ as $n \to \infty$ and $f(x_n) \to \mu$ as $n \to \infty$.

This stronger theorem is a little harder to prove and will not be proved here. If you chose in Section 7.2 to read the optional material on subsequences, then you can reach the theorem by clicking on the icon ⬛ . If you have not read the material on subsequences, then you can still read the proof of this theorem but the proof will be a little longer and can be reached by clicking on the icon ⬛ .

8.5   Some Facts About Limits of Functions

In this section we study some elementary properties of limits of functions. Some of these facts are analogous to theorems about sequences that we studied in Chapter 7, and although these can be deduced directly from the definition of this kind of limit, we shall deduce them much more quickly by making use of the corresponding facts about limits of sequences and the relationship between the two kinds of limits that we observed in Theorem 8.4.1. In this sense we can think such theorems as coming to us "for free".

8.5.1   The Uniqueness of Limits of Functions

Suppose that $S \subseteq \mathbf{R}$, and that $f : S \to \mathbf{R}$. Suppose that $f(x) \to \lambda$ as $x \to a$. Then the number λ is the only possible limit value of the function f at a.

Proof. Using our understanding that if $a \in \mathbf{R}$, then a is a limit point of S; that if $a = \infty$, then S is unbounded above; and that if $a = -\infty$, then S is unbounded below, choose a sequence (x_n) in the set $S \setminus \{a\}$ such that $x_n \to a$ as $n \to \infty$.

If λ and μ are extended real numbers that are both limit values of the function f at a, then it follows from Theorem 8.4.1 that $f(x_n) \to \lambda$ as $n \to \infty$ and also $f(x_n) \to \mu$ as $n \to \infty$. By the uniqueness theorem for limits of sequences (Theorem 7.4.1) we know that $\lambda = \mu$. Thus the function f can have not more than one limit value at the number a. ∎

8.5.2   Limit Notation for Functions

Suppose that S is a set of real numbers, that $f : S \to \mathbf{R}$, and that $f(x) \to \lambda$ as $x \to a$. Since we know that the limit value λ is unique, we can give it a name. We write it as

$$\lim_{x \to a} f(x).$$

Of course, the symbol x plays no special role in the latter expression. We could just as well have written

$$\lambda = \lim_{t \to a} f(t)$$

or even

$$\lambda = \lim_a f.$$

If λ is the limit of f from the left at a, then we write

$$\lambda = \lim_{x \to a-} f(x) = \lim_{a-} f = f(a-),$$

and we employ a similar notation for limits from the right.

8.5.3   The Sandwich Theorem for Limits of Functions

Suppose that f, g, and h are functions defined on a given set S of real numbers and that a is an extended real number. Suppose that the inequality

$$f(x) \le g(x) \le h(x)$$

holds for every number $x \in S \setminus \{a\}$. Suppose finally that the limits

$$\lim_{x \to a} f(x) \quad and \quad \lim_{x \to a} h(x)$$

exist and are equal to each other. Then the function g also has a limit at a and we have

$$\lim_{x \to a} f(x) = \lim_{x \to a} g(x) = \lim_{x \to a} h(x).$$

Proof. Suppose that (x_n) is a sequence in the set $S \setminus \{a\}$ and that $x_n \to a$ as $n \to \infty$. We deduce from Theorem 8.4.1 that

$$\lim_{n \to \infty} f(x_n) = \lim_{n \to \infty} h(x_n).$$

Since the inequality

$$f(x_n) \le g(x_n) \le h(x_n)$$

holds for every positive integer n, we deduce from the sandwich rule for sequences (Theorem 7.4.6) that

$$\lim_{n \to \infty} f(x_n) = \lim_{n \to \infty} g(x_n) = \lim_{n \to \infty} h(x_n).$$

Using Theorem 8.4.1 once again we deduce that

$$\lim_{x \to a} f(x) = \lim_{x \to a} g(x) = \lim_{x \to a} h(x).$$

8.5.4   Algebraic Rules for Limits of Functions

Suppose that f and g are functions defined on a set S of real numbers and that $f(x) \to \lambda$ as $x \to a$ and $g(x) \to \mu$ as $x \to a$. Then each of the following conditions holds as long as the appropriate limit value is defined:

1. $f(x) + g(x) \to \lambda + \mu$ *as $x \to a$.*
2. $f(x) - g(x) \to \lambda - \mu$ *as $x \to a$.*
3. $f(x)g(x) \to \lambda\mu$ *as $x \to a$.*
4. $f(x)/g(x) \to \lambda/\mu$ *as $x \to a$. The sense in which this fourth assertion is made is that there exists a neighborhood U of the number a such that $g(x) \neq 0$ whenever $x \in U \cap S \setminus \{a\}$ and that we have restricted the function f/g to the set $U \cap S \setminus \{a\}$.*

Proof. Suppose that (x_n) is a sequence in the set $S \setminus \{a\}$ and that $x_n \to a$ as $n \to \infty$. We deduce from Theorem 8.4.1 that

$$
\begin{aligned}
f(x_n) + g(x_n) &\to \lambda + \mu \\
f(x_n) - g(x_n) &\to \lambda - \mu \\
f(x_n)g(x_n) &\to \lambda\mu
\end{aligned}
$$

as $n \to \infty$. Now we need to explain why, if the ratio λ/μ is defined, the function f/g is defined at every number x in some neighborhood of the number a.In order for the ratio λ/μ to exist we must have $\mu \neq 0$, and therefore the condition $g(x) \to \mu$ as $x \to a$ guarantees that there is a neighborhood U of a such that $g(x) \neq 0$ whenever $x \in U \cap S \setminus \{a\}$. The function f/g is defined at every number x in the set $U \cap S \setminus \{a\}$. Since $x_n \in U \cap S \setminus \{a\}$ for all sufficiently large n, we know that

$$
f(x_n)/g(x_n) \to \lambda/\mu.
$$

All of the limit rules for functions therefore follow at once from the corresponding rules for sequences seen in Subsection 7.5.3.

8.6   The Composition Theorem for Limits

8.6.1   Introduction to the Composition Theorem

Suppose that f is a function from a set A of real numbers into a set B of real numbers and that g is a function from B into a set C of real numbers. As we know from Subsection 4.3.11, the composition $g \circ f$ of the functions f and g is defined by the equation

$$
(g \circ f)(x) = g(f(x))
$$

for every number $x \in A$. Suppose now that a, b, and c are extended real numbers; that $f(x) \to b$ as $x \to a$; and that $g(y) \to c$ as $y \to b$.See Figure 8.5. A natural

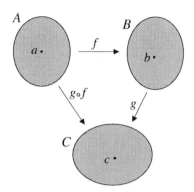

Figure 8.5

question to ask is whether $(g \circ f)(x) \to c$ as $x \to a$.At first sight, the answer seems to be *yes*. One might argue that if a number x in the set $A \setminus \{a\}$ is sufficiently close to a, then $f(x)$ will be a member of the set B lying close to b and so $g(f(x))$ ought to be close to c. However, this reasoning is not valid because the fact that $g(y) \to c$ as $y \to b$ does not guarantee that we can make $g(y)$ close to c for *all* numbers y in the set B that are sufficiently close to b.What the condition that $g(y) \to c$ as $y \to b$ actually says is that we can make $g(y)$ close to c for all numbers y in the set $B \setminus \{b\}$ that are sufficiently close to b.

Unfortunately, if x is a number in the set $A \setminus \{a\}$, there is no reason to expect that the number $f(x)$ should belong to the set $B \setminus \{b\}$. It is quite possible to have $f(x) = b$, and this possibility compels us to state the composition theorem more carefully. Before we state the composition theorem we shall look at a simple example that shows what can go wrong with the composition theorem if we don't approach it carefully enough.

8.6.2   Illustrating the Limitations of the Composition Theorem

In this example we shall take

$$A = B = C = \mathbf{R}$$

and we shall define $f(x) = 1$ for every number x and

$$g(y) = \begin{cases} 0 & \text{if} \quad y \neq 1 \\ 2 & \text{if} \quad y = 1. \end{cases}$$

We see that

$$\lim_{x \to 3} f(x) = 1$$

and

$$\lim_{y \to 1} g(y) = 0.$$

However, since $g(f(x)) = 2$ for every number x, we have

$$\lim_{x \to 3} g(f(x)) = 2 \neq 0.$$

8.6.3   Statement of the Composition Theorem for Limits

Suppose that f is a function from a set A of real numbers into a set B of real numbers and that g is a function from B into a set C of real numbers. Suppose that a, b, and c are extended real numbers; that $f(x) \to b$ as $x \to a$; and that $g(y) \to c$ as $y \to b$.Then the assertion

$$g(f(x)) \to c \quad as \quad x \to a$$

will be true if and only if at least one of the following two conditions holds:

1. *There exists a neighborhood U of a such that the inequality $f(x) \neq b$ holds for every number x in the set $A \cap U \setminus \{a\}$.*
2. *The number b lies in the set B and $g(b) = c$.*

Part 1 of the Proof.   We assume that condition 1 holds and we want to show that $g(f(x)) \to c$ as $x \to a$. Using condition 1 we choose a neighborhood U_1 of a such that for every number x in the set $U_1 \cap A \setminus \{a\}$ we have $f(x) \neq b$. See Figure 8.6.

To prove that $g(f(x)) \to c$ as $x \to a$, suppose that W is any neighborhood of the number c and, using the fact that $g(y) \to c$ as $y \to b$, choose a neighborhood V of b such that the condition $g(y) \in W$ will be satisfied for every number y in the set $V \cap B \setminus \{b\}$.

Having chosen this neighborhood V of b, we now use the fact that $f(x) \to b$ as $x \to a$ to choose a neighborhood U_2 of a such that the condition $f(x) \in V$ holds for every number x in the set $A \cap U_2 \setminus \{a\}$.

We define

$$U = U_1 \cap U_2$$

and we observe that U is a neighborhood of the number a and that the condition $g(f(x)) \in W$ holds for every number x in the set $U \cap A \setminus \{a\}$.

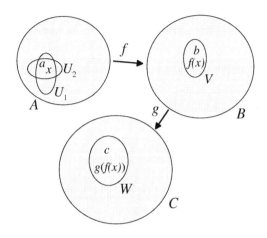

Figure 8.6

Part 2 of the Proof. We assume that condition 2 holds, and we want to show that $g\left(f(x)\right) \to c$ as $x \to a$. Suppose that W is any neighborhood of the number c and, using the fact that $g(y) \to c$ as $y \to b$, choose a neighborhood V of b such that the condition $g(y) \in W$ will be satisfied for every number y in the set $V \cap B \setminus \{b\}$. Since $c = g(b)$, the condition $g(y) \in W$ also holds when $y = b$. Therefore the condition $g(y) \in W$ holds for every number y in the set $V \cap B$.

Using the fact that $f(x) \to b$ as $x \to a$, we now choose a neighborhood U of a such that the condition $f(x) \in V$ holds for every number x in the set $U \cap A \setminus \{a\}$. We observe that the condition $g\left(f(x)\right) \in W$ holds for every number x in the set $U \cap A \setminus \{a\}$.

Part 3 of the Proof. We assume that neither condition 1 nor condition 2 holds. This time we need to explain why the assertion that $g\left(f(x)\right) \to c$ as $x \to a$ is false.

Since condition 1 is false, there must exist numbers x in the set A for which $f(x) = b$ and therefore $b \in B$. We can now deduce from the fact that condition 2 is false that $g(b) \neq c$, and, using the fact that $g(b) \neq c$, we choose a neighborhood W of the number c that does not contain the number $g(b)$. Now, to see that the assertion $g\left(f(x)\right) \to c$ as $x \to a$ is false, we observe that if U is any neighborhood of the number a, then, since condition 1 is false, there exist numbers x in the set $U \cap A \setminus \{a\}$ for which $f(a) = b$ and, for any such number x, we have

$$g\left(f(x)\right) = g(b) \notin W.$$

8.7   Continuity

8.7.1   Introduction to the Concept of Continuity

When we defined the notion $f(x) \to \lambda$ as $x \to a$ in Section 8.1 we were careful to point out that the number $f(a)$ plays no role at all in the definition. As you may recall, the notion $f(x) \to \lambda$ as $x \to a$ does not even require the number a to belong to the domain of the function f.

There are times, however, when the number a does belong to the domain of the function f and when the limit value λ of f at a is exactly equal to the number $f(a)$. This extra good behavior distinguishes between the concept of continuity that we are about to study in the present section and the notion of a limit that we have been studying up until now. We begin with the precise definition.

8.7.2   Continuity of a Function at a Given Number

Suppose that S is a set of real numbers, that $f : S \to \mathbf{R}$ and that $a \in S$. We say that the function f is **continuous** at the number a if for every neighborhood V of the number $f(a)$ there exists a neighborhood U of the number a such that the condition $f(x) \in V$ holds for every number $x \in U \cap S$.

Note that this definition does **not** require the number a to be a limit point of the domain S of the function f, but it does require a to belong to S.

8.7.3   Epsilon-Delta Condition for Continuity

By analogy with the theorems on limits that we saw in Subsection 8.1.4 we have the following equivalent forms of the definition of continuity of a function:

Suppose that S is a set of real numbers, that $f : S \to \mathbf{R}$, and that $a \in S$. Then the following three conditions are equivalent:

1. *The function f is continuous at the number a.*
2. *For every number $\varepsilon > 0$ there exists a number $\delta > 0$ such that the condition*

$$f(x) \in (f(a) - \varepsilon, f(a) + \varepsilon)$$

   holds for every number x in the set $(a - \delta, a + \delta) \cap S$.
3. *For every number $\varepsilon > 0$ there exists a number $\delta > 0$ such that the inequality*

$$|f(x) - f(a)| < \varepsilon$$

   holds for every number x in the set S that satisfies the inequality

$$|x - a| < \delta.$$

Proof. The proof of this theorem is similar to the proof of the analogous assertion for limits that we saw in Subsection 8.1.4 and will be left as an exercise. ◪

8.7.4 Continuity of a Function on a Set

Suppose that S is a set of real numbers and that $f : S \to \mathbf{R}$. We say that the function f is **continuous** on the set S if, for every number $x \in S$, the function f is continuous at x.

8.7.5 The Relationship Between Limits and Continuity

In this subsection we make a more careful comparison between the concept of a limit and the concept of continuity than we made in the introduction.

Suppose that S is a set of real numbers, that $f : S \to \mathbf{R}$, and that $a \in S$.

1. *If the number a is not a limit point of the set S, then the function f is automatically continuous at a.*
2. *If the number a is a limit point of the set S, then the following two conditions are equivalent:*
   (a) *The function f is continuous at the number a.*
   (b) *We have $f(x) \to f(a)$ as $x \to a$.*

Proof of Part 1. Suppose that V is a neighborhood of the number $f(a)$. Using the fact that a is not a limit point of the set S, we choose a neighborhood U of a such that the set $U \cap S \setminus \{a\}$ is empty. Then for every number x in the set $U \cap S$ we have

$$f(x) = f(a) \in V.$$

Proof of Part 2: Condition a Implies Condition b. We assume that the number a is a limit point of the set S and that f is continuous at a. To prove that $f(x) \to f(a)$ as $x \to a$, suppose that V is a neighborhood of the number $f(a)$. Using the fact that f is continuous at a, choose a neighborhood U of a such that the condition $f(x) \in V$ holds for every number x in the set $U \cap S$. Then the condition $f(x) \in V$ certainly holds for every number x in the set $U \cap S \setminus \{a\}$ and so $f(x) \to f(a)$ as $x \to a$. ∎

Proof of Part 2: Condition b Implies Condition a. We assume that the number a is a limit point of the set S and that $f(x) \to f(a)$ as $x \to a$. To prove that f is continuous at the number a, suppose that V is a neighborhood of the number $f(a)$. Using the fact that $f(x) \to f(a)$ as $x \to a$, we choose a neighborhood U of a such that the condition $f(x) \in V$ holds for every number x in the set $U \cap S \setminus \{a\}$. Since the condition $f(x) \in V$ is also true when $x = a$ we know that $f(x) \in V$ for every number x in the set $U \cap S$, and so we have shown that f is continuous at the number a. ∎

8.7.6   The Relationship Between Limits of Sequences and Continuity of Functions

By analogy with the relationship between limits of sequences and limits of functions that we saw in Section 8.4, we have the following relationship between limits of sequences and the concept of continuity:

Suppose that S is a set of real numbers, that $f : S \to \mathbf{R}$ and that $a \in S$. The following conditions are equivalent:

1. *The function f is continuous at the number a.*
2. *Given any sequence (x_n) in the set S that converges to the number a we have $f(x_n) \to f(a)$ as $n \to \infty$.*
3. *Given any sequence (x_n) in the set S that has the number a as a partial limit, the number $f(a)$ will be a partial limit of the sequence $(f(x_n))$.*

Proof that Condition 1 Implies Condition 2. We assume that the function f is continuous at a and that (x_n) is a sequence in S that converges to a, and we need to show that $f(x_n) \to f(a)$ as $n \to \infty$. Suppose that $\varepsilon > 0$.

Using the fact that f is continuous at the number a we choose a number $\delta > 0$ such that the inequality $|f(x) - f(a)| < \varepsilon$ holds for every number x in the set S that satisfies the inequality $|x - a| < \delta$. Now we use the fact that $x_n \to a$ as $n \to \infty$ to choose an integer N such that the inequality $|x_n - a| < \delta$ holds whenever $n \geq N$. We see that the inequality $|f(x_n) - f(a)| < \varepsilon$ holds whenever $n \geq N$.

Proof that Condition 1 Implies Condition 3. We leave this proof as an exercise.

Proof that if Condition 1 Is False, then Both Conditions 2 and 3 Are False. We assume that condition 1 is false, in other words, that f fails to be continuous at the number a. Choose a number $\varepsilon > 0$ such that it is impossible to find a number $\delta > 0$ such that the inequality $|f(x) - f(a)| < \varepsilon$ holds for every number x in the set S that satisfies the inequality $|x - a| < \delta$.

What we know about this number ε is that, if δ is any positive number, then there must exist a number x in the set S that satisfies the inequality $|x - a| < \delta$ even though $|f(x) - f(a)| \geq \varepsilon$. For every positive integer n we use the fact that $1/n > 0$ to choose a number $x_n \in S$ such that $|x_n - a| < 1/n$ and $|f(x_n) - f(a)| \geq \varepsilon$. In this way we have found a sequence (x_n) in the set S that converges to a and the fact that $|f(x_n) - f(a)| \geq \varepsilon$ for every n prevents the number $f(a)$ from being a partial limit of the sequence $(f(x_n))$. Therefore both of the conditions 2 and 3 are false. ∎

8.7.7 Algebraic Combinations of Continuous Functions

Suppose that f and g are functions from a set S of real numbers into $\mathbf{R}$, that $a \in \mathbf{R}$, and that both of the functions f and g are continuous at the number a. Then the functions $f + g$, and $f - g$, and fg are also continuous at the number a and, in the event that $g(x) \neq 0$ for every $x \in S$, the function f/g is also continuous at the number a.

This theorem follows simply from the algebraic rules for limits of sequences that we saw in Theorem 7.5.2 and the relationship between continuity and limits of sequences that we saw in Theorem 8.7.6. We leave the proof as an exercise.

8.7.8 Polynomials and Rational Functions

We deduce at once from the preceding theorem that all polynomials are continuous and that rational functions are continuous at every number at which their denominators are not zero.

8.7.9 The Composition Theorem for Continuity

Suppose that A and B are sets of real numbers, that $f : A \to B$, and that $g : B \to R$. Suppose that f is continuous at a number $x \in A$ and that g is continuous at the number $f(x)$. Then the composition function $g \circ f$ is continuous at the number x.

Proof. Suppose that (x_n) is a sequence in the set A that converges to the number x. Since f is continuous at the number x, we know from Theorem 8.7.6 that $f(x_n) \to f(x)$ as $n \to \infty$. Using Theorem 8.7.6 again and the fact that the function g is continuous at the number $f(x)$, we deduce that $g(f(x_n)) \to g(f(x))$ as $n \to \infty$, and using Theorem 8.7.6 once more we deduce that the function $g \circ f$ is continuous at the number x. ∎

Notice how much simpler the composition theorem for continuity is than the corresponding theorem, Theorem 8.6.3, for limits of functions.

8.7.10 Some Exercises on Continuity

1. Given that

$$f(x) = \frac{x - 1}{x^2 + 3}$$

for every number x, prove that the function f is continuous at the number 2.

2. Given that

$$f(x) = \begin{cases} x \sin \frac{1}{x} & \text{if } x \neq 0 \\ 0 & \text{if } x = 0, \end{cases}$$

prove that the function f is continuous at the number 0. Hint: Use the fact that $|f(x)| \leq |x|$ for every number x and use the sandwich theorem (Theorem 8.5.3). The graph of this function is illustrated in Figure 8.7.

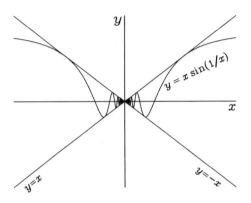

Figure 8.7

3. Given that f is the ruler function defined in Example 8 of Subsection 8.1.5, explain why f is continuous at every irrational number in the interval $[0, 1]$ and discontinuous at every rational number in $[0, 1]$.

4. Suppose that f and g are functions from a given set S of real numbers into $\mathbf{R}$ and that the inequality

$$|f(t) - f(x)| \leq |g(t) - g(x)|$$

holds for all numbers t and x in S.Prove that f must be continuous at every number at which the function g is continuous.

5. Given that f is a continuous function from a set S into $\mathbf{R}$, prove that the function $|f|$ is also continuous from S into $\mathbf{R}$.

6. Suppose that a and b are real numbers, that $a < b$, and that

$$f : [a, b] \to \mathbf{R}.$$

Prove that the following conditions are equivalent:

(a) The function f is continuous at the number a.

(b) For every number $\varepsilon > 0$ there exists a number $\delta > 0$ such that for every number x in the interval $[a, b]$ that satisfies the inequality $x - a < \delta$ we have $|f(x) - f(a)| < \varepsilon$.

7. Given that f is continuous on a closed set H and that (x_n) is a convergent sequence in the set H, prove that the sequence $(f(x_n))$ is also convergent.

Prove that this assertion is false if we omit the assumption that H is closed.

8. Prove that if a set S has no limit points, then every function $f : S \to \mathbf{R}$ is continuous on S.

9. Prove that if S is a set of real numbers and if no limit point of S belongs to the set S, then every function $f : S \to \mathbf{R}$ is continuous on S.

10. Suppose that f and g are functions from a set S to $\mathbf{R}$ and that f is continuous at a given number a at which the function g fails to be continuous.

   (a) What can we say about the continuity of the function $f + g$ at the number a?

   (b) What can we say about the continuity of the function fg at the number a?

   (c) What can we say about the continuity of the function fg at the number a if $f(a) = 0$ and g is a bounded function?

   (d) What can be said about the continuity of the function fg if $f(a) \neq 0$?

11. Give an example of two functions f and g that are both discontinuous at a given number a such that their sum $f + g$ is continuous at a.

12. ▇ Given that f is a continuous function from a closed set H into $\mathbf{R}$ and that $a \in H$, prove that the set

$$E = \{x \in H \mid f(x) = f(a)\}$$

   is closed. Hint: Consider the behavior of a convergent sequence in the set E.

13. Given that f and g are continuous functions from $\mathbf{R}$ to $\mathbf{R}$ and that

$$E = \{x \in \mathbf{R} \mid f(x) = g(x)\},$$

   prove that the set E must be closed.

14. Given that f and g are continuous functions from $\mathbf{R}$ to $\mathbf{R}$ and that $f(x) = g(x)$ for every rational number x, prove that $f = g$.

15. ▇ Given that $f : \mathbf{Z}^+ \to \mathbf{R}$, that $f(1) = 1$, and that the equation

$$f(x + t) = f(x) + f(t)$$

   holds for all positive integers x and t, prove that $f(x) = x$ for every positive integer x.

16. Given that $f : \mathbf{Z} \to \mathbf{R}$, that $f(1) = 1$, and that the equation

$$f(x + t) = f(x) + f(t)$$

   holds for all integers x and t, prove that $f(x) = x$ for every integer x.

17. ▇ Given that $f : \mathbf{Q} \to \mathbf{R}$, that $f(1) = 1$, and that the equation

$$f(x + t) = f(x) + f(t)$$

holds for all rational numbers x and t, prove that $f(x) = x$ for every rational number x.

18. Given that f is a continuous function from $\mathbf{R}$ to $\mathbf{R}$, that $f(1) = 1$, and that the equation

$$f(x + t) = f(x) + f(t)$$

holds for all rational numbers x and t, prove that $f(x) = x$ for every real number x.

19. ⬛ Given that f is an increasing function from $\mathbf{R}$ to $\mathbf{R}$, that $f(1) = 1$, and that the equation

$$f(x + t) = f(x) + f(t)$$

holds for all rational numbers x and t, prove that $f(x) = x$ for every real number x.[28]

Some additional exercises on continuity can be found by clicking on the icon ⬛ .

8.8   The Distance Function of a Set

This interesting and useful example of a continuous function can be found by clicking on the icon ⬛ .

8.9   The Behavior of Continuous Functions on Closed Bounded Sets

One of the most important properties of continuous functions in the study of differential calculus is the fact that every continuous function f on a closed bounded interval $[a, b]$ must have both a maximum value and a minimum value. In other words, if f is continuous on $[a, b]$, then there must exist numbers c and d in $[a, b]$ such that

$$f(c) \leq f(x) \leq f(d)$$

---

[28]   Notice how the fact that this function f is increasing makes it automatically continuous. This exercise is actually the simplest in a long line of interesting theorems that say, roughly speaking, that a function f satisfying the identity

$$f(x + y) = f(x) + f(y)$$

for all numbers x and y will either be continuous or it will be very badly behaved. See Boas [6] for some more detailed results of this type. Some abstract results of this type can be found in Lewin [18] .

for every $x \in [a, b]$. See Figure 8.8. The main purpose of this section is to

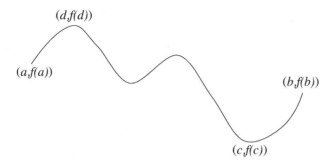

Figure 8.8

provide a proof of this assertion. In doing so, we shall extend the theorem slightly and prove that every function that is continuous on a nonempty, closed, bounded set of real numbers must have both a maximum and a minimum value. There are two important ingredients in the proof of this important fact:

1. In Theorem 6.4.2 we observed that every nonempty closed bounded set of real numbers has both a largest and a smallest member.
2. In Theorem 7.9.1 we learned that a set H of real numbers is closed and bounded if and only if every sequence in H has a partial limit in H.

8.9.1   Continuous Image of a Closed Bounded Set

Suppose that H is a closed bounded set of real numbers and that f is a function that is continuous on the set H. Then the range of the function f is also closed and bounded.

Proof. Suppose that (y_n) is a sequence in the set $f[H]$. For each n we choose a number x_n in the set H such that $y_n = f(x_n)$, and, in so doing, we have chosen a sequence in the set H. Using the fact that H is closed and bounded, we now choose a partial limit x of the sequence (x_n) such that $x \in H$. We now deduce from Theorem 8.7.6 that the number $f(x)$ is a partial limit of the sequence (y_n). Since every sequence in the set $f[H]$ has a partial limit in $f[H]$, this set must be closed and bounded. ■

8.9.2   Maxima and Minima of a Continuous Function

Suppose that f is a continuous function on a nonempty closed bounded set H. Then the function f has both a maximum and a minimum value.

Proof. The result follows at once from Theorem 6.4.2 and the fact that the set $f[H]$ is closed and bounded. ■

8.9.3   Some Exercises on Continuous Functions on Closed Bounded Sets

1. Give an example of a function f that is continuous on a closed set H such that the range $f[H]$ of the function f fails to be closed.
2. Give an example of a function f that is continuous on a closed set H such that the range $f[H]$ of the function f fails to be bounded.
3. Give an example of a function f that is continuous on a bounded set H such that the range $f[H]$ of the function f fails to be closed.
4. Give an example of a function f that is continuous on a bounded set H such that the range $f[H]$ of the function f fails to be bounded.
5. Prove that if a set H of real numbers is unbounded above and $f(x) = x$ for every number x in H, then f is a continuous function on H and f fails to have a maximum.
6. ▇ Prove that if H is a set of real numbers and a number a is close to H but does not belong to H, and if we define

$$f(x) = \frac{1}{|x - a|}$$

for every $x \in H$, then f is a continuous function on H but f has no maximum.

8.10   The Behavior of Continuous Functions on Intervals

8.10.1   Introduction to Functions on Intervals

The concept of an interval was first introduced in Example 5 of Subsection 4.2.9. Then in Section 5.8.1 we encountered a criterion for determining whether or not a given set is an interval. In this section we shall use that criterion in order to study the behavior of functions that are continuous on intervals. We shall see, among other things, that if a function is continuous on an interval, then its range is also an interval.

We shall also have something to say about one-one functions. In our introduction to one-one functions seen in Subsection 4.3.8, we saw the example that was illustrated in Figure 4.6 and that warns us that not every one-one function is monotone. We can obtain another example of this type by defining

$$f(x) = \begin{cases} x & \text{if } 0 \leq x < 1 \\ 3 - x & \text{if } 1 \leq x \leq 2 \end{cases}$$

as illustrated in Figure 8.9. This example shows that even if both the domain and

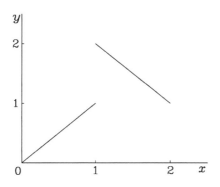

Figure 8.9

the range of a given one-one function are intervals, the function does not have to be monotone. As you can see, the function in this example is a one-one function from the interval $[0, 2]$ onto the interval $[0, 2]$, but it is not monotone and it is not continuous at the number 1.

On the other hand, we shall see in Theorem 8.10.7 that if a function is one-one and continuous on an interval, then it must be strictly monotone. Then in Theorem 8.10.9 we shall see that a monotone function defined on an interval is continuous if and only if its range is also an interval.

8.10.2   Bolzano's Intermediate Value Theorem

Suppose that f is a continuous function on an interval $[a, b]$, and that w is a number that lies between $f(a)$ and $f(b)$. Then there is at least one number c in the interval $[a, b]$ for which $f(c) = w$. This theorem is illustrated in Figure 8.10.

Proof. We shall assume, without loss of generality, that

$$f(a) < w < f(b).$$

We begin by defining

$$S = \{x \in [a, b] \mid f(x) < w\}.$$

The set S is nonempty because $a \in S$ and is bounded above because it is included in the interval $[a, b]$. We define $c = \sup S$.

For each positive integer n we use the fact that the number $c - 1/n$ is not an

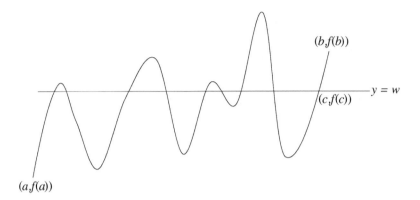

Figure 8.10

upper bound of S to choose a member x_n of S such that

$$c - \frac{1}{n} < x_n \leq c.$$

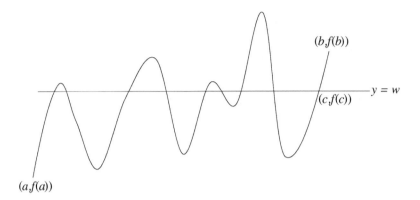

Since $x_n \to c$ as $n \to \infty$ and since $f(x_n) < w$ for every n, we have

$$f(c) = \lim_{n \to \infty} f(x_n) \leq w.$$

From the fact that $f(c) \leq w$ and $f(b) > w$, we deduce that $c \neq b$.Therefore, starting at a sufficiently large positive integer n, the sequence of numbers $c + 1/n$ is a sequence in the interval $[a, b]$ that converges to c.Since none of the numbers $c + 1/n$ can belong to the set S, we have

$$f(c) = \lim_{n \to \infty} f\left(c + \frac{1}{n}\right) \geq w,$$

which completes the proof that $f(c) = w$. ∎

8.10.3   Existence of nth Roots

Suppose that w is a positive number and that n is a positive integer. Then there exists a unique positive number x such that $w = x^n$.

Proof. We define

$$f(t) = t^n$$

for every number $t \geq 0$.Since this function f is strictly increasing, there cannot be more than one number x for which $w = f(x)$.Therefore, to complete the proof, all we need to do is show that there is at least one number x such that $w = f(x)$, and we shall obtain this fact from Bolzano's intermediate value theorem. We define $a = 0$ and $b = 1 + w$. Since f is continuous on the interval $[a, b]$ and since

$$f(a) = 0 < w < w + 1 \leq (w + 1)^n = f(b),$$

the Bolzano intermediate value theorem guarantees the existence of a number $x \in [a, b]$ such that $f(x) = w$. ∎

8.10.4   Continuous Image of an Interval

Suppose that f is a continuous function on an interval S. Then the range of f is also an interval.

Proof. In order to prove that the set $f[S]$ is an interval we shall use Theorem 5.8.1. Suppose that p and q are members of the set $f[S]$ and that w is a number satisfying the inequality

$$p < w < q.$$

Choose numbers a and b in the interval S such that $f(a) = p$ and $f(b) = q$. In the event that $a < b$, the fact that $w \in f[S]$ follows at once from the Bolzano intermediate value theorem applied to the function f on the interval $[a, b]$. In the event that $b < a$, we obtain $w \in f[S]$ by applying the Bolzano intermediate value theorem to f on the interval $[b, a]$. ∎

8.10.5   The Concept of a Switch

The concept of a switch that is defined in this subsection is useful when we want to prove that a given function is monotone.

Suppose that f is a function defined on a set S of real numbers. An ordered triple (a, b, c) of numbers that belong to S is said to be a **switch** of the function f if $a < b < c$ and either one of the following conditions holds:

1. $f(a) < f(b)$ and $f(b) > f(c)$.
2. $f(a) > f(b)$ and $f(b) < f(c)$.

Figure 8.11 illustrates the two ways in which a switch can occur.

8.10.6   Existence of a Switch when a Function Is not Monotone

Suppose that S is a set of real numbers, that f is a function defined on S and that the function f is not monotone. Then the function f has a switch.

Proof. Using the fact that the function f is not monotone, we choose numbers a and b in the set S such that $a < b$ and $f(a) < f(b)$ and choose numbers c and d in the set S such that $c < d$ and $f(c) > f(d)$. To obtain a contradiction we

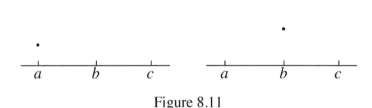

Figure 8.11

assume that the function f has no switch. We now break the proof down into the six cases that are illustrated in Figure 8.12.

$$
\begin{array}{cccc}
a & b & c & d \\
\hline
& a < b \leq c < d &
\end{array}
\qquad
\begin{array}{cccc}
a & c & b & d \\
\hline
& a \leq c < b \leq d &
\end{array}
$$

$$
\begin{array}{cccc}
c & a & b & d \\
\hline
& c < a < b \leq d &
\end{array}
\qquad
\begin{array}{cccc}
a & c & d & b \\
\hline
& a \leq c < d \leq b &
\end{array}
$$

$$
\begin{array}{cccc}
c & a & d & b \\
\hline
& c < a \leq d \leq b &
\end{array}
\qquad
\begin{array}{cccc}
c & d & a & b \\
\hline
& c < d < a < b &
\end{array}
$$

Figure 8.12

For example, in the first case, since $f(a) < f(b)$ and f has no switch, we must have $f(b) \leq f(c)$. Therefore $f(a) < f(c)$ and, because f has no switch, we must have $f(c) \leq f(d)$, which contradicts the choice of the numbers c and d. The other five cases can be handled similarly. ∎

8.10.7   Monotonicity of One-One Functions

Every one-one function that is continuous on an interval must be strictly monotone.

Proof. Suppose that f is a one-one continuous function on an interval S. We shall show that f is monotone. Once we have shown that f is monotone, it will follow at once from the fact that f is one-one that f is strictly monotone. In order to show that f is monotone, all we have to do is show that f does not have a switch.

To obtain a contradiction, assume that f does have a switch in S. Choose a switch (a, b, c) of the function f. We know that one of the following two conditions occurs:

1. $f(a) < f(b)$ and $f(b) > f(c)$.
2. $f(a) > f(b)$ and $f(b) < f(c)$.

We shall assume that the first of these conditions occurs and leave consideration of the second one as an exercise. [■] Choose a number α that is less than $f(b)$ but greater than both of the two numbers $f(a)$ and $f(c)$, as illustrated in Figure 8.13. We now apply Bolzano's intermediate value theorem (Theorem

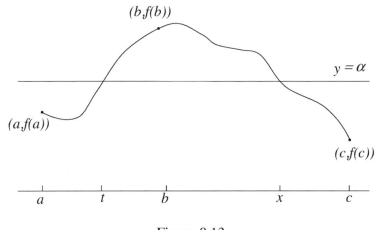

Figure 8.13

8.10.2) to choose a number t between a and b and a number x between b and c such that

$$f(t) = \alpha = f(x).$$

This choice of numbers t and x contradicts our assumption that the function f is one-one. ∎

8.10.8   One-Sided Limits of a Monotone Function

In this subsection we shall learn that a monotone function always has one-sided limits and that it is continuous at a number if and only if it does not "jump" there. This concept is illustrated in Figure 8.14.

Suppose that f is an increasing function on an set S and that a is a real number.

1. *If a is a limit point of the set $S \cap (-\infty, a)$, then the function f has a (possibly infinite) limit from the left at a and*

$$\lim_{x \to a-} f(x) = \sup \{ f(x) \mid x \in S \text{ and } x < a \}.$$

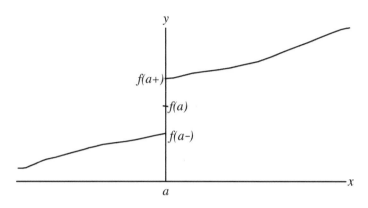

Figure 8.14

2. *If a is a limit point of the set $S \cap (a, \infty)$, then the function f has a (possibly infinite) limit from the right at a and*

$$\lim_{x \to a+} f(x) = \inf \left\{ f(x) \mid x \in S \text{ and } x > a \right\}.$$

3. *If a is a limit point of both of the sets $S \cap (-\infty, a)$ and $S \cap (a, \infty)$ and $a \in S$, then we have*

$$\lim_{x \to a-} f(x) \leq f(a) \leq \lim_{x \to a+} f(x),$$

and f will be continuous at the number a if and only if

$$\lim_{x \to a-} f(x) = \lim_{x \to a+} f(x).$$

Proof. We shall prove part 1 and leave part 2 as an exercise. 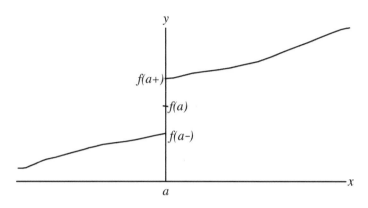 Part 3 follows at once from parts 1 and 2.

To prove part 1 we assume that the number a is a limit point of the set $S \cap (-\infty, a)$. We now define

$$\lambda = \sup \left\{ f(x) \mid x \in S \text{ and } x < a \right\}$$

with the understanding that λ may be equal to ∞ and what we need to show is that $f(x) \to \lambda$ as $x \to a-$. Since $f(x) \leq \lambda$ whenever $x \in S$ and $x < a$, in order to show that $f(x) \to \lambda$ as $x \to a-$, we need to show that for every number $w < \lambda$ the inequality $f(x) > w$ will hold whenever $x \in S$ and $x < a$ and x is sufficiently close to a. Suppose that $w < \lambda$. Using the fact that w is not an upper bound of the set

$$\{f(x) \mid x \in S \text{ and } x < a\},$$

we choose a number $c \in S$ such that $c < a$ and $f(c) > w$.

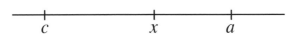

Then for every number x in the interval (c, a) we have

$$w < f(c) \leq f(x),$$

and so $f(x) \to \lambda$ as $x \to a-$. ■

8.10.9   Monotone Continuous Functions

Suppose that f is a monotone function defined on an interval S. Then the following conditions are equivalent:

1. *The function f is continuous on S.*
2. *The range of f is an interval.*

Proof. We know from Bolzano's theorem (Theorem 8.10.2) that if f is continuous on S, then the range of f is an interval. On the other hand, if f fails to be continuous at any number $a \in S$, then it follows at once from Theorem 8.10.8 that the range of f cannot be an interval. ■

8.11   Inverse Function Theorems for Continuity

8.11.1   Introduction to Inverse Function Theorems

As we saw in Subsection 4.3.12, if f is a one-one function from a set A onto a set B, then the inverse function f^{-1} is a one-one function from the set B onto the set A and for every $x \in A$ we have

$$f^{-1}(f(x)) = x.$$

In the event that A and B are sets of real numbers and f is continuous on the set A, a reasonable question to ask is whether the function f^{-1} has to be continuous on the set B. The answer is *no!* However, there are some important situations in which the inverse function of a continuous one-one function will automatically be continuous, and the description of any one of these is known as an inverse function theorem for continuity. In Section 9.3.8 we shall also see an inverse function theorem for derivatives.

8.11.2   An Example of a Discontinuous Inverse Function

With an eye on Figure 8.15 we define

$$f(x) = \begin{cases} x & \text{if } 0 \le x \le 1 \\ x - 1 & \text{if } 2 < x \le 3. \end{cases}$$

The function f is a one-one continuous function from the set $[0,1] \cup (2,3]$ onto

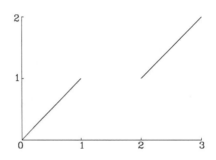

Figure 8.15

the interval $[0,2]$.Notice that the domain of this function f is not an interval and it is not a closed set.

8.11.3   Inverse Function Theorem for Functions on Closed Bounded Sets

Suppose that f is a one-one continuous function from a closed bounded set H onto a set K.Then the function f^{-1} is continuous from K onto H.

Proof. The proof will make use of the connection between continuity and limits of sequences that appears in Theorem 8.7.6. Suppose that $y \in K$ and, to prove that the function f^{-1} is continuous at the number y, suppose that (y_n) is a sequence in the set K that converges to y. We need to show that

$$f^{-1}(y_n) \to f^{-1}(y)$$

as $n \to \infty$. We define $x = f^{-1}(y)$, and, for each positive integer n, we define

$$x_n = f^{-1}(y_n).$$

So what we need to show is that $x_n \to x$ as $n \to \infty$, and in order to achieve this goal we shall make use of Theorem 7.9.3. We shall establish the condition $x_n \to x$ as $n \to \infty$ by showing that the number x is the only partial limit of the sequence (x_n). Suppose that t is any partial limit of the sequence (x_n). Since H is closed and bounded, we know that $t \in H$, and so it follows from Theorem

8.7.6 that $f(t)$ is a partial limit of the sequence (y_n). But the only partial limit of the convergent sequence (y_n) is its limit y, and so

$$f(t) = y = f(x);$$

and it follows from the fact that f is one-one that $t = x$. Thus x is the only partial limit of the sequence (x_n), as we have promised, and $x_n \to x$ as $n \to \infty$. ∎

8.11.4   Inverse Function Theorem for Functions on Intervals

Suppose that f is a one-one continuous function from an interval A onto a set B. Then the function f^{-1} is continuous from B onto A.

Proof. Since the set A is an interval, we know from Theorem 8.10.4 that the set B is an interval. We know from Theorem 8.10.7 that the function f must be strictly monotone. Since f^{-1} is a strictly monotone function from the interval B onto the interval A, it follows from Theorem 8.10.9 that f^{-1} is continuous on B. ∎

8.11.5   Exercises on Continuity of Functions on Intervals

1. ▦ Given that S is a set of positive numbers and that $f(x) = \sqrt{x}$ for all $x \in S$, prove that f is a one-one continuous function on S. Prove that S is an interval if and only if the set $f[S]$ is an interval.

2. ▦ Prove that there are three real numbers x satisfying the equation

$$x^3 - 4x - 2 = 0.$$

3. Is it true that, if a set S of real numbers is not an interval, then there must exist a one-one continuous function on S whose inverse function fails to be continuous?

4. ▦ Is it true that, if a set S of real numbers is not an interval and is not closed, then there must exist a one-one continuous function on S whose inverse function fails to be continuous?

5. Is it true that, if a set S of real numbers is not an interval and is not bounded, then there must exist a one-one continuous function on S whose inverse function fails to be continuous?

6. ▦ Prove that if f is a continuous function from the interval $[0, 1]$ into $[0, 1]$, then there must be at least one number $x \in [0, 1]$ such that $f(x) = x$. This assertion is the one-dimensional form of the **Brouwer fixed point theorem**.

8.12   Uniform Continuity

In this section we introduce a notion called **uniform continuity** that resembles continuity but is a little stronger. We can think of the condition of uniform continuity as saying that a given function is continuous at "about the same rate" throughout its domain. The notion of uniform continuity will be a useful tool when we come to the theory of integration in Chapter 11.

8.12.1   Introduction to Uniform Continuity

Suppose that f is a function defined on a set S of real numbers. We begin our discussion of uniform continuity by comparing the following two sets of conditions:

1. (a) For every number $x \in S$ the function f is continuous at x.
    (b) For every number $x \in S$ and for every number $\varepsilon > 0$ there exists a number $\delta > 0$ such that the inequality

    $$|f(t) - f(x)| < \varepsilon$$

    holds for every number t in the set S for which $|t - x| < \delta$.
    (c) For every number $\varepsilon > 0$ and for every number $x \in S$ there exists a number $\delta > 0$ such that the inequality

    $$|f(t) - f(x)| < \varepsilon$$

    holds for every number t in the set S for which $|t - x| < \delta$.
2. (a) For every number $\varepsilon > 0$ there exists a number $\delta > 0$ such that for every number $x \in S$ the inequality

    $$|f(t) - f(x)| < \varepsilon$$

    holds for every number t in the set S for which $|t - x| < \delta$.
    (b) For every number $\varepsilon > 0$ there exists a number $\delta > 0$ such that the inequality

    $$|f(t) - f(x)| < \varepsilon$$

    holds for every number $x \in S$ and every number t in the set S for which $|t - x| < \delta$.

The three conditions 1a, 1b, and 1c are obviously equivalent to one another, and what each of them says is that the function f is continuous on the set S.The conditions 2a and 2b are also obviously equivalent to one another, but they are not equivalent to the conditions 1a, 1b, and 1c. To help sort out the distinction between the two concepts, you may wish to make a review of Subsection 2.1.4.

Both sets of conditions require an inequality of the form

$$|f(t) - f(x)| < \varepsilon$$

to hold whenever the numbers t and x are close enough to each other to satisfy an inequality of the form $|t - x| < \delta$, but the two sets of conditions differ from one another in the way in which they introduce the number δ.

In the conditions 1a, 1b, and 1c, the number δ comes forward to answer the challenge posed by a specific number $\varepsilon > 0$ and a specific number $x \in S$. Both ε and x must be in our hands before we can say that there exists a number δ such that for every number $t \in S$ satisfying the inequality $|t - x| < \delta$, we have $|f(t) - f(x)| < \varepsilon$.

However, in the conditions 2a and 2b, the number δ responds to the challenge posed by the number ε acting alone. Once ε has been prescribed, the number δ comes forward to say that, for all numbers t and x that belong to the set S and satisfy the inequality $|t - x| < \delta$, we have $|f(t) - f(x)| < \varepsilon$.

8.12.2   Definition of Uniform Continuity

Suppose that f is a function defined on a set S of real numbers. We say that f is **uniformly continuous** on the set S if for every number $\varepsilon > 0$ it is possible to find a number $\delta > 0$ such that the inequality

$$|f(t) - f(x)| < \varepsilon$$

holds for all numbers t and x in the set S that satisfy the inequality $|t - x| < \delta$.

8.12.3   Failure of Uniform Continuity

Suppose that f is a function defined on a set S of real numbers. The assertion that the function f *fails* to be uniformly continuous on S denies that for every number $\varepsilon > 0$ it is possible to find a number $\delta > 0$ such that the inequality

$$|f(t) - f(x)| < \varepsilon$$

will hold for all numbers t and x in the set S that satisfy the inequality $|t - x| < \delta$. Denying that a condition holds for every number $\varepsilon > 0$ is the same as saying that there exists at least one number $\varepsilon > 0$ for which the condition does not hold.

Therefore, the assertion that f fails to be uniformly continuous on the set S says that it is possible to find a number $\varepsilon > 0$ such that for every positive number δ there exist numbers t and x in the set S such that $|t - x| < \delta$ and

$$|f(t) - f(x)| \geq \varepsilon.$$

8.12.4   The Relationship Between Limits of Sequences and Uniform Continuity of Functions

One of the tools that we used in our study of continuity was the relationship given in Subsection 8.7.6 between continuity and limits of sequences. We shall now explore a similar and very useful relationship between uniform continuity and limits of sequences.[29]

Suppose that f is a function from a set S of real numbers into $\mathbf{R}$. Then the following conditions are equivalent:

1. *If the function f is uniformly continuous on S, then, given any two sequences (x_n) and (t_n) in S such that $t_n - x_n \to 0$ as $n \to \infty$, we have $f(t_n) - f(x_n) \to 0$ as $n \to \infty$.*

2. *If the function f fails to be uniformly continuous on S, then it is possible to find two sequences (x_n) and (t_n) in S and a number $\varepsilon > 0$ such that $t_n - x_n \to 0$ as $n \to \infty$ and $|f(t_n) - f(x_n)| \geq \varepsilon$ for every n.*

Proof: To prove part 1 of the theorem, we assume that f is uniformly continuous on S. Suppose that (t_n) and (x_n) are sequences in S and that $t_n - x_n \to 0$ as $n \to \infty$. In order to show that $f(t_n) - f(x_n) \to 0$ as $n \to \infty$, suppose that $\varepsilon > 0$.

Using the fact that f is uniformly continuous on S we choose a number $\delta > 0$ such that whenever t and x are members of S and $|t - x| < \delta$ we have $|f(t) - f(x)| < \varepsilon$. Now, using the fact that $t_n - x_n \to 0$ as $n \to 0$ we choose an integer N such that the inequality

$$|t_n - x_n| < \delta$$

holds whenever $n \geq N$. Then whenever $n \geq N$ we have

$$|f(t_n) - f(x_n)| < \varepsilon.$$

Now, to prove part 2 of the theorem, we assume that f fails to be uniformly continuous on the set S. Using the remarks that appeared in Subsection 8.12.3, we choose a number $\varepsilon > 0$ such that for every positive number δ there must exist numbers t and x in S for which $|t - x| < \delta$ even though

$$|f(t) - f(x)| \geq \varepsilon.$$

For each positive integer n we use the fact that $1/n > 0$ to choose numbers that

---

[29]   The author is indebted to Sean Ellermeyer for his suggestion that the present material be elevated from its former status of *exercise* to a theorem in the text.

we shall call t_n and x_n in the set S such that

$$|t_n - x_n| < \frac{1}{n}$$

even though

$$|f(t_n) - f(x_n)| \geq \varepsilon.$$

In this way we have found two sequences (t_n) and (x_n) in S such that $t_n - x_n \to 0$ and

$$|f(t_n) - f(x_n)| \geq \varepsilon$$

for every n. ∎

8.12.5   Some Examples and Observations Illustrating Uniform Continuity

1. Suppose that $f(x) = x^2$ for $x \in [0, 1]$. As we know, this polynomial is continuous on $[0, 1]$, but we shall now go a little further and show that f is actually uniformly continuous on $[0, 1]$. Our proof will be based on the observation that if t and x are any numbers in the interval $[0, 1]$, then

    $$|f(t) - f(x)| = \left|t^2 - x^2\right| = |t - x|\,|t + x| \leq 2\,|t - x|.$$

    Now suppose that $\varepsilon > 0$ and define $\delta = \varepsilon/2$. Whenever t and x are numbers in the interval $[0, 1]$ and $|t - x| < \delta$ we have

    $$|f(t) - f(x)| \leq 2\,|t - x| < 2\left(\frac{\varepsilon}{2}\right) = \varepsilon.$$

2. Suppose that $f(x) = x^2$ for every real number x. Although this polynomial f is continuous on the set $\mathbf{R}$, it is not uniformly continuous. To show that f is not uniformly continuous on $\mathbf{R}$ we define $x_n = n$ and $t_n = n + 1/n$ for every positive integer n. We observe that, even though $t_n - x_n \to 0$ as $n \to \infty$, we have

    $$f(t_n) - f(x_n) = \left(n + \frac{1}{n}\right)^2 - n^2 = 2 + \frac{1}{n^2}$$

    for each n, and so we cannot have $f(t_n) - f(x_n) \to 0$ as $n \to \infty$. Intuitively we can say that the reason why this function f fails to be uniformly continuous on $\mathbf{R}$ is that the graph becomes too steep as we move far away from the origin. See Figure 8.16.

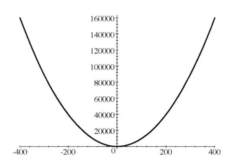

Figure 8.16

3. In this example we define

$$f(x) = \sin\frac{1}{x}$$

for every number $x > 0$. The graph of this function is illustrated in Figure 8.17. Although f is continuous on the interval $(0, \infty)$, it is not uniformly

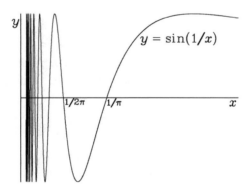

Figure 8.17

continuous. To show that f fails to be uniformly continuous we define

$$x_n = \frac{1}{n\pi}$$

$$t_n = \frac{1}{n\pi + \frac{\pi}{2}}$$

for every positive integer n. We see at once that $t_n - x_n \to 0$ as $n \to \infty$, but for each n we have

$$|f(t_n) - f(x_n)| = \left|\sin\left(n\pi + \frac{\pi}{2}\right) - \sin n\pi\right| = 1,$$

and so we do not have $f(t_n) - f(x_n) \to 0$ as $n \to \infty$.

4. Suppose that f is any function from the set $\mathbf{Z}$ of integers into $\mathbf{R}$. We shall see that f is automatically uniformly continuous on $\mathbf{Z}$. Suppose that $\varepsilon > 0$ and define $\delta = 1$. Whenever x and t are integers and $|t - x| < \delta$ we have $t = x$, and so

$$|f(t) - f(x)| = 0 < \varepsilon.$$

5. In this example we define

$$S = \left\{\frac{1}{n} \mid n \in \mathbf{Z}^+\right\}$$

and for every positive integer n we define

$$f\left(\frac{1}{n}\right) = (-1)^n.$$

Since no member of the set S is a limit point of S, we see at once from Theorem 8.7.5 that f is continuous on S. However, if we define

$$t_n = \frac{1}{2n}$$

and

$$x_n = \frac{1}{2n+1}$$

for every positive integer n, we see that $t_n - x_n \to 0$ as $n \to \infty$ but $f(t_n) - f(x_n)$ does not approach 0 as $n \to \infty$.

8.12.6   The Principal Theorem on Uniform Continuity

Suppose that f is a continuous function on a closed bounded set S. Then f is uniformly continuous on S.

Proof.   To obtain a contradiction, we suppose that f fails to be uniformly continuous on S. Using the relationship, Theorem 8.12.4, between uniform continuity and convergence of sequences, we choose two sequences (t_n) and (x_n) in S such that $t_n - x_n \to 0$ as $n \to \infty$ and such that

$$|f(t_n) - f(x_n)| \geq \varepsilon$$

for every n. Using the fact that the sequence (x_n) is a sequence in the closed bounded set S and Theorem 7.9.1 on existence of partial limits, we choose a partial limit x of (x_n) such that $x \in S$.

Now we use the fact that the function f is continuous at the number x to choose a number $\delta > 0$ such that whenever $t \in S$ and $|t - x| < \delta$ we have

$$|f(t) - f(x)| < \frac{\varepsilon}{2}.$$

The next step is to find a value of n for which both of the numbers t_n and x_n belong to the interval $(x - \delta, x + \delta)$.

		x_n	t_n	
$x - \delta$	$x - \delta/2$	x	$x + \delta/2$	$x + \delta$

For this purpose we use the fact $t_n - x_n \to 0$ as $n \to \infty$ to choose an integer N such that the inequality

$$|t_n - x_n| < \frac{\delta}{2}$$

holds whenever $n \geq N$. Now, using the fact that the sequence (x_n) is frequently in the interval $(x - \delta/2, x + \delta/2)$, we choose an integer $n \geq N$ such that

$$x_n \in \left(x - \frac{\delta}{2}, x + \frac{\delta}{2} \right).$$

We observe that

$$
\begin{aligned}
|t_n - x| &= |t_n - x_n + x_n - x| \\
&\leq |t_n - x_n| + |x_n - x| < \frac{\delta}{2} + \frac{\delta}{2} = \delta.
\end{aligned}
$$

Now that we have found a value of n for which both of the numbers t_n and x_n belong to the interval $(x - \delta, x + \delta)$, we observe that

$$
\begin{aligned}
|f(t_n) - f(x_n)| &= |f(t_n) - f(x) + f(x) - f(x_n)| \\
&\leq |f(t_n) - f(x)| + |f(x) - f(x_n)| < \frac{\varepsilon}{2} + \frac{\varepsilon}{2} = \varepsilon,
\end{aligned}
$$

which contradicts the way in which the numbers t_n and x_n were chosen. ∎

8.12.7   Exercises on Uniform Continuity

1. Is it true that if S is an unbounded set of real numbers and $f(x) = x^2$ for every number $x \in S$, then the function f fails to be uniformly continuous?

2. Given that

$$f(x) = \begin{cases} 1 & \text{if } 0 \le x < 2 \\ 0 & \text{if } 2 < x \le 3, \end{cases}$$

prove that f is continuous but not uniformly continuous on the set $[0, 2) \cup (2, 3]$.

3. Given that $f(x) = \sin(x^2)$ for all real numbers x, prove that f is not uniformly continuous on the set $\mathbf{R}$. See Figure 8.18.

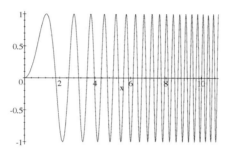

Figure 8.18

4.   Ask *Scientific Notebook* to make some 2D plots of the function f defined by the equation

$$f(x) = \sin(x \log x)$$

for $x > 0$. Plot the function on each of the intervals $[0, 50]$, $[50, 100]$, $[100, 150]$, and $[150, 200]$. Revise your plot and increase its sample size if it appears to contain errors. Why do these graphs suggest that f fails to be uniformly continuous on the interval $(0, \infty)$? Prove that this function does, indeed, fail to be uniformly continuous.

5. (a) A function f is said to be **Lipschitzian** on a set S if there exists a number k such that the inequality

$$|f(t) - f(x)| \le k|t - x|$$

holds for all numbers t and x in S. Prove that every Lipschitzian function is uniformly continuous.

    (b) Given that $f(x) = \sqrt{x}$ for all $x \in [0, 1]$, prove that f is uniformly continuous but not Lipschitzian on $[0, 1]$.

6. (a) Suppose that f is uniformly continuous on a set S, that (x_n) is a

sequence in the set S, and that (x_n) has a partial limit $x \in \mathbf{R}$. Prove that it is impossible to have $f(x_n) \to \infty$ as $n \to \infty$.

(b) Did you assume that $x \in S$ in part a? If you did, go back and do the exercise again. You have no information that $x \in S$. If you didn't assume that $x \in S$, you can sit this question out.

(c) Suppose that f is uniformly continuous on a bounded set S and that (x_n) is a sequence in S. Prove that it is impossible to have $f(x_n) \to \infty$ as $n \to \infty$.

(d) ▤ Prove that if f is uniformly continuous on a bounded set S, then the function f is bounded.

7. (a) ▤ Given that S is a set of real numbers, that $a \in \overline{S} \setminus S$ and that

$$f(x) = \frac{1}{x - a}$$

for all $x \in S$, prove that f is continuous on S but not uniformly continuous.

(b) Given that S is a set of real numbers and that S fails to be closed, prove that there exists a continuous function on S that fails to be uniformly continuous on S.

(c) Is it true that if S is an unbounded set of real numbers, then there exists a continuous function on S that fails to be uniformly continuous on S?

8. Is it true that the composition of a uniformly continuous function with a uniformly continuous function is uniformly continuous?

9. (a) Suppose that f is uniformly continuous on a set S and that (x_n) is a convergent sequence in S. Prove that the sequence $(f(x_n))$ cannot have more than one partial limit.

(b) In part a, did you assume that the limit of the sequence (x_n) belongs to S? If so, go back and do the problem again.

(c) Prove that if f is uniformly continuous on a set S and (x_n) is a convergent sequence in S, then the sequence $(f(x_n))$ is also convergent. Do *not* assume that the limit of (x_n) belongs to S.

(d) Suppose that f is uniformly continuous on a set S, that x is a real number and that (x_n) and (t_n) are sequences in S that converge to the number x. Prove that

$$\lim_{n \to \infty} f(x_n) = \lim_{n \to \infty} f(t_n).$$

(e) Suppose that f is uniformly continuous on a set S and that $x \in \overline{S} \setminus S$. Explain how we can use part d to extend the definition of the function f to the number x in such a way that f is continuous on the set $S \cup \{x\}$.

(f) Prove that if f is uniformly continuous on a set S, then it is possible to extend f to the closure $\overline{S}$ of S in such a way that f is uniformly continuous on $\overline{S}$.

10. Suppose that f is a continuous function on a bounded set S. Prove that the following two conditions are equivalent:

(a) The function f is uniformly continuous on S.

(b) It is possible to extend f to a continuous function on the set $\overline{S}$.

11. Given that f is a function defined on a set S of real numbers, prove that the following conditions are equivalent:

(a) The function f fails to be uniformly continuous on the set S.

(b) There exists a number $\varepsilon > 0$ and there exist two sequences (t_n) and (x_n) in S such that $t_n - x_n \to 0$ as $n \to \infty$ and

$$|f(x_n) - f(t_n)| \geq \varepsilon$$

for every n.

Chapter 9
Differentiation

This chapter begins the study of differential calculus. You will find many topics here that you have seen before and that you will now have the opportunity to place on a sound logical footing.

9.1 Introduction to the Concept of a Derivative

9.1.1 The Derivative as Seen in Elementary Calculus

From your courses in elementary calculus you know that if f is a given function and x is a given number, then the **derivative** of the function f at the number x is the number $f'(x)$ defined by the equation

$$f'(x) = \lim_{t \to x} \frac{f(t) - f(x)}{t - x}$$

as long as the latter limit exists. Another way in which this limit is traditionally written is

$$\lim_{h \to 0} \frac{f(x + h) - f(x)}{h}.$$

These two ways of describing the limit are equivalent to one another.

As you know, the limit

$$\lim_{t \to x} \frac{f(t) - f(x)}{t - x}$$

is motivated by looking at a figure like Figure 9.1. If $(x, f(x))$ and $(t, f(t))$ are two different points on the graph of a given function f, then the number

$$\frac{f(t) - f(x)}{t - x}$$

is the slope of the line segment that runs from $(x, f(x))$ to $(t, f(t))$. The limit of this expression as t approaches x gives us a meaning to the notion of *slope* of the graph of f at the point $(x, f(x))$, a notion that has many important applications. The applications of differentiation stem from the following key facts:

1. If the derivative of a function is zero everywhere in an interval, then the function must be constant in that interval.

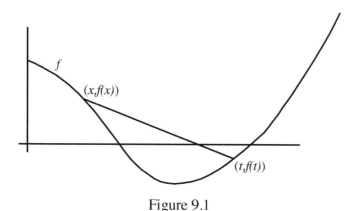

Figure 9.1

2. If the derivative of a function is positive everywhere in an interval, then the function must be strictly increasing in that interval.
3. If the derivative of a function is negative everywhere in an interval, then the function must be strictly decreasing in that interval.

On the face of it, no properties of the derivative can be simpler than these. For example, if the derivative of a function is positive in an interval, then the slope of the graph must be positive and so the graph must be rising from left to right. But looks can be deceiving. These three facts about differentiation are anything but simple. They depend upon the full force of the work that we have done so carefully on the preceding chapters, and, in depending upon that material, they depend upon the completeness of the number system $\mathbf{R}$.

9.1.2   What Happens in an Incomplete Number System?

In order to appreciate the way in which the three basic properties of derivatives listed in Subsection 9.1.1 depend upon the completeness of the number system $\mathbf{R}$, we shall imagine, for the moment, that we live in the days of Pythagoras and that the only numbers we know about are the rational numbers.

In this Pythagorean world, we can still define limits, we can still talk about continuous functions, and we can still define derivatives. What we can't do is prove any theorem that depends upon the existence of suprema and infima. We have lost the Cantor intersection theorem, the theorems on the existence of partial limits of sequences, and, most important of all, we have lost the theorems on continuous functions that appeared in Sections 8.9 and 8.10. As you will see in this chapter, those theorems play a fundamental role when we prove the properties of derivatives listed in Subsection 9.1.1. Without a complete number system, the apparently obvious facts listed in Subsection 9.1.1 are actually false.

To see how they can fail, we consider the function f defined by the equation

$$f(x) = \begin{cases} 1 & \text{if } x < 0 \text{ or } x^2 < 2 \\ 6 & \text{if } x > 0 \text{ and } x^2 > 2. \end{cases}$$

The graph of this function is illustrated in Figure 9.2. Since our Pythagorean

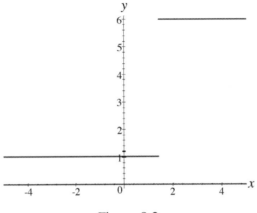

Figure 9.2

concept of a number, which includes rational numbers only, omits the number $\sqrt{2}$, we can assert that $f'(x) = 0$ for every "number" x even though the function f is not constant. In a similar way we can exhibit a function whose derivative is constantly equal to 1 even though the function fails to be increasing.

Thus, in the Pythagorean world, a function can have an everywhere zero derivative without being constant, it can have an everywhere positive derivative without being increasing, and it can have an everywhere negative derivative without being decreasing. We now leave the Pythagorean world and return to the complete number system $\mathbf{R}$ of the present day. Our sojourn there has taught us that the key facts about differentiation are not trivial and that, when we prove them, we shall have to make use of the completeness of the number system.

9.2   Derivatives and Differentiability

9.2.1   Definition of a Derivative

Suppose that f is a function defined on a set S of real numbers and suppose that x is a number that belongs to S and is also a limit point of S. The **derivative**

$f'(x)$ of the function f at the number x is the limit

$$\lim_{t \to x} \frac{f(t) - f(x)}{t - x}$$

as long as the latter limit exists.

In the event that the latter limit exists (and is finite), we say that the function f is **differentiable** at the number x. If f is differentiable at every number $x \in S$, then we say that f is **differentiable on** the set S and, in this event, the function f' whose value at every number $x \in S$ is $f'(x)$ is a function with domain S.

If f is differentiable on a set S and if the function f' is differentiable at a number $x \in S$, then the value of the derivative of f' at the number x is written as $f''(x)$ and is called the **second derivative** of the function f at the number x. The third and subsequent derivatives are defined similarly. If n is a positive integer, then the nth derivative of f is written as $f^{(n)}$. In keeping with this notation, we write $f^{(0)}$ for the function f itself.

9.2.2 Derivative of a Function Defined on an Interval

The general definition of a derivative that we have just seen provides us with three important special cases involving functions defined on an interval:

1. Suppose that f is a function defined on an interval S, that $x \in S$, and that x is not an endpoint of the interval S.

In this case the limit

$$\lim_{t \to x} \frac{f(t) - f(x)}{t - x}$$

is a regular two-sided limit and $f'(x)$ is called the **two-sided derivative** of f at the number x.

2. Suppose that f is a function defined on an interval S, that $x \in S$, and that x is the left endpoint of S.

In this case the number $f'(x)$ is the **derivative from the right** of the function f at the number x.

3. Suppose that f is a function defined on an interval S, that $x \in S$, and that x is the right endpoint of S.

In this case the number $f'(x)$ is the **derivative from the left** of the function f at the number x.

9.2.3   Some Examples of Derivatives

1. Suppose that c is a given real number and that $f(x) = c$ for every number x. Then, given any number x we have

$$\lim_{t \to x} \frac{f(t) - f(x)}{t - x} = \lim_{t \to x} \frac{c - c}{t - x} = \lim_{t \to x} 0 = 0.$$

Therefore the derivative of a constant function must be zero. Do not confuse this simple fact with the substantially more difficult fact that we mentioned in Subsection 9.1.1, that a function with a zero derivative must be constant.

2. Suppose that $f(x) = x$ for every number x. Then, given any number x we have

$$f'(x) = \lim_{t \to x} \frac{f(t) - f(x)}{t - x} = \lim_{t \to x} \frac{t - x}{t - x} = 1.$$

3. Suppose that n is a positive integer and that $f(x) = x^n$ for every number x. Then, given any number x we have

$$
\begin{aligned}
f'(x) &= \lim_{t \to x} \frac{t^n - x^n}{t - x} \\
&= \lim_{t \to x} \frac{(t - x)\left(t^{n-1} + t^{n-2}x + t^{n-3}x^2 + \cdots + tx^{n-2} + x^{n-1}\right)}{t - x} \\
&= \lim_{t \to x} \left(t^{n-1} + t^{n-2}x + t^{n-3}x^2 + \cdots + tx^{n-2} + x^{n-1}\right) = nx^{n-1}.
\end{aligned}
$$

4. In the preceding example we saw that if n is a positive integer and $f(x) = x^n$ for every number x, then $f'(x) = nx^{n-1}$ for every number x. We now extend this statement to include the case in which n is a negative integer. Suppose that n is a negative integer and that $f(x) = x^n$ for every number $x \neq 0$. We write $m = -n$ and note that m is a positive integer. Now, given any number $x \neq 0$, we have

$$
\begin{aligned}
f'(x) &= \lim_{t \to x} \frac{t^n - x^n}{t - x} = \lim_{t \to x} \frac{\frac{1}{t^m} - \frac{1}{x^m}}{t - x} \\
&= -\lim_{t \to x} \frac{t^m - x^m}{t^m x^m (t - x)} = -\frac{1}{x^m x^m} m x^{m-1} = nx^{n-1}.
\end{aligned}
$$

5. In this example we suppose that r is a positive rational number and that $f(x) = x^r$ for all numbers $x > 0$. We shall show that if x is any positive

number, then $f'(x) = rx^{r-1}$. We begin by choosing positive integers m and n such that $r = m/n$. Now, given any positive number x we have

$$
\begin{aligned}
f'(x) &= \lim_{t \to x} \frac{t^r - x^r}{t - x} = \lim_{t \to x} \frac{t^{m/n} - x^{m/n}}{t - x} \\
&= \lim_{t \to x} \frac{\left(t^{1/n}\right)^m - \left(x^{1/n}\right)^m}{\left(t^{1/n}\right)^n - \left(x^{1/n}\right)^n} \\
&= \lim_{t \to x} \frac{\left(t^{1/n} - x^{1/n}\right)\left(\left(t^{1/n}\right)^{m-1} + \left(t^{1/n}\right)^{m-2}\left(x^{1/n}\right) + \cdots + \left(x^{1/n}\right)^{m-1}\right)}{\left(t^{1/n} - x^{1/n}\right)\left(\left(t^{1/n}\right)^{n-1} + \left(t^{1/n}\right)^{n-2}\left(x^{1/n}\right) + \cdots + \left(x^{1/n}\right)^{n-1}\right)} \\
&= \lim_{t \to x} \frac{\left(t^{1/n}\right)^{m-1} + \left(t^{1/n}\right)^{m-2}\left(x^{1/n}\right) + \cdots + \left(x^{1/n}\right)^{m-1}}{\left(t^{1/n}\right)^{n-1} + \left(t^{1/n}\right)^{n-2}\left(x^{1/n}\right) + \cdots + \left(x^{1/n}\right)^{n-1}} \\
&= \frac{mx^{(m-1)/n}}{nx^{(n-1)/n}} = rx^{r-1}.
\end{aligned}
$$

6. In this example we suppose that r is a negative rational number and once again we take $f(x) = x^r$ for $x > 0$. By using the technique of Example 4, one may show that $f'(x) = rx^{r-1}$ for every number $x > 0$. We leave the details as an exercise. 📝

7. In this example we take

$$
f(x) = \begin{cases} x^2 & \text{if } x \text{ is rational} \\ 0 & \text{if } x \text{ is irrational.} \end{cases}
$$

For every number $t \neq 0$ we have

$$
\left| \frac{f(t) - f(0)}{t - 0} \right| = \left| \frac{f(t)}{t} \right| \leq |t|,
$$

and so it follows from the sandwich theorem (Theorem 8.5.3) that $f'(0) = 0$.

8. In this example we take

$$
f(x) = \begin{cases} x & \text{if } x \text{ is rational} \\ 0 & \text{if } x \text{ is irrational.} \end{cases}
$$

For every number $t \neq 0$ we have

$$\frac{f(t) - f(0)}{t - 0} = \begin{cases} 1 & \text{if } t \text{ is rational} \\ 0 & \text{if } t \text{ is irrational,} \end{cases}$$

and it is clear that the limit

$$\lim_{t \to x} \frac{f(t) - f(0)}{t - 0}$$

fails to exist.

9. In this example we define

$$f(x) = \begin{cases} x \sin \frac{1}{x} & \text{if } x \neq 0 \\ 0 & \text{if } x = 0. \end{cases}$$

For every number $t \neq 0$ we have

$$\frac{f(t) - f(0)}{t - 0} = \frac{t \sin \frac{1}{t} - 0}{t} = \sin \frac{1}{t},$$

and so it follows from Example 6 of Subsection 8.1.5 that $f'(0)$ does not exist. The graph of the function f is illustrated in Figure 9.3.

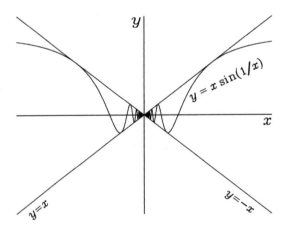

Figure 9.3

9.3   Some Elementary Properties of Derivatives

9.3.1   Differentiability and Continuity

Suppose that a function f is differentiable at a number x. Then f is continuous at x.

Proof. If S is the domain of the function f, then for every number $t \in S \setminus \{x\}$ we have

$$f(t) = \frac{f(t) - f(x)}{t - x}(t - x) + f(x)$$

and the latter expression approaches

$$f'(x) \times 0 + f(x) = f(x)$$

as $t \to x$. ∎

9.3.2   The Sum Rule

Suppose that f and g are functions defined on a set S and that both f and g are differentiable at a number x. Then the function $f + g$ is also differentiable at the number x and we have

$$(f + g)'(x) = f'(x) + g'(x).$$

Proof. For every number $t \in S \setminus \{x\}$ we have

$$\frac{(f + g)(t) - (f + g)(x)}{t - x} = \frac{f(t) - f(x)}{t - x} + \frac{g(t) - g(x)}{t - x}$$

and the latter expression approaches $f'(x) + g'(x)$ as $t \to x$. ∎

9.3.3   The Constant Multiple Rule

This rule will become obsolete as soon as we have stated the product rule in the next subsection, but it is worth stating anyway:

Suppose that f is a function defined on a set S, that c is a given number, and that the function f is differentiable at a number $x \in S$. Then the function cf is also differentiable at the number x and we have

$$(cf)'(x) = cf'(x).$$

We leave the proof of this rule as an exercise.

9.3.4   The Product Rule

Suppose that f and g are functions defined on a set S and that both f and g are differentiable at a number x. Then the function fg is also differentiable at the *number x and we have*

$$(fg)'(x) = f'(x)g(x) + f(x)g'(x).$$

Proof. For every number $t \in S \setminus \{x\}$ we have

$$\frac{f(t)g(t) - f(x)g(x)}{t - x} = \frac{f(t)g(t) - f(x)g(t) + f(x)g(t) - f(x)g(x)}{t - x}$$

$$= \left(\frac{f(t) - f(x)}{t - x}\right) g(t) + f(x) \left(\frac{g(t) - g(x)}{t - x}\right).$$

Since Theorem 9.3.1 guarantees that $g(t) \to g(x)$ as $t \to x$, we have

$$\lim_{t \to x} \frac{f(t)g(t) - f(x)g(x)}{t - x} = f'(x)g(x) + f(x)g'(x).$$

9.3.5   The Quotient Rule

Suppose that f and g are functions defined on a set S, that both f and g are differentiable at a number x, and that $g(x) \neq 0$. Then the function $\dfrac{f}{g}$ is also differentiable at the number x and we have

$$\left(\frac{f}{g}\right)'(x) = \frac{f'(x)g(x) - f(x)g'(x)}{(g(x))^2}.$$

Proof. Using the fact that the function g is continuous at the number x, we choose a number $\delta > 0$ such that $g(t) \neq 0$ whenever $t \in S$ and $|t - x| < \delta$. Now whenever $t \in S \setminus \{x\}$ and $|t - x| < \delta$ we have

$$\frac{\frac{f(t)}{g(t)} - \frac{f(x)}{g(x)}}{t - x} = \frac{f(t)g(x) - f(x)g(t)}{(t - x)\, g(t)g(x)}$$

$$= \frac{f(t)g(x) - f(x)g(x) - f(x)g(t) + f(x)g(x)}{(t - x)\, g(t)g(x)}$$

$$= \frac{\left(\frac{f(t) - f(x)}{t - x}\right) g(x) - f(x) \left(\frac{g(t) - g(x)}{t - x}\right)}{g(t)g(x)}$$

and the latter expression approaches

$$\frac{f'(x)g(x) - f(x)g'(x)}{(g(x))^2}$$

as $t \rightarrow x$. ∎

9.3.6   Differentiability and Limits of Sequences

Suppose that f is a function defined on a set S, that $x \in S$, and that the number x is a limit point of S. Suppose that w is a given real number. Then the following conditions are equivalent:

1. *The function f is differentiable at the number x and $f'(x) = w$.*
2. *For every sequence (x_n) in the set $S \setminus \{x\}$ that converges to x we have*

$$\frac{f(x_n) - f(x)}{x_n - x} \rightarrow w$$

as $n \rightarrow \infty$.

Proof. This theorem follows at once from the relationship between limits of sequences and limits of functions that we saw in Section 8.4. ∎

9.3.7   The Chain Rule

By analogy with the composition theorem for limits that we saw in Section 8.6 and the composition theorem for continuity that we saw in Subsection 8.7.9, the chain rule is the composition theorem for derivatives.

Suppose that f is a function defined on a set S, that $f(x)$ lies in a given set T for every number $x \in S$, and that g is a function defined on T. Suppose that $x \in S$, that f is differentiable at the number x, and that g is differentiable at the number $f(x)$. See Figure 9.4.

Then the composition function $g \circ f$ is continuous at the number x and we have

$$(g \circ f)'(x) = g'(f(x)) f'(x).$$

Proof. We begin by defining a function ϕ on the set T as follows:

$$\phi(y) = \begin{cases} \frac{g(y) - g(f(x))}{y - f(x)} & \text{if } y \in T \setminus \{f(x)\} \\ g'(f(x)) & \text{if } y = f(x). \end{cases}$$

Since $\phi(y) \rightarrow g'(f(x))$ as $y \rightarrow f(x)$, we see that the function ϕ is continuous at the number $f(x)$. Since f is continuous at the number x, it follows from the composition theorem for continuity (Theorem 8.7.9) that

$$\lim_{s \rightarrow x} \phi(f(s)) = \phi(f(x)) = g'(f(x)).$$

Now for every number s in the set $S \setminus \{x\}$, regardless of whether $f(s) = f(x)$

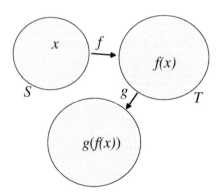

Figure 9.4

or $f(s) \neq f(x)$,

$$\frac{g\left(f(s)\right) - g\left(f(x)\right)}{s - x} = \phi\left(f(s)\right)\left(\frac{f(s) - f(x)}{s - x}\right)$$

and so it follows that

$$\lim_{s \to x} \frac{g\left(f(s)\right) - g\left(f(x)\right)}{s - x} = \lim_{s \to x} \phi\left(f(s)\right)\left(\frac{f(s) - f(x)}{s - x}\right) = g'\left(f(x)\right) f'(x).$$

9.3.8   Differentiation of Inverse Functions, Motivation

Suppose that f is a one-one continuous function on an interval S and that g is the inverse function of f. If both of the functions f and g are differentiable at each number in their domains, then, since

$$g\left(f(x)\right) = x$$

for every number $x \in S$, it follows from the chain rule that

$$g'\left(f(x)\right) f'(x) = 1$$

for all $x \in S$. Thus

$$g'\left(f(x)\right) = \frac{1}{f'(x)}.$$

In the next subsection we shall go one step further and observe that if f is a one-one continuous function on an interval and g is its inverse function, then the function g is automatically differentiable at the number $f(x)$ whenever f is differentiable at x and $f'(x) \neq 0$.

9.3.9   Theorem on Differentiation of Inverse Functions

The theorem contained in this subsection depends upon the results about continuous functions on intervals that we saw in Section 8.10. In particular, we need to know that a continuous one-one function on an interval has to be strictly monotone and that the inverse function of such a function is also continuous.

Suppose that f is a one-one continuous function on an interval S and that g is the inverse function of f. Suppose that f is differentiable at a given number $x \in S$ and that $f'(x) \neq 0$. Then the function g must be differentiable at the number $f(x)$ and we have

$$g'(f(x)) = \frac{1}{f'(x)}.$$

Proof. We write the range of the function f as T. Note that T is an interval and that g is a continuous function from T onto S. We shall prove the theorem by using Theorem 9.3.6. Suppose that (y_n) is a sequence in $T \setminus \{f(x)\}$ and that $y_n \to f(x)$ as $n \to \infty$. We need to show that

$$\frac{g(y_n) - g(f(x))}{y_n - f(x)} \to \frac{1}{f'(x)}$$

as $n \to \infty$. For each positive integer n we define $x_n = g(y_n)$. In this way we have defined a sequence (x_n) in the set $S \setminus \{x\}$, and, since the function g is continuous at the number $f(x)$, we have

$$\lim_{n \to \infty} x_n = \lim_{n \to \infty} g(y_n) = g(f(x)) = x.$$

Therefore, since f is differentiable at the number x we have

$$\lim_{n \to \infty} \frac{g(y_n) - g(f(x))}{y_n - f(x)} = \lim_{n \to \infty} \frac{x_n - x}{f(x_n) - f(x)}$$
$$= \lim_{n \to \infty} \left(\frac{f(x_n) - f(x)}{x_n - x} \right)^{-1} = \frac{1}{f'(x)},$$

as promised. ∎

9.3.10   Exercises on Derivatives

1. Given that $f(x) = |x|$ for every number x, prove that $f'(0)$ does not exist.
2. ⬛ Given that $f(x) = |x|$ for all $x \in [-2, -1] \cup [0, 1]$, prove that $f'(0)$ does exist.

3. Given that $f(x) = x\,|x|$ for every number x, determine whether or not $f'(0)$ exists.

4. This exercise concerns the function f defined by the equation

$$f(x) = \begin{cases} x^2 \sin \frac{1}{x} & \text{if } x \neq 0 \\ 0 & \text{if } x = 0. \end{cases}$$

You should assume all of the standard formulas for the derivatives of the functions sin and cos.

(a) Ⓝ Ask *Scientific Notebook* to make a 2D plot of the expression $x^2 \sin \frac{1}{x}$ on the interval $[-.2, .2]$ and then drag each of the expressions x^2 and $-x^2$ into your plot. Revise the plot and give the components different colors.

(b) Prove that the function f is differentiable on $\mathbf{R}$ but that the function f' is not continuous at the number 0.

5. This exercise concerns the function f defined by the equation

$$f(x) = \begin{cases} x^3 \sin \frac{1}{x} & \text{if } x \neq 0 \\ 0 & \text{if } x = 0. \end{cases}$$

(a) Ⓝ Ask *Scientific Notebook* to make a 2D plot of the expression $x^3 \sin \frac{1}{x}$ on the interval $[-.05, .05]$ and then drag each of the expressions x^3 and $-x^3$ into your plot. Revise the plot and give the components different colors.

(b) Prove that the function f' is continuous at the number 0 but does not have a derivative there.

6. Suppose that f is a function defined on an open interval (a, b) and that $x \in (a, b)$.

(a) ▦ Prove that if $f'(x)$ exists, then

$$f'(x) = \lim_{h \to 0} \frac{f(x+h) - f(x)}{h}.$$

(b) Prove that if the limit

$$\lim_{h \to 0} \frac{f(x+h) - f(x)}{h}$$

exists, then $f'(x)$ exists and is equal to this limit.

(c) ▇ Prove that if $f'(x)$ exists, then

$$f'(x) = \lim_{h \to 0} \frac{f(x+h) - f(x-h)}{2h}.$$

(d) Prove that if $f'(x)$ exists, then

$$f'(x) = \lim_{t \to x} \left(\lim_{u \to x} \frac{f(t) - f(u)}{t - u} \right).$$

9.4   The Mean Value Theorem

The mean value theorem that we shall study in this section is the bridge to the fundamental facts about derivatives that we listed in Subsection 9.1.1. This theorem depends upon the completeness of the real number system.

The mean value theorem is deduced from an important result called **Rolle's theorem**, which tells us that if a function satisfies certain conditions, then its derivative has to be zero somewhere. By making use of a simple technique known as Fermat's theorem, we shall show that the derivative of a given function must have a zero value somewhere by considering the numbers where the function takes a maximum or minimum value. Therefore, the key to Rolle's theorem is Theorem 8.9.2, which guarantees the existence of maxima and minima, and it is by making use of that theorem that we are invoking the completeness of the number system $\mathbf{R}$.

9.4.1   Fermat's Theorem

Suppose that f is a function defined on an interval S and that f is differentiable at a number $x \in S$.

1. *If x is not the right endpoint of S and $f'(x) > 0$, then f cannot have a maximum value at the number x.*
2. *If x is not the right endpoint of S and $f'(x) < 0$, then f cannot have a minimum value at the number x.*
3. *If x is not the left endpoint of S and $f'(x) < 0$, then f cannot have a maximum value at the number x.*
4. *If x is not the left endpoint of S and $f'(x) > 0$, then f cannot have a minimum value at the number x.*
5. *If x is not an endpoint of S and f has either a maximum or a minimum value at x, then we must have $f'(x) = 0$.*

Proof. We shall prove part 1 and leave the other parts as exercises. Suppose that

x is not the right endpoint of the interval S and that $f'(x) > 0$. See Figure 9.5. Using the fact that the interval $(0, \infty)$ is a neighborhood of the number $f'(x)$,

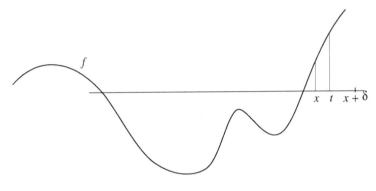

Figure 9.5

we choose a number $\delta > 0$ such that whenever $t \in S$ and $|x - t| < \delta$ we have

$$\frac{f(t) - f(x)}{t - x} > 0.$$

Using the fact that x is not the right endpoint of the interval S, we choose a number $t \in S$ such that $x < t < x + \delta$. Since $t - x > 0$ and

$$\frac{f(t) - f(x)}{t - x} > 0,$$

we see that $f(t) > f(x)$ and so $f(x)$ cannot be the maximum value of the function f. ■

9.4.2   Rolle's Theorem

Suppose that a and b are real numbers and that $a < b$. Suppose that f is a function that is continuous on the interval $[a, b]$, that f is differentiable[30] at every number x in the interval (a, b), and that $f(a) = f(b)$. Then there is at least one number $c \in (a, b)$ such that $f'(c) = 0$.

Proof. We illustrate this proof in Figure 9.6. If the number $f(a)$ is both the maximum and the minimum value of the function f, then f has to be constant on $[a, b]$ and $f'(x) = 0$ for every number $x \in [a, b]$. Otherwise, either the maximum or the minimum value of f must fail to occur at at a, and must therefore occur at

---

[30]   There is some overlap in these assumptions. If f is differentiable on (a, b), then f is automatically continuous on (a, b). To ensure that f is continuous on $[a, b]$, all we need to assume is that f is continuous from the right at a and from the left at b.

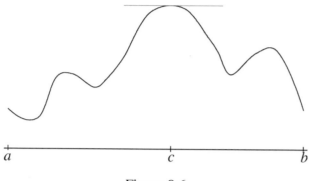

Figure 9.6

a number $c \neq a$. Since $f(a) = f(b)$, it is clear that $c \neq b$, and it follows from Fermat's theorem (Theorem 9.4.1) that $f'(c) = 0$. ∎

9.4.3   The Mean Value Theorem

Suppose that a and b are real numbers and that $a < b$. Suppose that f is a function that is continuous on the interval $[a, b]$, and that f is differentiable at every number x in the interval (a, b). Then there is at least one number $c \in (a, b)$ such that

$$f'(c) = \frac{f(b) - f(a)}{b - a}. \tag{9.1}$$

This theorem is illustrated in Figure 9.7. The theorem tells us that there must be at

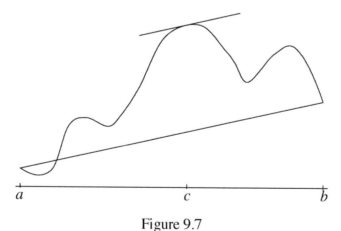

Figure 9.7

least one point $(c, f(c))$ on the graph of f where the slope of the graph is equal to the slope of the straight line that runs from the point $(a, f(a))$ to the point

$(b, f(b))$. Note that if $f(a) = f(b)$, then Equation (9.1) becomes $f'(c) = 0$, and so the mean value theorem reduces to Rolle's theorem.

Proof. For every number $x \in [a, b]$ we define

$$h(x) = f(x) - f(a) - \frac{f(b) - f(a)}{b - a} (x - a).$$

Like f, the function h is continuous on the interval $[a, b]$ and, like f, the function h is differentiable at every number $x \in (a, b)$. As a matter of fact, if $x \in (a, b)$, then

$$h'(x) = f'(x) - \frac{f(b) - f(a)}{b - a}. \tag{9.2}$$

However, unlike the function f, the function h takes the same value at the two numbers a and b because

$$h(a) = h(b) = 0.$$

Using Rolle's theorem we choose a number $c \in (a, b)$ such that $h'(c) = 0$ and, substituting $x = c$ in Equation (9.2), we obtain

$$f'(c) = \frac{f(b) - f(a)}{b - a}. \blacksquare$$

9.4.4   The Cauchy Mean Value Theorem

In this subsection we look at a slightly more general form of the mean value theorem that is known as the **Cauchy mean value theorem.**

Suppose that f and g are continuous functions on an interval $[a, b]$ that are differentiable at every number $x \in (a, b)$. Suppose that $g'(x) \neq 0$ for every $x \in (a, b)$. Then there exists a number $c \in (a, b)$ such that

$$\frac{f'(c)}{g'(c)} = \frac{f(b) - f(a)}{g(b) - g(a)}.$$

Proof. We note first that, because $g'(x) \neq 0$ for all $x \in (a, b)$, Rolle's theorem guarantees that $g(a) \neq g(b)$. We now define

$$h(x) = f(x) - f(a) - \left(\frac{f(b) - f(a)}{g(b) - g(a)} \right) (g(x) - g(a))$$

for every number $x \in [a, b]$. The function h is continuous on the interval $[a, b]$, and for every number $x \in (a, b)$ we have

$$h'(x) = f'(x) - \left(\frac{f(b) - f(a)}{g(b) - g(a)} \right) g'(x).$$

Furthermore, $h(a) = h(b) = 0$. Using Rolle's theorem we choose a number $c \in (a, b)$ such that $h'(c) = 0$ and we observe that

$$\frac{f'(c)}{g'(c)} = \frac{f(b) - f(a)}{g(b) - g(a)}. \quad \blacksquare$$

9.4.5   Exercises on the Mean Value Theorem

1. (a) Given that f is a function defined on an interval S and that $f'(x) = 0$ for every $x \in S$, prove that f must be constant on S.
   (b) ▤ Given that f is a function defined on an interval S and that $f'(x) > 0$ for every $x \in S$, prove that f must be strictly increasing on S.
   (c) Given that f is a function defined on an interval S and that $f'(x) < 0$ for every $x \in S$, prove that f must be strictly decreasing on S.

2. ▤ Suppose that f and g are functions defined on an interval S and that $f'(x) = g'(x)$ for every number $x \in S$. Prove that there exists a real number c such that the equation

$$f(x) = g(x) + c$$

   holds for every number $x \in S$.

3. Suppose that f is continuous on an interval $[a, b]$ and differentiable on the interval (a, b) and that $f(a) = f(b)$. Suppose that $a < c < b$ and that $f'(x) > 0$ when $a < x < c$ and $f'(x) < 0$ when $c < x < b$. Prove that $f(c)$ is the maximum value of the function f.

4. Given that f is a strictly increasing differentiable function on an interval S, is it true that $f'(x)$ must be positive for every $x \in S$?

5. ▤ Prove that if f is a differentiable function on an interval S and $f'(x) \neq 0$ for every $x \in S$, then the function f must be one-one.

6. Given that f is differentiable on an interval S and that the function f' is bounded on S, prove that f must be Lipschitzian (see Exercise 5a of Subsection 8.12.7) on S.

7. Given that f is a function defined on an interval S and that the inequality

$$|f(t) - f(x)| \leq |t - x|^2$$

   holds for all numbers t and x in S, prove that f must be constant.

8. Suppose that f and g are functions defined on $\mathbf{R}$ and that $f'(x) = g(x)$ and $g'(x) = -f(x)$ for every real number x.

   (a) Prove that $f''(x) = -f(x)$ for every number x.

   (b) Prove that the function $f^2 + g^2$ is constant.[31]

9. **H** Given that f is differentiable on the interval $(0, \infty)$ and that $f'(x) \to \lambda$ as $x \to \infty$, prove that

$$f(x + 1) - f(x) \to \lambda$$

   as $x \to \infty$.

10. Given that f is continuous on $[a, b]$ and differentiable on (a, b), and that $f'(x)$ approaches a limit $w \in \mathbf{R}$ as $x \to a$, prove that f must be differentiable at the number a and that $f'(a) = w$.

11. **H** Prove that if f is differentiable on an interval $[a, b]$ and $f'(a) < 0$ and $f'(b) > 0$, then there must be at least one number $c \in (a, b)$ for which $f'(c) = 0$.

12. Prove that if f is differentiable on an interval S, then the range of the function f' must be an interval.

13. **H** Suppose that f is differentiable on the interval $[0, \infty)$, that $f(0) = 0$ and that f' is increasing on $[0, \infty)$. Prove that if

$$g(x) = \frac{f(x)}{x}$$

   for all $x > 0$, then the function g is increasing on $(0, \infty)$.

14. Suppose that f is defined on an open interval S and that $f''(x) < 0$ for every number $x \in S$. Suppose that $a \in S$. Prove that if $x \in S$ and $x > a$, then

$$f(x) < f(a) + (x - a) f'(a).$$

   In other words, explain why, to the right of the point $(a, f(a))$, the graph of f lies below the tangent line to the graph at $(a, f(a))$.

15. Suppose that f is defined on an open interval S and that $f''(x) < 0$ for every number $x \in S$. Suppose that a and b belong to S and that $a < b$. Suppose that

$$g(x) = f(x) - f(a) - \left(\frac{f(b) - f(a)}{b - a} \right) (x - a)$$

   for all $x \in [a, b]$.

---

[31] Be careful to distinguish the notation f^2, which means ff, from the notation $f^{(2)}$ that stands for the second derivative of f.

(a) Prove that there exists a number $c \in (a, b)$ such that $g'(c) = 0$.

(b) Prove that the function g' is strictly decreasing on the interval $[a, b]$.

(c) Prove that the function g is strictly increasing on $[a, c]$ and strictly decreasing on $[c, b]$.

(d) Prove that $g(x) > 0$ for every $x \in (a, b)$.

(e) Prove that the straight line segment that joins the points $(a, f(a))$ and $(b, f(b))$ lies under the part of the graph of f that lies between the two points $(a, f(a))$ and $(b, f(b))$.

16. A function f defined on an interval S is said to be **convex** on S if, whenever a, x and b belong to S and $a < x < b$, we have

$$\frac{f(x) - f(a)}{x - a} \leq \frac{f(b) - f(x)}{b - x}.$$

Prove that if f is differentiable on an interval S, then f is convex on S if and only if the function f' is increasing.

17. ▣ Prove that if f is a convex function on an open interval S, then f must be continuous on S.

18. Ⓝ   By clicking on the icon ▣ you can reach some exercises that introduce Newton's method for approximating roots of an equation.

9.5   Taylor Polynomials

In this section we discuss a method by which a given function f can be approximated by polynomials that have the same behavior as f at a given number c. We begin with a brief discussion of polynomials.

9.5.1   Some Observations About Polynomials

As you may know, a function f is said to be a **polynomial** if it is possible to find a nonnegative integer n and numbers $a_0, a_1, a_2, \cdots, a_n$ such that the equation

$$f(x) = a_0 + a_1 x + a_2 x^2 + \cdots + a_n x^n = \sum_{j=0}^{n} a_j x^j \tag{9.3}$$

holds for every real number x. Our first observation is that if f is such a polynomial, then the coefficients $a_0, a_1, \cdots, a_n$ are uniquely determined by the behavior of the function f at the number 0.

We begin by putting $x = 0$ in Equation (9.3) to obtain $f(0) = a_0$. Now we differentiate both sides of Equation (9.3) to obtain the identity

$$f'(x) = 1a_1 + 2a_2 x^1 + 3a_3 x^2 + \cdots + n a_n x^{n-1}, \qquad (9.4)$$

and by putting $x = 0$ in Equation (9.4) we obtain $f'(0) = 1a_1$. Differentiating once more we obtain the identity

$$f''(x) = 2 \times 1a_2 + 3 \times 2a_3 x^1 + \cdots + n(n-1) a_n x^{n-2},$$

and by putting $x = 0$ we obtain $f''(0) = 2!a_2$. Continuing in this way, we see that if j is a nonnegative integer, then

$$a_j = \frac{f^{(j)}(0)}{j!}.$$

We conclude that a polynomial f can be expressed in the form shown in Equation (9.3) in only one way and that if x is any real number, we have

$$f(x) = \sum_{j=0}^{n} \frac{f^{(j)}(0)}{j!} x^j.$$

The largest value of j for which the coefficient $f^{(j)}(0)/j!$ is nonzero is called the **degree** of the polynomial f.

This technique for finding the coefficients of a polynomial can be applied a little more generally. Suppose that f is a polynomial of degree n having the form

$$f(x) = a_0 + a_1 x + a_2 x^2 + \cdots + a_n x^n$$

for every number x, and suppose that c is any real number. Given any number x we have

$$f(x) = a_0 + a_1 \left((x - c) + c\right) + a_2 \left((x - c) + c\right)^2 + \cdots + a_n \left((x - c) + c\right)^n,$$

and, by expanding each of these terms in powers of $(x - c)$, we see that there are numbers $b_0, b_1, \cdots, b_n$ such that the equation

$$f(x) = b_0 + b_1 (x - c) + b_2 (x - c)^2 + \cdots + b_n (x - c)^n$$

holds for every number x. By substituting $x = c$ we obtain $f(c) = b_0$, and, by differentiating repeatedly and substituting $x = c$, we obtain the equation

$$b_j = \frac{f^{(j)}(c)}{j!}$$

for every nonnegative integer j. Thus if x is any number, we have

$$f(x) = \sum_{j=0}^{n} \frac{f^{(j)}(c)}{j!} (x - c)^j .$$

9.5.2   Taylor Polynomials of a Function

Suppose that f is a given function that has derivatives of all orders at a given number c. For each nonnegative integer n, the nth **Taylor polynomial** of the function f centered at the number c is the polynomial p_n that is defined by the equation

$$\begin{aligned}
p_n(x) &= \sum_{j=0}^{n} \frac{f^{(j)}(c)}{j!} (x - c)^j \\
&= f(c) + \frac{f'(c)}{1!} (x - c) + f^{(2)}(c)\frac{(x - c)^2}{2!} + \cdots + f^{(n)}(c)\frac{(x - c)^n}{n!}
\end{aligned}$$

for every number x. The nth Taylor polynomial of a function centered at 0 is also called the nth **Maclaurin polynomial** of the function.

The important property of the Taylor polynomial p_n is that it is an exact fit to the function f at the number c in the sense that

$$\begin{aligned}
p_n(c) &= f(c) \\
p'(c) &= f'(c) \\
&\vdots \\
p^{(n)}(c) &= f^{(n)}(c).
\end{aligned}$$

From the discussion that appears in Subsection 9.5.1 we see that if the function f is itself a polynomial of degree not exceeding n, then the nth Taylor polynomial of f is precisely the function f.

9.5.3   Some Examples of Taylor Polynomials

Some of the examples in this subsection are phrased as exercises that are designed to be done while reading the on-screen version of the text.

1. Suppose that

$$f(x) = 2 - 4x + 3x^2 + 7x^3 + 5x^4$$

and that p_n is the nth Taylor polynomial of f centered at 0. Then for every

number x we have $p_0(x) = 2$,

$$p_3(x) = 2 - 4x + 3x^2 + 7x^3,$$

and $p_n(x) = f(x)$ whenever $n \geq 4$.

If q_n is the nth Taylor polynomial of f centered at 1, then for every number x we have $q_0(x) = 13$,

$$q_3(x) = 13 + 43(x - 1) + 54(x - 1)^2 + 27(x - 1)^3,$$

and $q_4(x) = f(x)$ whenever $n \geq 4$.

2. 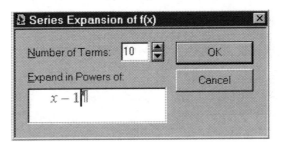 In this example we shall show how to use *Scientific Notebook* to evaluate the ninth Taylor polynomial centered at 1 of the function f defined by the equation

$$f(x) = \frac{x^2 + 1}{\sqrt{x^2 + x + 1}}$$

for every number x. If you are reading the on-screen version of this book, point at the expression

$$\frac{x^2 + 1}{\sqrt{x^2 + x + 1}},$$

open the Compute menu and click on Power Series. Fill in the dialogue box as shown in Figure 9.8. You will obtain the equation

Figure 9.8

$$\frac{x^2+1}{\sqrt{x^2+x+1}} = \left(\tfrac{2}{3}\sqrt{3}\right) + \left(\tfrac{1}{3}\sqrt{3}\right)(x-1) + \left(\tfrac{5}{36}\sqrt{3}\right)(x-1)^2 +$$
$$\left(-\tfrac{5}{72}\sqrt{3}\right)(x-1)^3 + \left(\tfrac{17}{576}\sqrt{3}\right)(x-1)^4 + \left(-\tfrac{11}{1152}\sqrt{3}\right)(x-1)^5 +$$
$$\left(\tfrac{49}{41\,472}\sqrt{3}\right)(x-1)^6 + \left(\tfrac{115}{82\,944}\sqrt{3}\right)(x-1)^7 + \left(-\tfrac{6205}{3981\,312}\sqrt{3}\right)(x-1)^8 +$$
$$\left(\tfrac{8251}{7962\,624}\sqrt{3}\right)(x-1)^9 + O\left((x-1)^{10}\right).$$

For now, we shall ignore the last term that appears as $O\left((x-1)^{10}\right)$. After deleting the last term we see that the ninth Taylor polynomial of the function f has the value

$$\left(\frac{2}{3}\sqrt{3}\right) + \left(\frac{1}{3}\sqrt{3}\right)(x-1) + \left(\frac{5}{36}\sqrt{3}\right)(x-1)^2 +$$

$$\left(-\frac{5}{72}\sqrt{3}\right)(x-1)^3 + \left(\frac{17}{576}\sqrt{3}\right)(x-1)^4 + \left(-\frac{11}{1152}\sqrt{3}\right)(x-1)^5 +$$

$$\left(\frac{49}{41\,472}\sqrt{3}\right)(x-1)^6 + \left(\frac{115}{82\,944}\sqrt{3}\right)(x-1)^7 +$$

$$\left(-\frac{6205}{3981\,312}\sqrt{3}\right)(x-1)^8 + \left(\frac{8251}{7962\,624}\sqrt{3}\right)(x-1)^9$$

at each number x.

3. Working on-screen, point at the expression

$$\frac{\arctan x}{\sqrt{1+x+x^2}}$$

and calculate its ninth Taylor polynomial centered at 1. Then highlight your answer and click on Evaluate Numerically to obtain a numeric form of this Taylor polynomial.

4. Working on-screen, sketch the graph

$$y = \frac{\arctan x}{\sqrt{1+x+x^2}}$$

and drag in the graphs of some of its Maclaurin polynomials. Do not make your domain too large and make your graphs in different colors so that you can see what you have drawn. Repeat the process using Taylor polynomials with center 1.

5. If you are reading the on-screen version of this text, you can see an animated sequence of graphs that illustrate the Maclaurin polynomials of the function f defined by the equation

$$f(x) = x\sin\left(x^2 + 3x + 1\right)$$

for every number x. To see the animation, click on the icon .

9.5.4 Some Exercises on Taylor Polynomials

1. Suppose that f is a polynomial whose degree does not exceed a given positive integer k and that $f^{(j)}(0) = 0$ for every $j = 0, 1, 2, \cdots, k$. Prove

that f is the constant function zero.

2.   ■   Suppose that f and g are two polynomials whose degrees do not exceed a given positive integer k and that $f^{(j)}(0) = g^{(j)}(0)$ for every $j = 0, 1, 2, \cdots, k$. Prove that $f(x) = g(x)$ for every number x.

3.   Prove that if f is a polynomial whose degree does not exceed a given positive integer k and n is an integer satisfying $n \geq k$, then the nth Taylor polynomial of f is f itself.

4.   Given a nonnegative integer n and that

$$f(x) = (1 + x)^n$$

for every number x, work out the nth Taylor polynomial of f and obtain a simple proof of the **binomial theorem**

$$(1 + x)^n = \sum_{j=0}^{n} \binom{n}{j} x^j.$$

As you may know, the expression $\binom{n}{j}$, which is called the n, j **binomial coefficient,** is defined by the equation

$$\binom{n}{j} = \frac{n(n-1)(n-2) \cdots (n-j+1)}{j!} = \frac{n!}{(n-j)!j!}$$

whenever n and j are integers and $0 \leq j \leq n$. We shall define binomial coefficients more generally In Example 3 of Subsection 12.7.9.

9.5.5   A Version of Rolle's Theorem for Higher Derivatives

Suppose that a and b are real numbers, that $a < b$, and that n is a nonnegative integer. Suppose that g is a function whose nth derivative $g^{(n)}$ is continuous on the interval $[a, b]$ and is differentiable at every number x in the interval (a, b), and that the following two conditions are satisfied:

1. *Whenever $0 \leq j \leq n$ we have $g^{(j)}(a) = 0$.*
2. *We have $g(b) = 0$.*

Then there is at least one number $c \in (a, b)$ such that $g^{(n+1)}(c) = 0$.

Proof. In the event that $n = 0$, the statement of this theorem reduces to Rolle's theorem. In the general case, we prove the theorem by applying Rolle's theorem $n + 1$ times. First we apply Rolle's theorem to the function g on the interval $[a, b]$ to choose a number c_1 such that $g'(c_1) = 0$. Next we use the fact that $g'(a) = g'(c_1) = 0$ and we apply Rolle's theorem to the function g' on the interval $[a, c_1]$ to choose a number c_2 such that $g''(c_2) = 0$.

Continuing in this way we construct a strictly decreasing sequence of numbers

$$c_1, c_2, c_3, \cdots, c_n, c_{n+1}$$

and we complete the proof by noting that if $c = c_{n+1}$, then $g^{(n+1)}(c) = 0$. ∎

9.5.6   A Version of the Mean Value Theorem for Higher Derivatives

Suppose that a and b are real numbers, that $a < b$, and that n is a nonnegative integer. Suppose that f is a function whose nth derivative $f^{(n)}$ is continuous on the interval $[a, b]$ and is differentiable at every number x in the interval (a, b).

Then there is at least one number $c \in (a, b)$ such that

$$f(b) = \sum_{j=0}^{n} \frac{f^{(j)}(a)}{j!} (b - a)^j + \frac{f^{(n+1)}(c)}{(n + 1)!} (b - a)^{n+1}.$$

Proof. We define p to be the nth Taylor polynomial of f centered at a. Now we need to show that there is a number $c \in (a, b)$ such that

$$f(b) = p(b) + \frac{f^{(n+1)}(c)}{(n + 1)!} (b - a)^{n+1}.$$

We begin by defining

$$g(x) = f(x) - p(x) - k (x - a)^{n+1}$$

for $x \in [a, b]$, where the number k is chosen in such a way that $g(b) = 0$. As a matter of fact,

$$k = \frac{f(b) - p(b)}{(b - a)^{n+1}}. \tag{9.5}$$

Now we apply Theorem 9.5.5 to the function g on the interval $[a, b]$ to choose a number $c \in (a, b)$ such that $g^{(n+1)}(c) = 0$. Since the degree of the polynomial p does not exceed n, we have

$$g^{(n+1)}(x) = f^{(n+1)}(x) - 0 - k (n + 1)!$$

for every $x \in (a, b)$, and substituting $x = c$ we obtain

$$0 = f^{(n+1)}(c) - k (n + 1)!. \tag{9.6}$$

Combining Equations (9.5) and (9.6) we obtain

$$f(b) = p(b) + \frac{f^{(n+1)}(c)}{(n+1)!} (b-a)^{n+1} .$$

9.6   Indeterminate Forms

9.6.1   Introduction to Indeterminate Forms

The algebraic rules for limits given in Theorem 8.5.4 tell us that if we have two limits

$$\lim_{x \to a} f(x) = \lambda \quad \text{and} \quad \lim_{x \to a} g(x) = \mu$$

then the conditions

$$
\begin{aligned}
\lim_{x \to a} (f(x) + g(x)) &= \lambda + \mu \\
\lim_{x \to a} (f(x) - g(x)) &= \lambda - \mu \\
\lim_{x \to a} (f(x)g(x)) &= \lambda\mu \\
\lim_{x \to a} \left(\frac{f(x)}{g(x)} \right) &= \frac{\lambda}{\mu}
\end{aligned}
$$

hold whenever the expressions on the right are defined. When the expressions on the right are not defined, the process of deciding whether or not the limits exist and what they are is considerably more difficult. Limits of this type are known as **indeterminate forms**.

In this section we shall discuss some techniques for evaluating indeterminate forms, among which is an important result that is known as L'Hôpital's rule.[32] As you probably know from your elementary calculus courses, many indeterminate forms can be evaluated very easily with the help of L'Hôpital's rule. But L'Hôpital's rule is not a silver bullet, and its importance is often overrated. We begin this section with some examples of indeterminate forms that are easier to evaluate without the help of L'Hôpital's rule.

9.6.2   Some Examples of Indeterminate Forms

1. The limit expression

$$\lim_{x \to \infty} \left(\sqrt{x^2 + x + 1} - \sqrt{x^2 - x + 1} \right)$$

---

[32]   Although first published by L'Hôpital, this theorem was in fact discovered by Johann Bernouli, who communicated it to L'Hôpital in a letter.

is indeterminate but we can see that the limit exists because, for every number $x \neq 0$, the expression $\sqrt{x^2 + x + 1} - \sqrt{x^2 - x + 1}$ equals

$$\frac{\left(\sqrt{x^2 + x + 1} - \sqrt{x^2 - x + 1}\right)\left(\sqrt{x^2 + x + 1} + \sqrt{x^2 - x + 1}\right)}{\sqrt{x^2 + x + 1} + \sqrt{x^2 - x + 1}}$$

$$= \frac{2x}{\sqrt{x^2 + x + 1} + \sqrt{x^2 - x + 1}} = \frac{2}{\sqrt{1 + \frac{1}{x} + \frac{1}{x^2}} + \sqrt{1 - \frac{1}{x} + \frac{1}{x^2}}}$$

and the latter expression approaches 1 as $x \to \infty$.

2. This example makes use of the function log that is introduced in Chapter 10.[33] From the fact that $\log' x = 1/x$ for all $x > 0$ we know that the function log is increasing. We deduce from Theorem 8.10.8 that

$$\lim_{x \to \infty} \log x = \sup \{\log x \mid x > 0\},$$

and, since the function log is unbounded above, we conclude that $\lim_{x \to \infty} \log x = \infty$. Therefore the limit expression

$$\lim_{x \to \infty} \frac{\log(x - 1)}{\log x}$$

is indeterminate. Now, given any number $x > 1$ we have

$$1 - \frac{\log(x - 1)}{\log x} = \frac{\log x - \log(x - 1)}{\log x} = \frac{\log\left(\frac{x}{x-1}\right)}{\log x}.$$

Since the numerator of the latter expression approaches $\log 1 = 0$ as $x \to \infty$ and $\log x \to \infty$ as $x \to \infty$, we can use the algebraic rules, Theorem 8.5.4, to obtain

$$\lim_{x \to \infty} \left(1 - \frac{\log(x - 1)}{\log x}\right) = 0$$

and we conclude that

$$\lim_{x \to \infty} \frac{\log(x - 1)}{\log x} = 1.$$

---

[33]   In this book we adopt the convention that the symbols log and ln mean the same thing. Both of them stand for the natural logarithm.

3. If q is any real number, then the limit expression

$$\lim_{x \to \infty} x \left(1 - \frac{(x-1)^q}{x^q} \right)$$

is indeterminate. To evaluate this limit we observe that if $x > 1$, we have

$$x \left(1 - \frac{(x-1)^q}{x^q} \right) = \frac{x^q - (x-1)^q}{x^{q-1}}.$$

We now define $f(x) = x^q$ for all $x > 0$, and, for any given number $x > 1$, we apply the mean value theorem to f on the interval $[x-1, x]$ to obtain a number $c \in (x-1, x)$ such that

$$x^q - (x-1)^q = \frac{f(x) - f(x-1)}{x - (x-1)} = f'(c) = qc^{q-1}.$$

Thus

$$\frac{x^q - (x-1)^q}{x^{q-1}} = q \left(\frac{c}{x} \right)^{q-1}.$$

If $q \geq 1$, then we have

$$q \left(\frac{x-1}{x} \right)^{q-1} \leq x \left(1 - \frac{(x-1)^q}{x^q} \right) \leq q \left(\frac{x}{x} \right)^{q-1},$$

and if $q < 1$, we have

$$q \left(\frac{x-1}{x} \right)^{q-1} \geq x \left(1 - \frac{(x-1)^q}{x^q} \right) \geq q \left(\frac{x}{x} \right)^{q-1}$$

for all $x > 1$. In either case, since the two outer expressions approach q as $x \to \infty$, it follows from the sandwich theorem (Theorem 8.5.3) that

$$\lim_{x \to \infty} x \left(1 - \frac{(x-1)^q}{x^q} \right) = q.$$

4. If q is any real number, then the limit expression

$$\lim_{x \to \infty} x \, (\log x) \left(1 - \frac{1}{x} - \frac{(x-1) \, (\log (x-1))^q}{x \, (\log x)^q} \right)$$

is indeterminate. To evaluate this limit we observe that if $x > 1$, we have

$$x \left(\log x\right) \left(1 - \frac{1}{x} - \frac{(x-1)\left(\log\left(x-1\right)\right)^q}{x\left(\log x\right)^q}\right)$$

$$= \frac{(x-1)\left[\left(\log x\right)^q - \left(\log\left(x-1\right)\right)^q\right]}{\left(\log x\right)^{q-1}}.$$

We now define $f(x) = (\log x)^q$ for all $x > 1$, and, for any given number $x > 2$, we apply the mean value theorem to f on the interval $[x - 1, x]$ to obtain a number $c \in (x - 1, x)$ such that

$$(\log x)^q - (\log\left(x-1\right))^q = \frac{f(x) - f(x-1)}{x - (x-1)} = f'(c) = \frac{q\left(\log c\right)^{q-1}}{c},$$

and so

$$\frac{(x-1)\left[\left(\log x\right)^q - \left(\log\left(x-1\right)\right)^q\right]}{\left(\log x\right)^{q-1}} = q \left(\frac{\log c}{\log x}\right)^{q-1} \left(\frac{x-1}{c}\right).$$

If $q \geq 1$, then we have

$$q \left(\frac{\log\left(x-1\right)}{\log x}\right)^{q-1} \left(\frac{x-1}{x}\right) \leq \frac{(x-1)\left[\left(\log x\right)^q - \left(\log\left(x-1\right)\right)^q\right]}{\left(\log x\right)^{q-1}} \leq q,$$

and if $q < 1$, we have

$$q \left(\frac{x-1}{x}\right) \leq \frac{(x-1)\left[\left(\log x\right)^q - \left(\log\left(x-1\right)\right)^q\right]}{\left(\log x\right)^{q-1}} \leq q \left(\frac{\log\left(x-1\right)}{\log x}\right)^{q-1}$$

for all $x > 2$. In either case, since the two outer expressions approach q as $x \to \infty$, it follows from the sandwich theorem (Theorem 8.5.3) that

$$\lim_{x\to\infty} x \left(\log x\right) \left(1 - \frac{1}{x} - \frac{(x-1)\left(\log\left(x-1\right)\right)^q}{x\left(\log x\right)^q}\right) = q.$$

9.6.3   L'Hôpital's rule

Suppose that f and g are differentiable on an interval S, that either $a \in S$ or a is an endpoint of S, and that the expression $f'(x)/g'(x)$ approaches a limit value λ as $x \to a$. Suppose that one of the following two conditions holds:

1. *$f(x) \to 0$ as $x \to a$ and $g(x) \to 0$ as $x \to a$.*
2. *$|g(x)| \to \infty$ as $x \to a$.*

Then we have

$$\lim_{x \to a} \frac{f(x)}{g(x)} = \lambda.$$

Note that both a and λ are permitted to be infinite. You can find a proof of L'Hôpital's rule by clicking on the icon  . In the following subsection we provide a proof of a special case of L'Hôpital's rule that contains many of the ingredients of the main theorem.

9.6.4   A Simple Form of L'Hôpital's Rule

Suppose that c and a are real numbers satisfying $c < a$, that f and g are differentiable functions on the interval (c, a), and that the expression $f'(x)/g'(x)$ approaches a limit value λ as $x \to a$. Suppose that

$$\lim_{x \to a} f(x) = \lim_{x \to a} g(x) = 0.$$

Then we have

$$\lim_{x \to a} \frac{f(x)}{g(x)} = \lambda.$$

Proof. The fact that $f'(x)/g'(x) \to \lambda$ as $x \to a$ guarantees that $g'(x) \neq 0$ whenever x is close enough to a. Choose a number $b < a$ such that $g'(x) \neq 0$ whenever $b < x < a$.

We define $f(a) = g(a) = 0$ and observe that the functions f and g are continuous on the interval $[b, a]$. Now, to prove that $f(x)/g(x) \to \lambda$ as $x \to a$, suppose that (x_n) is a sequence in the interval (b, a) and that $x_n \to a$ as $n \to \infty$. We need to show that $f(x_n)/g(x_n) \to \lambda$ as $n \to \infty$. For each n, we apply the Cauchy mean value theorem (Theorem 9.4.4) to the functions f and g on the interval $[x_n, .a]$ to choose a number $t_n \in (x_n, a)$ such that

$$\frac{f'(t_n)}{g'(t_n)} = \frac{f(x_n) - f(a)}{g(x_n) - g(a)}.$$

Since $t_n \to a$ as $n \to \infty$, we see that

$$\frac{f(x_n)}{g(x_n)} = \frac{f(x_n) - f(a)}{g(x_n) - g(a)} = \frac{f'(t_n)}{g'(t_n)} \to \lambda$$

as $n \to \infty$, as required. ∎

9.6.5   Some Exercises on Indeterminate Forms

1. **H** **N** Evaluate each of the following limits. In each case, use *Scientific Notebook* to verify that your limit value is correct.

$$\lim_{x \to 0} \left(\frac{x - \sin x}{x^3} \right) \qquad\qquad \lim_{x \to 0} \left(\frac{\tan x - x}{x - \sin x} \right)$$

$$\lim_{x \to 0} \left(\frac{\tan x - \sin x}{x^3} \right) \qquad\qquad \lim_{x \to 0} \left(\frac{\log (1 + x)}{x} \right)$$

$$\lim_{x \to 0} \left((1 + x)^{1/x} \right) \qquad\qquad \lim_{x \to 0} \left(\frac{e - (1 + x)^{1/x}}{x} \right)$$

$$\lim_{x \to \infty} \left(\frac{x^{100}}{\exp\left[(\log x)^2 \right]} \right) \qquad\qquad \lim_{x \to \pi/2} (\sin x)^{\tan x}$$

$$\lim_{x \to \infty} \left(x^{\frac{\log x}{x}} \right) \qquad\qquad \lim_{x \to \infty} \frac{x^{\log x}}{(x + 1)^{\log(x+1)}}$$

$$\lim_{x \to 0} \left(\frac{e^x - \log (1 + x) - 1}{x^2 (x + 2)} \right) \qquad\qquad \lim_{x \to \infty} \left(\sqrt[4]{x^4 - 5x^3 + 8x^2 - 2x + 1} - x \right)$$

2. **H** Given that $\alpha > 0$, evaluate the limit

$$\lim_{x \to \infty} x \left(\frac{(2x + 2)^\alpha - (2x + 1)^\alpha}{(2x + 2)^\alpha} \right).$$

3. **N** Evaluate the limits

$$\lim_{x \to \infty} \left((\log (x + 1))^\alpha - (\log x)^\alpha \right)$$

$$\lim_{x \to \infty} \left(\frac{\log (x + 1)}{\log x} \right)^x \qquad\qquad \lim_{x \to \infty} \left(\frac{\log (x + 1)}{\log x} \right)^{x \log x}.$$

For the latter two limits, check the limit value with *Scientific Notebook*.

4. (a) Prove that if $0 < a < b$ and $x \geq 1$, then

$$xa^{x-1} \leq \frac{b^x - a^x}{b - a} \leq xb^{x-1}.$$

(b) ▉ Prove that

$$\lim_{x\to\infty} \left((\log (x+1))^{\log x} - (\log x)^{\log x} \right) = \infty.$$

(c) Prove that

$$\lim_{x\to\infty} \left((\log (x+1))^{\log \log x} - (\log x)^{\log \log x} \right) = 0.$$

5. ▉ Prove that if q is any given number, then

$$\lim_{x\to\infty} x^q \left(1 - \frac{q \log x}{x} \right)^x = 1.$$

6. ▉ Suppose that we want to evaluate the limit

$$\lim_{x\to\infty} \frac{3}{1+x+x^2}.$$

Would it be correct to use L'Hôpital's rule by taking $f(x) = 3$ and $g(x) = 1 + x + x^2$ for every x and then arguing that

$$\lim_{x\to\infty} \frac{3}{1+x+x^2} = \lim_{x\to\infty} \frac{f(x)}{g(x)} = \lim_{x\to\infty} \frac{f'(x)}{g'(x)} = \lim_{x\to\infty} \frac{0}{1+2x} = 0?$$

The answer is certainly correct, but is the reasoning correct? Have we made a valid use of L'Hôpital's rule?

7. Should L'Hôpital's rule be used to evaluate the limits

$$\lim_{x\to\infty} \frac{e^{\sin x}}{x} \quad \text{and} \quad \lim_{x\to\infty} \frac{x e^{\sin x}}{\log x}?$$

Chapter 10
The Exponential and Logarithmic Functions

10.1   The Purpose of This Chapter

Although the exponential and logarithmic functions have appeared fairly extensively in the preceding chapters, their appearance was restricted to illustrative examples. They were never part of the central narrative and were never used in the development of the theory. However, as you continue your reading of this text, the role played by these functions will begin to change, and, eventually, they will become part of the fabric of what we are studying.

The purpose of this chapter is to place the exponential and logarithmic functions on a sound footing. Up until now we have not seen a precise definition of expressions of the type a^x or $\log_a x$. If you look back at the axioms for the real number system $\mathbf{R}$ that appeared in Section 5.2, you will see that they contain nothing at all about exponential expressions. Thus, we need to supply a definition for the expression a^x. We need to establish the familiar identities such as

$$a^t a^x = a^{t+x}.$$

We need to show that if a is any positive number that is unequal to 1, and if

$$f(x) = a^x$$

for every real number x, then the function f is a strictly monotone differentiable function from $\mathbf{R}$ onto the interval $(0, \infty)$. It will then follow that the inverse function of this function f, which is written as $\log_a$, is a strictly monotone differentiable function from $(0, \infty)$ onto $\mathbf{R}$. As you know, the function $\log_a$ is called the logarithm with base a. Finally, we need to show that there is a positive number e that has the special property that if we define $\exp(x) = e^x$ for every number x, then

$$\exp'(x) = \exp x$$

for every number x and

$$\log_e'(x) = \frac{1}{x}$$

for every positive number x.

Some further remarks that are mainly of interest to instructors can be reached by clicking on the icon [image] .

If you wish, you may regard this chapter as being optional. If you decide to skip this chapter, then you should assume the familiar properties of the functions exp and log as they are listed in the preceding paragraphs.

10.2   Integer Exponents

This section is being provided for the sake of completeness. It should be read very rapidly and, perhaps, parts of it should be skipped altogether.

10.2.1   Positive Integer Exponents

If a is any number and n is any positive integer, then the expression a^n is, as usual, defined to be the product of a with itself n times. If a and b are any real numbers and m and n are positive integers, then the following rules hold:

1. $a^1 = a$.
2. $a^m a^n = a^{m+n}$.
3. $a^n b^n = (ab)^n$.
4. $(a^m)^n = a^{mn}$.
5. If we define $f(x) = x^n$ for every real number $x \geq 0$, then the function f is a strictly increasing continuous function from the interval $[0, \infty)$ onto $[0, \infty)$.

We recall from Subsection 7.7.2 that if $c > 1$ and if we define $x_n = c^n$ for every positive integer n, then the geometric sequence (x_n) is strictly increasing and $x_n \to \infty$ as $n \to \infty$. On the other hand, if $-1 < c < 1$, then the sequence (x_n) converges to zero.

10.2.2   General Integer Exponents

Suppose that a is any nonzero number and that n is an integer. In the event that $n > 0$, the expression a^n has already been defined. We now define $a^0 = 1$ and, in the event that $n < 0$, we define

$$a^n = \frac{1}{a^{-n}}.$$

We see easily that if a and b are nonzero numbers and m and n are any integers, then the first four equations that appear in Subsection 10.2.1 must hold. The fifth statement also has an analog for negative exponents: If n is a negative integer, and if we define $f(x) = x^n$ for every real number $x > 0$, then the function f is a strictly decreasing continuous function from the interval $(0, \infty)$ onto $(0, \infty)$.

10.3 Rational Exponents

In this section we introduce rational exponents and we extend the rules for exponents that we saw for integers in Section 10.2 to include rational exponents. We begin by looking at an elementary fact about radicals upon which the definition of rational exponents is based.

10.3.1 An Elementary Fact about Radicals

Suppose that $a > 0$; that $m, n, p,$ and q are integers; and that the integers n and q are positive. Suppose finally that $m/n = p/q$. Then we must have

$$\left(\sqrt[n]{a}\right)^m = \sqrt[n]{a^m} = \left(\sqrt[q]{a}\right)^p = \sqrt[q]{(a^p)}.$$

Proof. In view of the fact that the function f in Condition 5 of Subsection 10.2.1 is one-one, we can obtain the desired result by showing that

$$\left(\left(\sqrt[n]{a}\right)^m\right)^{nq} = \left(\sqrt[n]{a^m}\right)^{nq} = \left(\left(\sqrt[q]{a}\right)^p\right)^{nq} = \left(\sqrt[q]{(a^p)}\right)^{nq}.$$

It is easy to see that each of the preceding four numbers is equal to a^{mq}. For example, remembering that $np = mq$, we see that

$$\left(\left(\sqrt[n]{a}\right)^m\right)^{nq} = \left(\sqrt[n]{a}\right)^{mnq} = \left(\left(\sqrt[n]{a}\right)^n\right)^{mq} = a^{mq}$$

and

$$\left(\sqrt[q]{(a^p)}\right)^{nq} = \left(\left(\sqrt[q]{(a^p)}\right)^q\right)^n = (a^p)^n = a^{pn} = a^{mq}.$$

We leave the remaining two identities as an exercise. ■

10.3.2 Defining Rational Exponents

As you know, there are certain instances in which an expression a^x can exist when $a < 0$ even if x is not an integer. For example,

$$(-1)^{1/3} = \sqrt[3]{-1} = -1.$$

However, we shall ignore these special cases, and, for the purpose of defining an expression of the form a^x when x is rational, we shall require the number a to be positive.

Given any positive number a and any rational number x we define a^x by choosing two integers m and n such that $n > 0$ and $x = m/n$ and defining

$$a^x = \left(\sqrt[n]{a}\right)^m = \sqrt[n]{(a^m)}.$$

In view of the preceding technical lemma, this definition of a^x is not dependent on our choice of integers m and n satisfying the condition $x = m/n$.

10.3.3   The Laws of Rational Exponents

Suppose that a and b are positive numbers and that t and x are any rational numbers. Then the following conditions hold:

1. $a^t a^x = a^{t+x}$.
2. $a^x b^x = (ab)^x$.
3. $\left(a^t\right)^x = a^{tx}$.
4. *If $a > 1$ and $x > 0$, then $a^x > 1$.*
5. *If $a > 1$ and $t < x$, then $a^t < a^x$.*

We leave the proof of this theorem as an exercise.

10.3.4   A Continuity Theorem for Constant Rational Exponents

Suppose that c is any rational number and that

$$f(x) = x^c$$

for every positive real number x. Then the function f is continuous on the interval $(0, \infty)$.

Proof. Choose integers m and n such that $n > 0$ and $c = m/n$. Since the function g defined by the equation

$$g(x) = x^n$$

for all $x > 0$ is strictly increasing and continuous from $(0, \infty)$ onto $(0, \infty)$, it follows from a form of the inverse function theorem (Theorem 8.11.4) that the inverse function g^{-1} of g is continuous. Since

$$f(x) = \left(g^{-1}(x)\right)^m$$

for every number $x > 0$, the continuity of f follows at once from the composition theorem for continuity (Theorem 8.7.9). ∎

10.3.5   The Behavior of nth Roots as $n \to \infty$

If a is any positive number, then

$$\lim_{n \to \infty} a^{1/n} = 1.$$

Proof.[34] We shall prove the theorem for the case $a \geq 1$. Once this has been

---

[34]   As you may have noticed, we studied much more interesting limits than this one in Section 9.6, and if we could use the techniques of that section here, the present theorem would be trivial. However, in Section 9.6, we made extensive use of exponential and logarithmic functions, and so we cannot use that material here as we develop the theory of those functions.

done, the case $a < 1$ will follow from the identity

$$a^{1/n} = \frac{1}{\left(a^{-1}\right)^{1/n}}.$$

Suppose that $a \geq 1$. From part 5 of Theorem 10.3.3 we know that the sequence $\left(a^{1/n}\right)$ is decreasing and it therefore follows from the monotone sequence theorem (Theorem 7.7.1) that this sequence converges to its infimum, which we shall write as b. Since the number 1 is a lower bound of the sequence $\left(a^{1/n}\right)$ we know that $b \geq 1$. However, if b were greater than 1, it would follow from Theorem 7.7.2 that $b^n \to \infty$ as $n \to \infty$, contradicting the fact that $b^n \leq a$ for every n. Thus $b = 1$ and the proof is complete. ■

10.3.6   Exponential Function with Rational Domain

Suppose that $a > 1$ and that

$$f(x) = a^x$$

for every rational number x.

1. *The function f is strictly increasing on the set $\mathbf{Q}$ of rational numbers.*
2. *If x is any real number, we have*

$$\sup\left\{f(s) \mid s \in \mathbf{Q} \text{ and } s < x\right\} = \inf\left\{f(t) \mid t \in \mathbf{Q} \text{ and } x < t\right\}.$$

3. *The function f is continuous on the set $\mathbf{Q}$ of rational numbers.*

Proof. The fact that f is strictly increasing follows at once from part 5 of Theorem 10.3.3. Now suppose that x is any real number. In view of one of the forms of the sandwich theorem (Theorem 7.4.7) we can prove the assertion 2 by finding sequences (s_n) and (t_n) of rational numbers such that $s_n < x < t_n$ for each n and such that

$$\lim_{n \to \infty} \left(f(t_n) - f(s_n)\right) = 0.$$

For this purpose we choose rational numbers s_n and t_n for each n such that

$$s_n < x < t_n \quad \text{and} \quad t_n - s_n < \frac{1}{n}.$$

For each n we have

$$0 < f(t_n) - f(s_n) = f(s_n)\left[f(t_n - s_n) - 1\right] \leq f(t_1)\left(a^{1/n} - 1\right).$$

Since the latter expression approaches 0 as $n \to \infty$, it follows that

$$\lim_{n \to \infty} \left(f(t_n) - f(s_n)\right) = 0.$$

We observe finally that, in view of Theorem 8.10.8, the fact that f is continuous at every rational number follows from assertion 2. ∎

10.4　Real Exponents

In this section we introduce real exponents and we extend the rules for exponents to include all real exponents. We shall also show that whenever a is a positive number unequal to 1, the exponential function base a is strictly monotone and continuous, and we shall introduce the function $\log_a$.

10.4.1　Definition of Real Exponents

Given any real number $a > 1$ and any number x we define

$$a^x = \sup \left\{ a^s \mid s \in \mathbf{Q} \text{ and } s < x \right\} = \inf \left\{ a^t \mid t \in \mathbf{Q} \text{ and } x < t \right\}.$$

In view of Theorem 10.3.6, this definition of a^x does not conflict with the meaning of a^x when x is rational.

If a is a positive real number and $a < 1$, we define

$$a^x = \left(\frac{1}{a} \right)^{-x}.$$

Finally, if $a = 1$, then we define $a^x = 1$ for every real number x.

10.4.2　Continuity of the Exponential Function

Suppose that a is any positive number unequal to 1 and that we have defined

$$f(x) = a^x$$

for every real number x. Then the function f is a strictly monotone continuous function from $\mathbf{R}$ onto the interval $(0, \infty)$.

Proof. Suppose first that $a > 1$. It is clear that f is a strictly increasing function on the set $\mathbf{R}$, and from Theorems 10.3.6 and 8.10.8 we see that f is continuous at every real number. Since the range of f is an interval and $\inf f = 0$ and $\sup f = \infty$, we conclude that the range of f is the interval $(0, \infty)$.

Now that we know that the theorem holds for $a > 1$, we can deduce it at once for the case $0 < a < 1$ by using the identity

$$a^x = \left(\frac{1}{a} \right)^{-x}. \quad ∎$$

10.4.3   Logarithms

Given any positive number $a \neq 1$, the function $\log_a$, which is called the logarithm with base a, is defined to be the inverse function of the function f that is defined by the equation

$$f(x) = a^x$$

for all $x \in \mathbf{R}$. Thus if $x = a^u$, then $u = \log_a x$.

From a form of the inverse function theorem (Theorem 8.11.4) we deduce that the function $\log_a$ is a strictly monotone continuous function from the interval $(0, \infty)$ onto $\mathbf{R}$.

10.4.4   The Laws of Real Exponents

Suppose that a and b are positive numbers and that t and x are any real numbers. Then the following conditions hold:

1. $a^t a^x = a^{t+x}$.
2. $a^x b^x = (ab)^x$.
3. $(a^t)^x = a^{tx}$.

Proof. Choose sequences (t_n) and (x_n) of rational numbers such that $t_n \to t$ and $x_n \to x$ as $n \to \infty$. From the continuity of the exponential function with base a we see that

$$a^t a^x = \lim_{n \to \infty} \left(a^{t_n} a^{x_n} \right) = \lim_{n \to \infty} \left(a^{t_n + x_n} \right) = a^{t+x}.$$

In the same way we see that

$$a^x b^x = \lim_{n \to \infty} \left(a^{x_n} b^{x_n} \right) = \lim_{n \to \infty} \left((ab)^{x_n} \right) = (ab)^x.$$

To prove the third assertion we observe first that, if n is any positive integer, then it follows from Theorem 10.3.4, the rules for rational exponents, and the continuity of the exponential function that

$$\left(a^t \right)^{x_n} = \lim_{m \to \infty} \left(a^{t_m} \right)^{x_n} = \lim_{m \to \infty} \left(a^{t_m x_n} \right) = a^{t x_n}.$$

Using the continuity of the exponential function again we obtain

$$\left(a^t \right)^x = \lim_{n \to \infty} \left(a^t \right)^{x_n} = \lim_{n \to \infty} \left(a^{t x_n} \right) = a^{tx}. \quad \blacksquare$$

10.4.5   The Laws of Logarithms

Now that we have established a proper foundation for the exponential function, the laws of logarithms can be proved by the same methods that are used in a precalculus course.

Suppose that a and b are positive numbers unequal to 1, that t and x are positive numbers and that c is any real number. Then the following conditions hold:

1. $\log_a (tx) = \log_a t + \log_a x$.

2. $\log_a \left(\dfrac{t}{x} \right) = \log_a t - \log_a x$.

3. $\log_a (x^c) = c \left(\log_a x \right)$

4. $\log_a x = \dfrac{\log_b x}{\log_b a}$.

5. $a^x = b^{x \log_b a}$.

You can reach a proof of these assertions by clicking on the icon .

10.5　Differentiating the Exponential Function: Intuitive Approach

A movie preview of this section can be seen by clicking on the link .

10.5.1　Introduction

If a is a positive number and $f(x) = a^x$ for every real number x, then the problem of finding the derivative $f'(x)$ of the function f at a given number x is the problem of finding the limit

$$\lim_{h \to 0} \frac{f(x + h) - f(x)}{h} = \lim_{h \to 0} \frac{a^{x+h} - a^x}{h} = \lim_{h \to 0} a^x \left(\frac{a^h - 1}{h} \right).$$

Thus the problem of differentiating the exponential function with base a boils down to the problem of finding the limit

$$\lim_{h \to 0} \frac{a^h - 1}{h}.$$

If we call this limit $\phi(a)$, then for every real number x we have

$$f'(x) = \lim_{h \to 0} a^x \left(\frac{a^h - 1}{h} \right) = a^x \lim_{h \to 0} \left(\frac{a^h - 1}{h} \right) = \phi(a)a^x.$$

In particular, $f'(0) = \phi(a)$ so that $\phi(a)$ is the slope of the graph $y = a^x$ at the point $(0, 1)$, as illustrated in Figure 10.1.

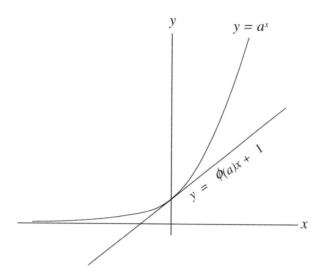

Figure 10.1

In this section we discuss the limit $\phi(a)$ intuitively. In the next section we shall see a rigorous proof that the limit exists, and we shall also see that somewhere between 2 and 3 there is a value of a for which the number $\phi(a) = 1$. This special value of a is called e. What is important about the number e is that if $f(x) = e^x$ for every number x, then given any number x we have

$$f'(x) = \phi(e)e^x = e^x.$$

The remainder of this section contains some important interactive material that should best be read in the on-screen version of this book.

10.5.2   A Numerical Approach to the Limit $\phi(a)$

The on-screen version of this book provides you with the opportunity of motivating the limit $\phi(a)$ by asking *Scientific Notebook* to evaluate the expression

$$\frac{a^h - 1}{h}$$

numerically for some chosen value of a and a variety of values of h that are close to 0. For example, the fact that

$$\frac{2^{.01} - 1}{.01} = .69556$$

tells us that $\phi(2)$ is approximately 0.696. Since

$$\frac{2^{.0001} - 1}{.0001} = .69317,$$

we see that a better approximation to $\phi(2)$ is 0.693. You will use the *formula* feature of *Scientific Notebook* to explore expressions of this type interactively.

10.5.3   A Graphical Approach to the Limit $\phi(a)$

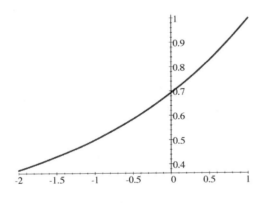   The on-screen version of this book provides you with the opportunity of drawing a graph of the form

$$y = \frac{a^x - 1}{x}$$

for a chosen value of a and then to watch the graph change automatically as the value of a changes. For example, when $a = 2$, the graph becomes

$$y = \frac{2^x - 1}{x}$$

and is illustrated in Figure 10.2. As you can see, this graph meets the y-axis just

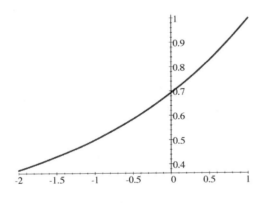

Figure 10.2

below the point $(0, 0.7)$ and illustrates the fact that the limit

$$\phi(2) = \lim_{x \to 0} \frac{2^x - 1}{x}$$

is a little below 0.7. By zooming into your graph and adjusting the value of a you will be able to see the behavior of the quantity $\phi(a)$ as a varies and you will be able to search for a value of a for which $\phi(a) = 1$.

Finally, you will be invited to click on the link [image] that will play a sound movie that further illustrates the interactive exercise in which you have engaged.

10.5.4   Another Graphical View of the Limit

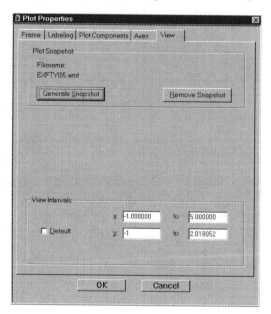 In the on-screen version of the text you will be invited to supply the definition

$$\phi(a) = \lim_{h \to 0} \frac{a^h - 1}{h}$$

to *Scientific Notebook* and then to draw the graph of the function ϕ. By choosing the domain interval as $[-1, 5]$ and overriding the default setting in the plot properties dialogue box to prevent y from being less than -1 (see Figure 10.3) you

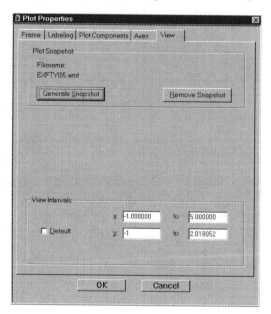

Figure 10.3

will make your graph appear as in Figure 10.4. This figure helps us understand that there must be a value of a between 2 and 3 for which $\phi(a) = 1$.

10.6   Differentiating the Exponential Function: Rigorous Approach

In this section we show that if a is a positive number, then the limit $\phi(a)$ defined in Section 10.5 does indeed exist and that there is precisely one positive number

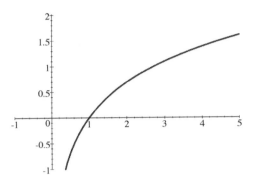

Figure 10.4

a for which $\phi(a) = 1$. In order to establish the existence of the limit

$$\lim_{h \to 0} \frac{a^h - 1}{h},$$

we shall show first that if we define

$$g(x) = \frac{a^x - 1}{x},$$

then the function g is increasing. The existence of the one-sided limits of g will then follow from Theorem 8.10.8, and all we shall have to do is compare the limits from the left and right at 0.

We begin with a technical lemma that will help us to prove that the function g is increasing.

10.6.1   A Useful Inequality

If $a > 0$ and n is a positive integer, then

$$\frac{a^{n+1} - 1}{n + 1} \geq \frac{a^n - 1}{n}.$$

Proof. Suppose that $a > 0$ and that n is a positive integer. The desired inequality is equivalent to the assertion that

$$na^{n+1} - (n + 1)\, a^n + 1 \geq 0.$$

For each positive integer n we define

$$x_n = na^{n+1} - (n + 1)\, a^n + 1.$$

We need to see that $x_n \geq 0$ for each n. For this purpose we note the obvious fact that $x_1 \geq 0$ and we can see that the sequence (x_n) is increasing by observing for each n that

$$x_{n+1} - x_n = (n+1)\, a^n \,(a-1)^2 \geq 0.$$

Therefore $x_n \geq 0$ for each n. ∎

10.6.2   An Extension of the Preceding Inequality

Suppose that $a > 0$, that t and x are positive rational numbers, and that $t \leq x$. Then we have

$$\frac{a^t - 1}{t} \leq \frac{a^x - 1}{x}.$$

Proof. Choose positive integers m, n, and q such that $t = m/q$ and $x = n/q$, and define $b = a^{1/q}$. The desired inequality is equivalent to the assertion that

$$\frac{b^m - 1}{m} \leq \frac{b^n - 1}{n},$$

which is clear, in view of Lemma 10.6.1. ∎

10.6.3   A Further Extension of the Inequality

Suppose that $a > 0$, that t and x are positive real numbers, and that $t \leq x$. Then we have

$$\frac{a^t - 1}{t} \leq \frac{a^x - 1}{x}.$$

Proof. Choose sequences (t_n) and (x_n) of rational numbers such that $0 < t_n < t$ and $x < x_n$ for each n and such that $t_n \to t$ and $x_n \to x$ as $n \to \infty$. We observe that

$$\frac{a^t - 1}{t} = \lim_{n \to \infty} \frac{a^{t_n} - 1}{t_n} \leq \lim_{n \to \infty} \frac{a^{x_n} - 1}{x_n} = \frac{a^x - 1}{x}. \quad ∎$$

10.6.4   Exponential Functions are Differentiable

Suppose that $a > 0$.

1. *The limit*

$$\phi(a) = \lim_{h \to 0} \frac{a^h - 1}{h}$$

   exists.

2. *If we define $f(x) = a^x$ for every real number x, then for every number x we have*

$$f'(x) = \phi(a)a^x.$$

3. *Unless $a = 1$ we have $\phi(a) \neq 0$.*

Proof. The existence of the one-sided limit

$$\lim_{h \to 0+} \frac{a^h - 1}{h}$$

follows at once from Theorems 10.6.3 and 8.10.8. We write this one-sided limit as $\phi(a)$. Now, given any number $h < 0$, since

$$\frac{a^h - 1}{h} = a^h \left(\frac{a^{-h} - 1}{-h} \right)$$

and since the exponential function f is continuous at 0 we see that

$$\lim_{h \to 0-} \frac{a^h - 1}{h} = \lim_{h \to 0-} a^h \left(\frac{a^{-h} - 1}{-h} \right) = 1\phi(a) = \phi(a).$$

Thus the number $\phi(a)$ is the two-sided limit:

$$\phi(a) = \lim_{h \to 0} \frac{a^h - 1}{h}$$

and we have proved assertion 1. Assertion 2 follows at once from assertion 1. Finally, if $a \neq 1$, then, because the function f is not constant, we see from assertion 2 that $\phi(a) \neq 0$. ∎

10.6.5   Natural Exponents and Logarithms
Now that we have shown that the exponential functions are differentiable, it is time to say a little more about the number $\phi(a)$ for any given positive number a. We shall begin by relating the number $\phi(a)$ to the number $\phi(2)$. If a is any positive number, then by using the chain rule to differentiate both sides of the equation

$$a^x = 2^{x \log_2 a}$$

we obtain

$$\phi(a)a^x = \phi(2) 2^{x \log_2 x} \log_2 a,$$

from which we conclude that

$$\phi(a) = \phi(2) \log_2 a.$$

Therefore the condition $\phi(a) = 1$ will hold if and only if $\phi(2) \log_2 a = 1$; in other words,

$$a = 2^{1/\phi(2)}.$$

Therefore there is precisely one positive number a for which $\phi(a) = 1$, and we define e to be this number. Since $\phi(e) = 1$ we know that if f is the exponential function with base e, then

$$f'(x) = 1e^x.$$

The exponential function with base e is called the **natural exponential function** and is written as exp. Thus

$$\exp'(x) = \exp(x)$$

for every real number x. The logarithm with base e is known as the natural logarithm and is written as log and also as ln . Note that if $x > 0$ and $t = \log x$, then, since $x = \exp(t)$, it follows from Theorem 9.3.9 that

$$\log' x = \frac{1}{\exp' t} = \frac{1}{\exp(t)} = \frac{1}{x}.$$

We have now derived all of the familiar identities on the calculus of the exponential and logarithmic functions that appear in an elementary calculus course.

10.6.6 Some Exercises on the Exponential and Logarithmic Functions

1. Given that a is a positive number and that $f(x) = a^x$ for every real number x, prove that

$$f'(x) = a^x \log a$$

for every number x. In other words, prove that the number we called $\phi(a)$ in the preceding sections is just $\log a$.

2. Given that a is a positive number and that $a \neq 1$, and given that $f(x) = \log_a x$ for every $x > 0$, prove that

$$f'(x) = \frac{1}{x (\log a)}$$

for every number $x > 0$.

3. Given that f and g are differentiable functions and that f is positive, use the fact that

$$f(x)^{g(x)} = \exp[g(x) (\log(f(x)))]$$

for each x to find a formula for the derivative of the function f^g.

4. Given that f is differentiable on $\mathbf{R}$, that $f(0) = 1$, and that $f' = f$, prove that the function $f/\exp$ is constant and then conclude that $f = \exp$.

5. Suppose that $f : \mathbf{R} \to \mathbf{R}$ and that $f''(x) = f(x)$ for every real number x.

   (a) Given that $g = f' + f$, prove that $g' = g$ and deduce that there exists a real number a such that $g'(x) = 2ae^x$ for every number x.

   (b) Given that $h(x) = f(x)e^x$ for every real number x, prove that the equation $h'(x) = 2ae^{2x}$ holds for every number x and deduce that there is there is a number b such that the equation

   $$f(x) = ae^x + be^{-x}$$

   holds for every real number x.

6. Suppose that $f : \mathbf{R} \to \mathbf{R}$ and that for all numbers t and x we have

   $$f(t + x) = f(t)f(x).$$

   (a) Prove that either $f(x) = 0$ for every real number x or $f(x) \neq 0$ for every real number x.

   (b) Prove that if f is not the constant zero, then $f(0) = 1$ and that $f(x) > 0$ for every number x.

   (c) Prove that if f is not the constant zero and if $a = f(1)$, then for every rational number x we have $f(x) = a^x$. Deduce that if f is continuous on the set $\mathbf{R}$, then the equation $f(x) = a^x$ holds for every real number x. Compare this exercise with the last few exercises in Subsection 8.7.10.

7. Suppose that α is a nonzero real number and that

   $$S = \{x \in \mathbf{R} \mid 1 + \alpha x > 0\}.$$

   (a) ▉ Prove that if

   $$g(x) = \alpha x - (1 + \alpha x) \log (1 + \alpha x)$$

   for all $x \in S$, then $g(x) < 0$ for every nonzero number $x \in S$.

   (b) ▉ Prove that if

   $$f(x) = \frac{\log (1 + \alpha x)}{x}$$

   for every nonzero number $x \in S$ and if $f(0) = \alpha$, then the function f is continuous and strictly decreasing on S. Deduce that the inequality $f(x) < \alpha$ holds for every positive number $x \in S$.

(c) **H** Prove that the inequality

$$(1 + \alpha x)^{1/x} < e^\alpha$$

holds for every positive number $x \in S$.

8. (a) By applying Theorem 9.5.6 to the function exp on the interval $[0, 1]$, show that if n is a positive integer, then there must exist a number $c \in (0, 1)$ such that

$$e = 1 + 1 + \frac{1}{2!} + \frac{1}{3!} + \cdots + \frac{1}{n!} + \frac{e^c}{(n+1)!}.$$

(b) Deduce that if n is a positive integer, then

$$0 < e - \sum_{j=0}^{n} \frac{1}{j!} < \frac{e}{(n+1)!}.$$

(c) By putting $n = 2$ in the latter inequality, prove that $e < 3$.

(d) Prove that if n is a positive integer, we have

$$0 < e - \sum_{j=0}^{n} \frac{1}{j!} < \frac{3}{(n+1)!}$$

and deduce that

$$\lim_{n \to \infty} \sum_{j=0}^{n} \frac{1}{j!} = e.$$

(e) **H** Prove that the number e is irrational.

9. **H** Prove that if (x_n) is a sequence of positive numbers and if

$$\lim_{n \to \infty} \frac{x_{n+1}}{x_n} = \alpha,$$

then

$$\lim_{n \to \infty} \sqrt[n]{x_n} = \alpha.$$

Deduce that

$$\lim_{n \to \infty} \frac{n}{\sqrt[n]{n!}} = e.$$

Chapter 11
The Riemann Integral

This chapter presents the theory of Riemann integration. If you would prefer to replace this chapter with a more extensive presentation that includes Riemann-Stieltjes integration, you can reach the alternative presentation from the on-screen version of this book by clicking on the icon ▉ .

11.1 Introduction to the Concept of an Integral

11.1.1 What Is an Integral?

When you studied elementary calculus, you acquired a good intuitive feel for the idea that an integral of the form

$$\int_a^b f(x)dx$$

is the "signed area" of the region illustrated in Figure 11.1 that lies between the lines $x = a$ and $x = b$, the x-axis, and the curve $y = f(x)$. As you know,

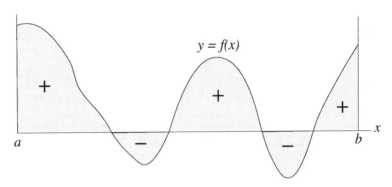

Figure 11.1

each portion of area above the x-axis is taken as positive and each portion of area below the x-axis is taken as negative.

However, even though you may have become quite skillful in the art of evaluating integrals and using them in mathematics, science, engineering, or economics, you have probably not yet seen a precise definition of an integral, and you have not seen how the theory of integration can be placed on a sound logical foundation. That is what we are going to do in this chapter. We shall explore the very meaning of the word *integral*. After a very brief look at the work of Eudoxus and Archimedes, we shall jump to the nineteenth century and study the theory of integration as it was developed by Cauchy, Darboux, and Riemann, and which has come to be known as Riemann integration. Perhaps we should mention that, although this theory of Riemann integration is logically sound and is able to serve many of the applications of mathematics to other disciplines, it is not all that mathematicians want it to be. There are several modern theories of integration that are more sophisticated and more satisfactory and which we shall mention briefly from time to time. But we shall leave their detailed study for a more advanced course.

11.1.2   What Is Area?

In view of the simple picture that we have just seen that describes an integral of the form $\int_a^b f(x)dx$ as being a combination of areas, you may, perhaps, wonder why we still have to provide a meaning for the idea of an integral. After all, if $\int_a^b f(x)dx$ stands for the signed area illustrated in Figure 11.1, then, on the face of it, it looks as though we have already given the integral a solid definition. But, in attempting to define an integral in terms of area we run into a snag: The definition requires us to have a prior understanding of the concept of area. Do we have such an understanding? The answer is *no!* The following philosophy sums up the approach to area that you may have found in your elementary mathematics courses:

> *Area is there! Area has a meaning that is absolute and that has nothing to do with what you or I may think it means. We don't need to define the area of a given region. We can see it.*

But this philosophy is quite wrong. We can't "see" area. All we can see is a piece of paper. Until we have given a clear definition of the concept of area, it is by no means clear what we mean by the area shown in Figure 11.1.

In the simplest situations, we can assign a meaning to the idea of area by using a concept of "paving". If, for example, we decide to measure length in feet, then the unit of area is a square foot, which is the area of a square with a side of 1 foot. A given plane region can be said to have an area of a certain number of square feet if we can "pave" that region with paving stones, each of which is a square with a side of 1 foot. Thus, for example, if we have a rectangle with a length of 4 feet and a width of 3 feet as illustrated in Figure 11.2, then, because the rectangle can be paved with 12 one-by-one squares, the rectangle has an area of 12 square feet. This example motivates the well-known formula for the area of

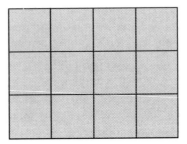

Figure 11.2

a rectangle:

$$\text{area of a rectangle} = \text{its length times its width.}$$

However, even for rectangles, this simple "paving" approach to area is too restrictive. Suppose, for example, that we have a rectangle with a length of $\sqrt{13} + 1$ and a width of $\sqrt{13} - 1$, as illustrated in Figure 11.3. If we believe that the area

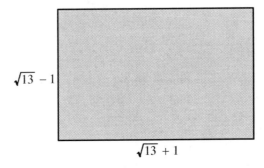

Figure 11.3

of this rectangle should be its length times its width, then this area must be

$$\left(\sqrt{13} + 1\right)\left(\sqrt{13} - 1\right) = 13 - 1 = 12.$$

But this rectangle cannot be paved with one-by-one squares. In fact, it cannot be paved with equal squares of *any* size. To see why such paving is impossible, suppose that we have managed, somehow, to pave the $\sqrt{13} + 1$ by $\sqrt{13} - 1$ rectangle with m rows and n columns of squares of a given fixed size, as illustrated in Figure 11.4, where m and n are integers. We see at once that this supposition

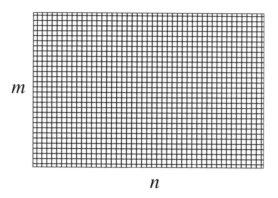

Figure 11.4

leads to a contradiction because it implies that

$$\frac{m}{n} = \frac{\sqrt{13} - 1}{\sqrt{13} + 1} = \frac{\left(\sqrt{13} - 1\right)\left(\sqrt{13} - 1\right)}{\left(\sqrt{13} + 1\right)\left(\sqrt{13} - 1\right)} = \frac{7}{6} - \frac{1}{6}\sqrt{13},$$

which is impossible because the latter number is irrational.

Thus the notion of area is more complicated than one might think. The simple "paving" approach fails even if we confine ourselves to rectangles which, from the standpoint of area, are the simplest of all geometric figures. If we extend our discussion to more complicated geometric figures, the problem becomes harder still. As you know from elementary calculus, the area of the region illustrated in Figure 11.5 is

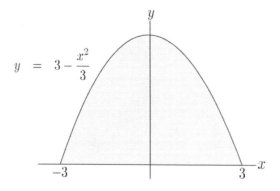

Figure 11.5

$$\int_{-3}^{3} \left(3 - \frac{x^2}{3} \right) dx = 12,$$

but there is no way in which this region with a curved boundary can be paved with squares.

For a slightly more extensive discussion of the area concept, click on the icon ![icon] .

11.1.3   The Simplest Type of Integral

In the light of the examples that appear in Subsection 11.1.2, we have to accept that the notion of area is anything but simple. We shall agree to *define* the area of a rectangle as its length times its width, but we are clearly still a long way from understanding what we mean by an integral, even of the easiest kind of function. Quadratic functions are out of the question, and even sloping straight lines are still beyond our reach. In fact, the only kind of function that seems to have an "obvious" integral at this stage is a function like the one illustrated in Figure 11.6. Functions of this type are known as **step functions** and will be introduced

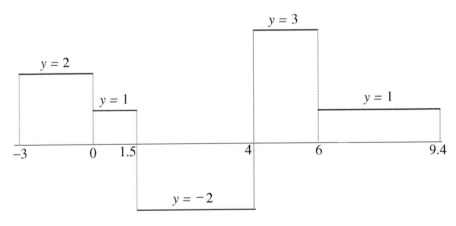

Figure 11.6

precisely in the next section. If we write the function whose graph appears in Figure 11.6 as f, then we should expect that

$$\int_{-3}^{9.4} f(x)dx = (3)(2) + (1.5)(1) + (2.5)(-2) + (2)(3) + (3.4)(1) = 11.9.$$

Notice that it shouldn't matter how the function f is defined at the four numbers where its graph jumps.

Notice also that the process of integrating a step function is *pure algebra*, because all we need are the operations of arithmetic: addition, subtraction, multiplication, and division. We do not require such notions as *limit, supremum,* or *infimum,* which constitute the bread and butter of calculus. These limit notions begin to appear, however, the moment we want to extend the theory of integration to a wider class of functions. There are several different ways of extending the theory to a wider class of functions, and each of these ways yields its own theory of integration. Thus, every theory of integration contains the theory of integration of step functions at its core and then extends this theory to a wider class of functions by some sort of limit process as a function is approximated ever more closely by step functions. The way in which different theories of integration differ from one another is precisely how the approximation of a given function by step functions takes place.

We therefore begin this chapter with some sections on the theory of integration of step functions. Hopefully, you should be able to proceed relatively quickly through these sections to Section 11.5, where we begin our actual study of the Riemann integral.

11.2 Partitions and Step Functions

11.2.1 Definition of a Partition

- Suppose that a and b are real numbers and that $a \leq b$. A **partition** $\mathcal{P}$ of the interval $[a, b]$ is a finite strictly increasing sequence $(x_0, x_1, \cdots, x_n)$ as illustrated in Figure 11.7, where n is a nonnegative integer, $x_0 = a$, and $x_n = b$. If $\mathcal{P}$ is the partition $(x_0, x_1, \cdots, x_n)$ of the interval $[a, b]$, then the

Figure 11.7

numbers $x_0, x_1, \cdots, x_n$ are called the **points**, and, for $j = 1, 2, \cdots, n$, the intervals $[x_{j-1}, x_j]$ and (x_{j-1}, x_j) are respectively called the **closed intervals** and the **open intervals** of the partition $\mathcal{P}$.
- A partition $\mathcal{Q}$ of an interval $[a, b]$ is said to be a **refinement** of a partition $\mathcal{P}$ if every point of $\mathcal{P}$ is also a point of $\mathcal{Q}$. If $\mathcal{P}$ and $\mathcal{Q}$ are any partitions of an interval $[a, b]$, then the **common refinement** is the partition whose points are the numbers that are either points of $\mathcal{P}$ or points of $\mathcal{Q}$. The common refinement of partitions $\mathcal{P}$ and $\mathcal{Q}$ is written as $\mathcal{P} \cup \mathcal{Q}$, even though the latter expression is not a union in quite the usual sense. Remember that a partition

is not merely the set of its points; it is a finite sequence running from its least point to its greatest.

- If $\mathcal{P}$ is the partition $(x_0, x_1, \cdots, x_n)$ of an interval $[a, b]$, then the largest of the lengths of the intervals of $\mathcal{P}$ is called the **mesh** of P and is written as $\|\mathcal{P}\|$.

- The intervals of a partition do not have to have the same length, but, when they do, we call the partition **regular**. If $\mathcal{P}$ is the partition $(x_0, x_1, \cdots, x_n)$ of $[a, b]$ and $\mathcal{P}$ is regular, then we say that $\mathcal{P}$ is the **regular n-partition** of the interval $[a, b]$, and in this case we have

$$x_j = a + \frac{j(b-a)}{n}$$

for each $j = 0, 1, \cdots, n$. Note that if $\mathcal{P}$ is the regular n-partition of $[a, b]$, then

$$\|\mathcal{P}\| = \frac{b-a}{n}.$$

11.2.2   Some Examples of Partitions

1. The regular 2-partition of $[0, 1]$ is the sequence $\left(0, \frac{1}{2}, 1\right)$. The regular 3-partition of $[0, 1]$ is the partition $\left(0, \frac{1}{3}, \frac{2}{3}, 1\right)$. The common refinement of these two partitions is the partition

$$\left(0, \frac{1}{3}, \frac{1}{2}, \frac{2}{3}, 1\right).$$

   The regular 6- partition of $[0, 1]$ is the partition

$$\left(0, \frac{1}{6}, \frac{1}{3}, \frac{1}{2}, \frac{2}{3}, \frac{5}{6}, 1\right).$$

   We illustrate these partitions in Figure 11.8.

2. If $a = b$, then the only partition of the interval $[a, b]$ is the regular 0-partition (a) that has one point and no intervals.

3. If n is a positive integer,, then the regular 2^{n+1}-partition of an interval is a refinement of the regular 2^n-partition and has one-half the mesh.

11.2.3   Definition of a Step Function

Suppose that

$$\mathcal{P} = (x_0, x_1, \cdots, x_n)$$

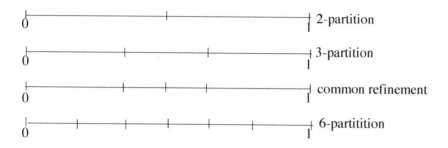

Figure 11.8

is a partition of a given interval $[a, b]$ and that f is a function defined on $[a, b]$. We say that the function f **steps within the partition** $\mathcal{P}$ if f is constant in each of the open intervals (x_{j-1}, x_j) of $\mathcal{P}$.

Note that a function f that steps within a partition $\mathcal{P}$ can take any values it likes at the points of $\mathcal{P}$. But between any two consecutive points of $\mathcal{P}$, the function f must be constant. Note also that if a function f steps within a given partition $\mathcal{P}$, then f will certainly step within any refinement of $\mathcal{P}$.

A function f defined on an interval $[a, b]$ is said to be a **step function on the interval** $[a, b]$ if it is possible to find a partition $\mathcal{P}$ of $[a, b]$ such that f steps within $\mathcal{P}$.

Thus if f is a step function on an interval $[a, b]$, then there exists a partition

$$\mathcal{P} = (x_0, x_1, \cdots, x_n)$$

of $[a, b]$, and, for each $j = 1, 2, \cdots, n$, there exists a number α_j such that f takes the constant value α_j on the open interval (x_{j-1}, x_j).

In addition to defining a step function on a particular interval, we also speak of a **step function**. A function f defined on the set $\mathbf{R}$ of real numbers is said to be a **step function** if it is possible to find an interval $[a, b]$ and a partition $\mathcal{P}$ of $[a, b]$ such that f steps within $\mathcal{P}$ and such that $f(x) = 0$ for every number

$x \in \mathbf{R} \setminus [a, b]$. For example, if

$$
f(x) = \begin{cases}
0 & \text{if } x < 1 \\
1 & \text{if } 1 \le x \le 2 \\
-1 & \text{if } 2 < x < 4 \\
2 & \text{if } 4 \le x < 5 \\
3 & \text{if } 5 \le x < 7 \\
0 & \text{if } x \ge 7,
\end{cases}
$$

then f is a step function. The graph of this function is illustrated in Figure 11.9.

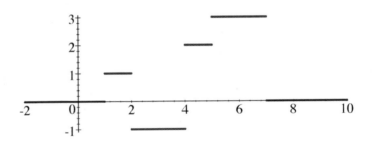

Figure 11.9

11.2.4   Some Exercises on Step Functions

1. ▦ True or false? If f is a step function on an interval $[a, b]$ and $[c, d]$ is a subinterval of $[a, b]$, then f is a step function on $[c, d]$.

2. ▦ True or false? If f is a step function, then, given any interval $[a, b]$, the function f is a step function on $[a, b]$.

3. ▦ Give an example of a step function on the interval $[0, 2]$ that does not step within any regular partition of $[0, 2]$.

4. Explain why a step function must always be bounded.

5. ▦ Prove that if f and g are step functions on an interval $[a, b]$, then so are their sum $f + g$ and their product fg.

6. Prove that if f and g are step functions, then so are their sum $f + g$ and their product fg.

7. Prove that a continuous step function on an interval must be constant on that interval.

11.3   Integration of Step Functions

11.3.1   The Sum of a Step Function over a Partition

Suppose that a function f defined on an interval $[a, b]$ steps within a partition

$$\mathcal{P} = (x_0, x_1, \cdots, x_n)$$

of $[a, b]$, taking the constant value α_j in each open interval (x_{j-1}, x_j). The **sum** $\Sigma(\mathcal{P}, f)$ of f over the partition $\mathcal{P}$ is defined by the equation

$$\Sigma(\mathcal{P}, f) = \sum_{j=1}^{n} \alpha_j (x_j - x_{j-1}).$$

So, for example, if f is the function defined on the interval $[-2, 10]$ whose graph is shown in Figure 11.9, and if

$$\mathcal{P} = (-2, 1, 2, 4, 5, 7, 10),$$

then $\Sigma(\mathcal{P}, f) =$

$$0(1 - (-2)) + 1(2 - 1) + (-1)(4 - 2) + 2(5 - 4) + 3(7 - 5) + 0(10 - 7).$$

Now suppose that

$$\mathcal{Q} = (-2, -1, 0, 1, 2, 3, 4, 5, 6, 7, 8, 9, 10).$$

The partition $\mathcal{Q}$ is a refinement of the partition $\mathcal{P}$ and, of course, f steps within $\mathcal{Q}$. Now write out the summation $\Sigma(\mathcal{Q}, f)$ and convince yourself that

$$\Sigma(\mathcal{Q}, f) = \Sigma(\mathcal{P}, f).$$

11.3.2   Summing over a Refinement of a Given Partition

Suppose that a function f defined on an interval $[a, b]$ steps within a partition

$$\mathcal{P} = (x_0, x_1, \cdots, x_n)$$

of $[a, b]$ and suppose that, for each $j = 1, 2, \cdots, n$, the constant value of f in the open interval (x_{j-1}, x_j) is called α_j. Now suppose that a new partition $\mathcal{Q}$ of $[a, b]$ is made by adding one extra point t. For some j we have

$$x_{j-1} < t < x_j,$$

and we illustrate this fact in the following figure.

Since all of the terms of the summation $\Sigma\left(\mathcal{Q}, f\right)$ are identical to terms in the summation $\Sigma\left(\mathcal{P}, f\right)$ except for the two terms that are drawn from the interval $\left(x_{j-1}x_j\right)$, we see that

$$\Sigma\left(\mathcal{Q}, f\right) - \Sigma\left(\mathcal{P}, f\right) = \alpha_j\left(t - x_{j-1}\right) + \alpha_j\left(x_j - t\right) - \alpha_j\left(x_j - x_{j-1}\right) = 0,$$

and so

$$\Sigma\left(\mathcal{Q}, f\right) = \Sigma\left(\mathcal{P}, f\right).$$

Thus if a refinement $\mathcal{Q}$ of the given partition $\mathcal{P}$ is made by adding just one new point, then the sums of f over $\mathcal{P}$ and $\mathcal{Q}$ are the same. We can now extend this observation to any refinement $\mathcal{Q}$ of the given partition $\mathcal{P}$. If $\mathcal{Q}$ is any refinement of $\mathcal{P}$, then, since $\mathcal{Q}$ can be obtained from $\mathcal{P}$ in finitely many steps, in which we add just one new point, and since the addition of one point does not affect the sum, we see again that

$$\Sigma\left(\mathcal{Q}, f\right) = \Sigma\left(\mathcal{P}, f\right).$$

11.3.3   Uniqueness of Sums over Partitions

Suppose that a function f defined on an interval $[a, b]$ steps within each of two partitions $\mathcal{P}$ and $\mathcal{Q}$ of $[a, b]$. By what we saw in Subsection 11.3.2 we know that

$$\Sigma\left(\mathcal{P}, f\right) = \Sigma\left(\mathcal{P} \cup \mathcal{Q}, f\right) = \Sigma\left(\mathcal{Q}, f\right).$$

In the light of this uniqueness property of sums over partitions, we can now make the following definition:

11.3.4   Integral of a Step Function

Suppose that f is a step function on an interval $[a, b]$. The **integral** of the function f over the interval $[a, b]$, which we write as $\int_a^b f$, is defined to be the sum $\Sigma\left(\mathcal{P}, f\right)$, where $\mathcal{P}$ is any partition of $[a, b]$ within which the function f steps.

An alternative notation for the symbol $\int_a^b f$ is $\int_a^b f(x)dx$.

11.3.5   Integral of a Constant Function

Suppose that f takes the constant value α on an interval $[a, b]$. If we define $\mathcal{P}$ to be the partition whose only points are a and b, then f steps within $\mathcal{P}$ and

$$\int_a^b f = \Sigma\left(\mathcal{P}, f\right) = \alpha\left(b - a\right).$$

Although the number α and the function whose constant value is α are not the same thing, we shall follow tradition and use the symbol α to denote this function. Thus

$$\int_a^b \alpha = \alpha\left(b - a\right).$$

Incidentally, if $a = b$, then any function f defined on the interval $[a, b]$ must be constant and we have

$$\int_a^a f = 0.$$

11.3.6   Linearity of the Integral of Step Functions

Suppose that f and g are step functions on an interval $[a, b]$ and that c is a given number.

1. *The function $f + g$ is a step function on $[a, b]$ and*

$$\int_a^b \left(f + g\right) = \int_a^b f + \int_a^b g.$$

2. *The function cf is a step function on $[a, b]$ and*

$$\int_a^b \left(cf\right) = c\int_a^b f.$$

Proof. Choose partitions $\mathcal{P}$ and $\mathcal{Q}$ of the interval $[a, b]$ such that f steps within $\mathcal{P}$ and g steps within $\mathcal{Q}$, and express the common refinement $\mathcal{P} \cup \mathcal{Q}$ of $\mathcal{P}$ and $\mathcal{Q}$ as

$$\mathcal{P} \cup \mathcal{Q} = \left(x_0, x_1, \cdots, x_n\right).$$

For each $j = 1, 2, \cdots, n$, we write the constant values of the functions f and g in the interval (x_{j-1}, x_j) as α_j and β_j, respectively. Since the function $f + g$ takes the constant value $\alpha_j + \beta_j$ in each interval (x_{j-1}, x_j), we know that $f + g$

is a step function and that

$$\int_a^b (f+g) \;=\; \sum_{j=1}^n \left(\alpha_j + \beta_j\right) \left(x_j - x_{j-1}\right)$$

$$= \sum_{j=1}^n \alpha_j \left(x_j - x_{j-1}\right) + \sum_{j=1}^n \beta_j \left(x_j - x_{j-1}\right) = \int_a^b f + \int_a^b g.$$

This proves the first part of the theorem. We leave the proof of the second part as an exercise. ▨ ■

11.3.7   Nonnegativity of the Integral of Step Functions

If f is a nonnegative step function on an interval $[a, b]$, then

$$\int_a^b f \geq 0.$$

We leave the proof of this assertion as an exercise. ▨

As a corollary to this nonnegativity property, we observe that if f and g are step functions on an interval $[a, b]$ and $f \leq g$, then, since $g - f$ is nonnegative, we have

$$\int_a^b g - \int_a^b f = \int_a^b (g - f) \geq 0.$$

and so

$$\int_a^b f \leq \int_a^b g.$$

11.3.8   Integral of an Absolute Value

Suppose that f is a step function on an interval $[a, b]$ that takes the constant value α_j in each of the open intervals (x_{j-1}, x_j) of a partition

$$\mathcal{P} = (x_0, x_1, \cdots, x_n).$$

Since the function $|f|$ takes the constant value $|\alpha_j|$ in each interval (x_{j-1}, x_j), we know that $|f|$ is also a step function and we have

$$\left| \int_a^b f \right| = \left| \sum_{j=1}^n \alpha_j \left(x_j - x_{j-1}\right) \right| \leq \sum_{j=1}^n |\alpha_j| \left(x_j - x_{j-1}\right) = \int_a^b |f|.$$

11.3.9   Additivity of the Integral of Step Functions

Suppose that $a \leq b \leq c$ and that f is a function defined on the interval $[a, c]$.

a   b   c

1. *The function f is a step function on the interval $[a, c]$ if and only if f is a step function on both of the intervals $[a, b]$ and $[b, c]$.*
2. *In the event that f is a step function on these intervals, we have*

$$\int_a^c f = \int_a^b f + \int_b^c f.$$

Proof. Suppose that f is a step function on the interval $[a, c]$ and choose a partition $\mathcal{P}$ of $[a, c]$ within which f steps. By refining the partition $\mathcal{P}$, if necessary, we can ensure that the number b is one of its points. Suppose that

$$\mathcal{P} = (x_0, x_1, \cdots, x_n)$$

and that $b = x_k$.

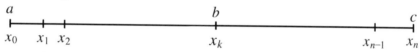

We now define partitions $\mathcal{P}_1$ and $\mathcal{P}_2$ of the intervals $[a, b]$ and $[b, c]$, respectively, as follows:

$$\mathcal{P}_1 = (x_0, x_1, \cdots, x_k) \quad \text{and} \quad \mathcal{P}_2 = (x_k, x_{k+1}, \cdots, x_n),$$

and we observe that

$$\int_a^c f = \Sigma\,(\mathcal{P}, f) = \Sigma\,(\mathcal{P}_1, f) + \Sigma\,(\mathcal{P}_2, f) = \int_a^b f + \int_b^c f.$$

It remains to show that if f is a step function on both of the intervals $[a, b]$ and $[b, c]$, then f is a step function on $[a, c]$. We leave the proof of this assertion as an exercise. ■

11.3.10   Integrating a Step Function on the Entire Line

Suppose that f is a step function. Recall that this means that there exists an interval $[a, b]$ such that f is a step function on the interval $[a, b]$ and such that $f(x) = 0$ for every $x \in \mathbf{R} \setminus [a, b]$. Our first task in this subsection is to show that if $[a, b]$ and $[c, d]$ are both intervals of this type, then

$$\int_a^b f = \int_c^d f.$$

To establish this equality, suppose that f is a step function on each of two intervals $[a, b]$ and $[c, d]$ and that $f(x) = 0$ whenever x lies outside of either one of these two intervals. Choose a number p that is less than both of the numbers a and c and choose a number q that is greater than both of the numbers b and d.

Using the additivity property, we observe that

$$\int_p^q f = \int_p^a f + \int_a^b f + \int_b^q f$$

$$= 0\,(a - p) + \int_a^b f + 0\,(q - b) = \int_a^b f,$$

and we see similarly that

$$\int_p^q f = \int_c^d f.$$

If f is a step function, then the integral of f on the entire line, which we write as $\int_{-\infty}^\infty f$, is defined to be

$$\int_a^b f,$$

where $[a, b]$ is any bounded interval outside of which f takes the constant value zero. In view of the observation that we made a moment ago, it makes no difference which such interval $[a, b]$ we choose. So, for example (see Figure 11.10), if

$$f(x) = \begin{cases} -3 & \text{if } -1 < x < 2 \\ 2 & \text{if } 2 \le x \le 5 \\ 0 & \text{if } x > 5 \\ 0 & \text{if } x \le -1 \end{cases}$$

then

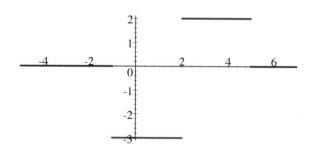

Figure 11.10

$$\int_{-\infty}^{\infty} f = \int_{-4}^{13} f = \int_{-1}^{5} f = (-3)(2-(-1)) + 2(5-2) = -3.$$

11.3.11   Exercises on Integration of Step Functions

1. ▣ Given that f is the function defined by the equation

$$f(x) = \begin{cases} 0 & \text{if } x < 1 \\ 1 & \text{if } 1 \leq x \leq 2 \\ -1 & \text{if } 2 < x < 4 \\ 2 & \text{if } 4 \leq x < 5 \\ 3 & \text{if } 5 \leq x < 7 \\ 0 & \text{if } x \geq 7 \end{cases}$$

whose graph appears in Figure 11.9, evaluate $\int_{-\infty}^{\infty} f$.

2. Prove that if f is a step function, then so is the function $|f|$ and we have

$$\left| \int_{-\infty}^{\infty} f \right| \leq \int_{-\infty}^{\infty} |f|.$$

3. Given that f is a function defined on $\mathbf{R}$ and that the set of numbers x for which $f(x) \neq 0$ is finite, explain why f must be a step function and why

$$\int_{-\infty}^{\infty} f = 0.$$

4. ▣ Given that f is a nonnegative step function and that

$$\int_{-\infty}^{\infty} f = 0,$$

prove that the set of numbers x for which $f(x) \neq 0$ must be finite.

5. Given that f and g are step functions and that c is a real number, prove that

$$\int_{-\infty}^{\infty} cf = c \int_{-\infty}^{\infty} f$$

and

$$\int_{-\infty}^{\infty} (f + g) = \int_{-\infty}^{\infty} f + \int_{-\infty}^{\infty} g.$$

11.4   Elementary Sets

11.4.1   Introduction to Elementary Sets

In this section we present an elementary discussion of the concept of the *length* of a set of real numbers. In the event that the set is an interval, the concept is quite simple because the obvious meaning of the length of an interval $[a, b]$ is the number $b - a$. However, even if a set is not an interval, it may be possible to assign a length to it. For example, if

$$E = [0, 1] \cup [2, 5],$$

then the natural meaning for the length of E is the number

$$m(E) = (1 - 0) + (5 - 2) = 4.$$

At a more advanced level of study than the one that we are undertaking here, the concept of length can be extended from these humble beginnings to yield a function m that is known as **one-dimensional Lebesgue measure** that was introduced by Henri Lebesgue in 1902 in his classic memoir, *Integrale longueur aire*. This function assigns to every set S of real numbers a nonnegative number $m(S)$ that is the natural meaning of the notion of *length* of the set S. The function m, or, more precisely, the restriction of the function m to a special family of sets known as the **Lebesgue measurable sets**, plays a fundamental role in the modern theories of integration. For some further information about the history of the measure concept, see Kline [16], pages 1040–1051, and you might possibly want to look at the work of Hawkins to which Kline refers on page 1051.

We, however, shall restrict our discussion of Lebesgue measure to a much smaller family of sets than the family of Lebesgue measurable sets. The sets upon which we shall focus our attention are called **elementary sets**, and it will turn out that a set of real numbers is elementary if and only if it can be expressed as the union of finitely many bounded intervals. Thus, for example, the set

$$E = [0, 1] \cup [2, 5]$$

that we discussed a moment ago is an elementary set. You will see why after we have given the precise definition in Subsection 11.4.3.

11.4.2   Characteristic Function of a Set

If S is a set of real numbers, then the **characteristic function** χ_S of the set S is defined by the equation

$$\chi_S(x) = \begin{cases} 1 & \text{if } x \in S \\ 0 & \text{if } x \in \mathbf{R} \setminus S. \end{cases}$$

11.4.3   Definition of an Elementary Set

A set E of real numbers is said to be an **elementary set** if its characteristic function χ_E is a step function.

Thus a set E of real numbers is elementary if and only if there exists an interval $[a, b]$ and a partition

$$\mathcal{P} = (x_0, x_1, \cdots, x_n)$$

of $[a, b]$ such that the function χ_E is zero at every number x that does not belong to the interval $[a, b]$ and the function χ_E steps within the partition $\mathcal{P}$.

More specifically, we can say that a set E of real numbers is elementary if and only if there exists an interval $[a, b]$ and a partition

$$\mathcal{P} = (x_0, x_1, \cdots, x_n)$$

of $[a, b]$ such that $E \subseteq [a, b]$ and for every $j = 1, 2, \cdots, n$ the open interval (x_{j-1}, x_j) is either a subset of E or is disjoint from E, depending on whether the constant value of χ_E in the interval (x_{j-1}, x_j) is 1 or 0.

11.4.4   Some Facts About Elementary Sets

1. *Every bounded interval is an elementary set.*
2. *The intersection of two elementary sets is an elementary set.*
3. *The union of two elementary sets is an elementary set.*
4. *The difference $A \setminus B$ of two elementary sets A and B is an elementary set. (See Subsection 4.2.7 for the definition of set difference.)*
5. *A set E is elementary if and only if it is the union of finitely many bounded intervals.*

Proof. Statement 1 is obvious. Now suppose that A and B are elementary sets. Since

$$\chi_{A \cap B} = (\chi_A)(\chi_B),$$

and since the product of two step functions must be a step function, the set $A \cap B$ is elementary. Since

$$\chi_{A \cup B} = \chi_A + \chi_B - (\chi_A)(\chi_B),$$

we see that the set $A \cup B$ must be elementary. Since

$$\chi_{A\setminus B} = \chi_A - (\chi_A)(\chi_B),$$

we see that $A \setminus B$ must be elementary.

We have thus completed the proof of statements 2, 3, and 4, and it is also clear at this stage that the union and intersection of finitely many elementary sets must be elementary. In particular, the union of finitely many bounded intervals must be an elementary set.

Suppose, finally, that E is an elementary set. We need to prove that E can be expressed as the union of finitely many bounded intervals. Choose an interval $[a, b]$ and a partition

$$\mathcal{P} = (x_0, x_1, \cdots, x_n)$$

of $[a, b]$ such that $E \subseteq [a, b]$ and for every $j = 1, 2, \cdots, n$ the open interval (x_{j-1}, x_j) is either a subset of E or is disjoint from E. Since E is the union of some (possibly none) of the intervals (x_{j-1}, x_j) and some (possibly none) of the points of $\mathcal{P}$, we see that E can be expressed as the union of finitely many bounded intervals. ∎

11.4.5   Motivating the Lebesgue Measure of an Elementary Set

In the introduction to this section we suggested that if

$$E = [0, 1] \cup [2, 5],$$

then the Lebesgue measure (length) of this set E should be given by

$$m(E) = (1 - 0) + (5 - 2).$$

Now look at the graph of the function χ_E that is illustrated in Figure 11.11. We

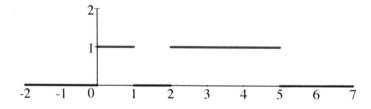

Figure 11.11

see at once that

$$\int_{-\infty}^{\infty} \chi_E = \int_0^1 1 + \int_2^5 1 = (1-0) + (5-2),$$

and this equation suggests the definition of the Lebesgue measure of an elementary set.

11.4.6   Definition of Lebesgue Measure of an Elementary Set

If E is an elementary set, then we define the **Lebesgue measure** $m(E)$ of the set E by the equation

$$m(E) = \int_{-\infty}^{\infty} \chi_E.$$

11.4.7   Some Properties of Lebesgue Measure

1. If a and b are numbers and $a \le b$, then

$$m([a,b]) = m([a,b)) = m((a,b]) = m((a,b)) = b - a.$$

2. If E and F are elementary sets, then it follows from the inequality

$$\chi_{E \cup F} \le \chi_E + \chi_F$$

   and nonnegativity (Theorem 11.3.7), that

$$m(E \cup F) \le m(E) + m(F).$$

3. If E and F are elementary sets that do not intersect with each other, then it follows from the equation

$$\chi_{E \cup F} = \chi_E + \chi_F$$

   and linearity that

$$m(E \cup F) = m(E) + m(F).$$

4. If E and F are elementary sets and $E \subseteq F$, then it follows from the fact that

$$F = E \cup (F \setminus E)$$

   that $m(F) = m(E) + m(F \setminus E)$ and therefore

$$m(F) - m(E) = m(F \setminus E).$$

5. If E and F are elementary sets and $E \subseteq F$, then it follows from the inequality $\chi_E \subseteq \chi_F$ and nonnegativity (Theorem 11.3.7) that $m(E) \leq m(F)$.

11.4.8   Approximation by Closed Sets and Open Sets

Suppose that a and b are real numbers and that $a \leq b$. Given any number $\varepsilon > 0$, if we define

$$H = \left[a + \frac{\varepsilon}{5}, b - \frac{\varepsilon}{5}\right]$$

(understood to be the empty set in the event that $a + \varepsilon/5 > b - \varepsilon/5$) and if we define

$$U = \left(a - \frac{\varepsilon}{5}, b + \frac{\varepsilon}{5}\right),$$

then H is a closed elementary set and U is an open elementary set. We see also that

$$H \subseteq [a, b] \subseteq U$$

and that $m(U \setminus H) < \varepsilon$.

For the moment we shall say that a given elementary set E *can be approximated* if for every number $\varepsilon > 0$ there exists a closed elementary subset H of E and an open elementary set $U \supseteq E$ such that $m(U \setminus H) < \varepsilon$. The preceding argument shows that every bounded interval can be approximated. We shall now show that if two elementary sets can be approximated, so can their union. From this fact it will follow that the union of finitely many elementary sets that can be approximated is an elementary set that can also be approximated. Since every elementary set is the union of finitely many bounded intervals, it will follow that every elementary set can be approximated.

Suppose then that E_1 and E_2 are elementary sets that can be approximated and suppose that $\varepsilon > 0$. Using the fact that E_1 and E_2 can be approximated, we choose closed elementary sets H_1 and H_2 and open elementary sets U_1 and U_2 such that

$$H_1 \subseteq E_1 \subseteq U_1 \quad \text{and} \quad H_2 \subseteq E_2 \subseteq U_2$$

and such that

$$m(U_1 \setminus H_1) < \frac{\varepsilon}{2} \quad \text{and} \quad m(U_2 \setminus H_2) < \frac{\varepsilon}{2}.$$

Since

$$H_1 \cup H_2 \subseteq E_1 \cup E_2 \subseteq U_1 \cup U_2$$

and

$$
\begin{aligned}
m\left((U_1 \cup U_2) \setminus (H_1 \cup H_2)\right) &\leq m\left((U_1 \setminus H_1) \cup (U_2 \setminus H_2)\right) \\
&\leq m\left(U_1 \setminus H_1\right) + m\left(U_2 \setminus H_2\right) < \varepsilon,
\end{aligned}
$$

we see that the set $E_1 \cup E_2$ can be approximated.

We have therefore proved the following properties of elementary sets:

1. *If E is any elementary set and if $\varepsilon > 0$, then there exists a closed elementary subset H of E and an open elementary set U that includes E such that $m(U \setminus H) < \varepsilon$.*

2. *If E is any elementary set and if $\varepsilon > 0$, then there exists a closed elementary subset H of E and an open elementary set U that includes E such that*

$$m(H) > m(E) - \varepsilon \quad and \quad m(U) < m(E) + \varepsilon.$$

Note that the second of these two assertions follows from the first one and the identity

$$m(U) - m(H) = m(U \setminus H).$$

11.4.9   Integration over an Elementary Set

If f is a step function and E is an elementary set, then the **integral of f over E**, which we write as $\int_E f$, is defined to be the integral

$$\int_{-\infty}^{\infty} f \chi_E.$$

Note that if E is an interval $[a, b]$, then, given any step function f, since

$$
f(x)\chi_E(x) = \begin{cases} f(x) & \text{if} \quad x \in [a, b] \\ 0 & \text{if} \quad x \in \mathbf{R} \setminus [a, b] \end{cases}
$$

we have

$$\int_E f = \int_a^b f.$$

In the same way we can see that if f is a step function and

$$E = [0, 1] \cup [2, 5],$$

then

$$\int_E f = \int_0^1 f + \int_2^5 f.$$

11.4.10   Some Properties of Integrals on Elementary Sets

1. If f is a step function and A and B are elementary sets that are disjoint from each other, then it follows from the equation

$$f\chi_{A\cup B} = f\chi_A + f\chi_B$$

that

$$\int_{-\infty}^{\infty} f\chi_{A\cup B} = \int_{-\infty}^{\infty} f\chi_A + \int_{-\infty}^{\infty} f\chi_B,$$

and so

$$\int_{A\cup B} f = \int_A f + \int_B f.$$

2. If f is a step function and E is an elementary set, and if, for some number α, we have $|f(x)| \leq \alpha$ for every $x \in E$, then it follows from nonnegativity (Theorem 11.3.7) and the inequality

$$|f\chi_E| \leq \alpha\chi_E$$

that

$$\left| \int_E f \right| = \left| \int_{-\infty}^{\infty} f\chi_E \right| \leq \int_{-\infty}^{\infty} |f\chi_E| \leq \int_{-\infty}^{\infty} \alpha\chi_E = \alpha \int_{-\infty}^{\infty} \chi_E = \alpha m(E).$$

11.4.11   Exercises on Elementary Sets

1. **H** Given that A and B are elementary sets, prove that

$$m(A \cup B) = m(A) + m(B) - m(A \cap B).$$

2. Prove that if E is an elementary set and $m(E) = 0$, then E must be finite.
3. Explain why the set of all rational numbers in the interval $[0, 1]$ is not elementary.
4. Prove that if E is an elementary subset of $[0, 1]$ and if every rational number in the interval belongs to E, then the set $[0, 1] \setminus E$ must be finite.
5. **H** Give an example of a set A of numbers such that, if E is any elementary subset of A, we have $m(E) = 0$ and if E is any elementary set that includes A, we have $m(E) \geq 1$.

6. ▥ Given that E is an elementary set that is not closed and that F is a closed elementary subset of E, prove that $m(E \setminus F) > 0$.

7. ▥ Given that f is a step function, that E is an elementary set, and that $f(x) = 0$ whenever $x \in \mathbf{R} \setminus E$, prove that

$$\int_E f = \int_{-\infty}^{\infty} f.$$

8. Given that f and g are step functions, that E is an elementary set, and that $f(x) \leq g(x)$ whenever $x \in E$, prove that

$$\int_E f \leq \int_E g.$$

9. Given that f is a nonnegative step function, that A and B are elementary sets, and that $A \subseteq B$, prove that

$$\int_A f \leq \int_B f.$$

10. ▥ Given that f is a step function and that E is an elementary set, prove that

$$\left| \int_E f \right| \leq \int_E |f|.$$

11. ▥ Given that A and B are elementary sets, prove that

$$\int_A \chi_B = \int_B \chi_A = m(A \cap B).$$

The on-screen version of this book contains a special group of exercises that are designed to be done as a special project. These exercises require you to have read some of the chapter on infinite series (Chapter 12). To reach this group of exercises, click on the icon ▣ .

11.5   Riemann Integrability and the Riemann Integral

11.5.1   Introduction to the Riemann Integral

▣ At the heart of the theory of Riemann integration lies a technique that was discovered by Eudoxus and then developed into a sophisticated theory by Archimedes, somewhere around the year 250 B.C.E. Archimedes calculated the areas and volumes of a number of geometric figures by approximating them both from inside and from outside by polygonal figures, as illustrated in Figure 11.12. He reasoned that if A is a plane region and if s and S are two polygons satisfying

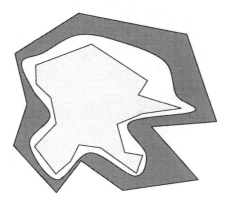

Figure 11.12

the condition $s \subseteq A \subseteq S$, then

$$\text{area}\,(s) \le \text{area}\,(A) \le \text{area}\,(S)$$

and therefore, given any number $\varepsilon > 0$, we can guarantee that both area (s) and area (S) must approximate area (A) with an error less than ε, simply by making

$$\text{area}\,(S) - \text{area}\,(s) < \varepsilon.$$

The analog of this idea for functions would be as follows: Suppose that f is a bounded function defined on an interval $[a, b]$ and that s and S are step functions on $[a, b]$ satisfying the inequality $s \le f \le S$. Then, whatever the symbol $\int_a^b f$ ought to mean, we should have

$$\int_a^b s \le \int_a^b f \le \int_a^b S,$$

and, given $\varepsilon > 0$, we can guarantee that the numbers $\int_a^b s$ and $\int_a^b S$ approximate $\int_a^b f$ with an error less than ε by making

$$\int_a^b S - \int_a^b s < \varepsilon.$$

For a function f that happens to be nonnegative, Figures 11.13 and 11.14 illustrate the numbers $\int_a^b s$ and $\int_a^b S$, respectively.

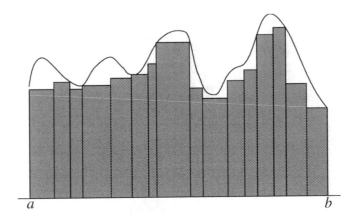

Figure 11.13

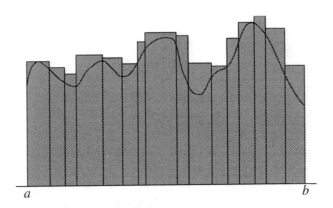

Figure 11.14

11.5.2   Definition of Integrability and the Integral

Suppose that f is a bounded function defined on an interval $[a, b]$. If s and S are step functions on the interval $[a, b]$ and

$$s \leq f \leq S,$$

then it follows from nonnegativity (Theorem 11.3.7) that

$$\int_a^b s \leq \int_a^b S.$$

Therefore

$$\sup \left\{ \int_a^b s \mid s \text{ is a step function and } s \leq f \right\}$$

$$\leq \inf \left\{ \int_a^b S \mid S \text{ is a step function and } f \leq S \right\}.$$

In the event that

$$\sup \left\{ \int_a^b s \mid s \text{ is a step function and } s \leq f \right\}$$

$$= \inf \left\{ \int_a^b S \mid S \text{ is a step function and } f \leq S \right\},$$

we say that the function f is **Riemann integrable** on the interval $[a, b]$ and we define

$$\int_a^b f = \sup \left\{ \int_a^b s \mid s \text{ is a step function and } s \leq f \right\}$$

$$= \inf \left\{ \int_a^b S \mid S \text{ is a step function and } f \leq S \right\}.$$

Usually we shall refer to a Riemann integrable function more briefly as an **integrable** function. Note that every step function is integrable and that the definition just given does not conflict with the definition of the integral of a step function that was given earlier.

11.5.3   A Necessary and Sufficient Condition for Integrability

Suppose that f is a bounded function on an interval $[a, b]$. Then the following conditions are equivalent:

1. *The function f is integrable.*
2. *There exist two sequences (s_n) and (S_n) of step functions such that*

$$s_n \leq f \leq S_n$$

for each n and such that

$$\lim_{n \to \infty} \int_a^b (S_n - s_n) = \lim_{n \to \infty} \left(\int_a^b S_n - \int_a^b s_n \right) = 0.$$

Furthermore, for any choice of sequences (s_n) and (S_n) satisfying the latter *condition we have*

$$\int_a^b f = \lim_{n\to\infty} \int_a^b s_n = \lim_{n\to\infty} \int_a^b S_n.$$

3. *There exist two sequences* (g_n) *and* (h_n) *of functions that are integrable on* $[a, b]$ *such that* $g_n \le f \le h_n$ *for each* n *and such that*

$$\lim_{n\to\infty} \left(\int_a^b h_n - \int_a^b g_n \right) = 0.$$

Proof. The equivalence of conditions 1 and 2 follows at once from Theorem 7.4.7, and condition 2 clearly implies condition 3. Now suppose that condition 3 holds and choose two sequences (g_n) and (h_n) of functions that are integrable on $[a, b]$ such that $g_n \le f \le h_n$ for each n and such that

$$\lim_{n\to\infty} \left(\int_a^b h_n - \int_a^b g_n \right) = 0.$$

For each n we use the fact that g_n and h_n are integrable to choose step functions $s_n \le g_n$ and $S_n \ge h_n$ such that

$$\int_a^b s_n > \int_a^b g_n - \frac{1}{n} \quad \text{and} \quad \int_a^b S_n < \int_a^b h_n + \frac{1}{n}.$$

Since

$$0 \le \int_a^b S_n - \int_a^b s_n < \int_a^b h_n - \int_a^b g_n + \frac{2}{n}$$

for each n, we conclude from the sandwich theorem that

$$\lim_{n\to\infty} \left(\int_a^b S_n - \int_a^b s_n \right) = 0,$$

and we have shown that condition 3 implies condition 2. ∎

11.5.4   Squeezing a Function on an Interval

Suppose that f is a bounded function on an interval $[a, b]$. A pair of sequences (s_n) and (S_n) of step functions is said to **squeeze** the function f on the interval $[a, b]$ if we have

$$s_n \le f \le S_n$$

for each n and we have

$$\lim_{n \to \infty} \int_a^b (S_n - s_n) = 0.$$

From this definition and the preceding theorem we can make the following conclusion:

Suppose that f is a bounded function on an interval $[a, b]$. Then the following conditions are equivalent:

1. *The function f is integrable.*
2. *There exists a pair of sequences (s_n) and (S_n) of step functions that squeezes f on $[a, b]$. Furthermore, for any such pair of sequences we have*

$$\int_a^b f = \lim_{n \to \infty} \int_a^b s_n = \lim_{n \to \infty} \int_a^b S_n.$$

11.6   Some Examples of Integrable and Nonintegrable Functions

11.6.1   A Linear Function

We can at last find the area of a triangle. In this example we shall consider the integral

$$\int_0^1 x\,dx.$$

We define $f(x) = x$ for $0 \le x \le 1$ and, for each positive integer n, we define $\mathcal{P}_n$ to be the regular n-partition of the interval $[0, 1]$. Thus

$$\mathcal{P}_n = \left(\frac{0}{n}, \frac{1}{n}, \cdots, \frac{n}{n} \right).$$

Now, for each n we define two step functions s_n and S_n by making

$$s_n(x) = S_n(x) = x$$

whenever x is a point of the partition $\mathcal{P}_n$ and, in each interval

$$\left(\frac{j-1}{n}, \frac{j}{n} \right)$$

of the partition $\mathcal{P}_n$, we make s_n and S_n take the constant values $(j-1)/n$ and

j/n, respectively. The graphs of the functions s_n, f, and S_n are illustrated for the case $n = 15$ in Figure 11.15. For each n we have

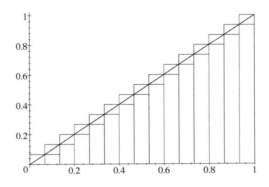

Figure 11.15

$$\int_0^1 (S_n - s_n) = \sum_{j=1}^n \left(\frac{j}{n} - \frac{j-1}{n} \right) \frac{1}{n} = \frac{1}{n},$$

and so

$$\lim_{n \to \infty} \int_0^1 (S_n - s_n) = 0.$$

We deduce from Theorem 11.5.3 that f is integrable on the interval $[0, 1]$ and that

$$\int_0^1 x\, dx = \lim_{n \to \infty} \int_0^1 S_n = \lim_{n \to \infty} \sum_{j=1}^n \frac{j}{n} \frac{1}{n} = \lim_{n \to \infty} \frac{1}{2} \left(1 + \frac{1}{n} \right) = \frac{1}{2}.$$

11.6.2   A Quadratic Function

In this example we take $f(x) = x^2$ for $0 \le x \le 1$ and, once again, we take $\mathcal{P}_n$ to be the regular n-partition of the interval $[0, 1]$ for each n. Now, for each n we define two step functions s_n and S_n by making

$$s_n(x) = S_n(x) = x^2$$

whenever x is a point of the partition $\mathcal{P}_n$ and, in each interval

$$\left(\frac{j-1}{n}, \frac{j}{n} \right)$$

of the partition $\mathcal{P}_n$, we make s_n and S_n take the constant values $(j-1)^2/n^2$ and j^2/n^2, respectively. The graphs of the functions s_n, f, and S_n are illustrated for the case $n = 15$ in Figure 11.16. For each n we have

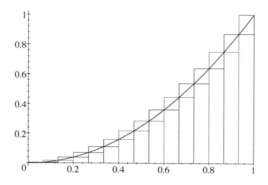

Figure 11.16

$$\int_0^1 (S_n - s_n) = \sum_{j=1}^n \left(\left(\frac{j}{n}\right)^2 - \left(\frac{j-1}{n}\right)^2 \right) \frac{1}{n} = \frac{1}{n},$$

and so

$$\lim_{n\to\infty} \int_0^1 (S_n - s_n) = 0.$$

We deduce from Theorem 11.5.3 that f is integrable on the interval $[0, 1]$ and that

$$\int_0^1 x^2 dx = \lim_{n\to\infty} \int_0^1 S_n = \lim_{n\to\infty} \sum_{j=1}^n \left(\frac{j}{n}\right)^2 \frac{1}{n} = \lim_{n\to\infty} \left(\frac{1}{3} + \frac{1}{2n} + \frac{1}{6n^2}\right) = \frac{1}{3}.$$

11.6.3   Monotone Functions Are Integrable

We shall confine our attention to increasing functions and leave the analogous case for decreasing functions as an exercise.

Suppose that f is an increasing function on an interval $[a, b]$. For each positive integer n we define $\mathcal{P}_n$ to be the regular n-partition of $[a, b]$. Thus, if for a given value of n we write

$$\mathcal{P}_n = (x_0, x_1, \cdots, x_n),$$

then for each $j = 0, 1, \cdots, n$ we have

$$x_j = a + \frac{j\,(b-a)}{n}.$$

Now, by analogy with the approach that we took in the two preceding examples, we define two step functions s_n and S_n by making

$$s_n(x) = S_n(x) = f(x)$$

whenever x is a point of the partition $\mathcal{P}_n$ and, in each interval (x_{j-1}, x_j) of the partition $\mathcal{P}_n$, we make s_n and S_n take the constant values $f(x_{j-1})$ and $f(x_j)$, respectively. The graphs of the functions s_n, f, and S_n are illustrated for the case $n = 16$ in Figure 11.17. For each n we have $s_n \leq f \leq S_n$ and

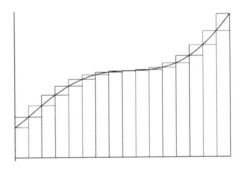

Figure 11.17

$$\int_a^b (S_n - s_n) \;=\; \sum_{j=1}^{n} (f(x_j) - f(x_{j-1})) \frac{b-a}{n}$$

$$= \frac{b-a}{n} \sum_{j=1}^{n} (f(x_j) - f(x_{j-1})) = \frac{(b-a)\,(f(b) - f(a))}{n},$$

and so

$$\lim_{n \to \infty} \int_a^b (S_n - s_n) = 0.$$

We deduce from Theorem 11.5.3 that f is integrable on the interval $[a, b]$. ∎

11.6.4   A Nonintegrable Function
In this example we look at a function f on the interval $[0, 1]$ that is too discontin-

uous to be integrable. We define

$$f(x) = \begin{cases} 1 & \text{if } x \in [0, 1] \cap \mathbf{Q} \\ 0 & \text{if } x \in [0, 1] \setminus \mathbf{Q}. \end{cases}$$

Note that this function is discontinuous at every number in the interval $[0, 1]$. Now, to see that f fails to be integrable, suppose that s is any step function on the interval $[0, 1]$ that satisfies the inequality $s \leq f$ and choose a partition

$$\mathcal{P} = (x_0, x_1, \cdots, x_n)$$

of $[0, 1]$ such that s steps within $\mathcal{P}$. If we call the constant value of s in each interval (x_{j-1}, x_j) by the name α_j, then, since there must exist irrational numbers in each interval (x_{j-1}, x_j), it follows from the inequality $s \leq f$ that $\alpha_j \leq 0$ for each j. Therefore

$$\int_0^1 s = \sum_{j=1}^n \alpha_j (x_j - x_{j-1}) \leq 0.$$

In a similar way we can show that if S is a step function on $[0, 1]$ that satisfies the inequality $f \leq S$, then

$$\int_0^1 S \geq 1,$$

and so we conclude that f is not integrable.

11.6.5   The Role of the Cantor Set in Riemann Integration

This optional topic can be reached from the on-screen version of the book by clicking on the icon ▓ .

11.6.6   Some Exercises on the Riemann Integral

1. ▓ Prove that the integral

$$\int_1^4 3x^2 dx$$

exists and has the value 63.

2. ▓ In this exercise we take $f(x) = \sqrt{x}$ for $x \in [0, 1]$. Given a positive integer n, we shall take $\mathcal{P}_n$ to be the partition of $[0, 1]$ defined by the equation

$$\mathcal{P}_n = \left(\frac{0^2}{n^2}, \frac{1^2}{n^2}, \frac{2^2}{n^2}, \cdots, \frac{n^2}{n^2} \right).$$

Prove that if we define a step function S_n on $[0, 1]$ by making

$$S_n(x) = \sqrt{x}$$

whenever x is a point of the partition $\mathcal{P}_n$ and giving S_n the constant value j/n in each interval

$$\left(\frac{(j-1)^2}{n^2}, \frac{j^2}{n^2} \right)$$

of the partition $\mathcal{P}_n$, then

$$\int_0^1 \sqrt{x}\,dx = \lim_{n\to\infty} \int_0^1 S_n = \frac{2}{3}.$$

3. Prove that

$$+ + \int_0^1 \sqrt[3]{x}\,dx = \frac{3}{4}.$$

11.7   Some Properties of the Riemann Integral

In this section we shall explore the analogs for Riemann integrals of the properties of linearity, nonnegativity, and additivity that we saw in Section 11.3 for integration of step functions. Each of these properties will be deduced using the corresponding statement for step functions.

11.7.1   Linearity of the Riemann Integral

Suppose that f and g are integrable functions on an interval $[a, b]$ and that c is a real number.

1. *The function $f + g$ is integrable on $[a, b]$ and*

$$\int_a^b (f + g) = \int_a^b f + \int_a^b g.$$

2. *The function cf is integrable on $[a, b]$ and*

$$\int_a^b cf = c \int_a^b f.$$

Proof. Using the fact that f and g are integrable on $[a, b]$, we choose a pair of sequences (s_n) and (S_n) that squeezes f on $[a, b]$ and a pair of sequences (s_n^*)

and (S_n^*) that squeezes g on $[a, b]$. Thus, for each n we have

$$s_n \le f \le S_n \quad \text{and} \quad s_n^* \le g \le S_n^*$$

and

$$\lim_{n \to \infty} \int_a^b (S_n - s_n) = \lim_{n \to \infty} \int_a^b (S_n^* - s_n^*) = 0.$$

Since

$$s_n + s_n^* \le f + g \le S_n + S_n^*$$

for each n and since

$$\lim_{n \to \infty} \int_a^b ((S_n + S_n^*) - (s_n + s_n^*)) = \lim_{n \to \infty} \int_a^b (S_n - s_n) + \lim_{n \to \infty} \int_a^b (S_n^* - s_n^*) = 0,$$

we know that the function $f + g$ is integrable. Finally

$$\int_a^b (f + g) = \lim_{n \to \infty} \int_a^b (S_n + S_n^*) = \lim_{n \to \infty} \left(\int_a^b S_n + \int_a^b S_n^* \right) = \int_a^b f + \int_a^b g.$$

Now, to prove the second assertion, we observe that if $c \ge 0$, we have

$$c s_n \le c f \le c S_n$$

for each n and if $c < 0$, we have

$$c S_n \le c f \le c s_n$$

for each n. So, in both of these cases, the pair of sequences $(c s_n)$ and $(c S_n)$ squeezes cf on $[a, b]$ and so cf is integrable on $[a, b]$. Finally,

$$\int_a^b cf = \lim_{n \to \infty} \int_a^b c s_n = \lim_{n \to \infty} c \int_a^b s_n = c \int_a^b f.$$

11.7.2   Nonnegativity of the Riemann Integral

Suppose that f and g are integrable functions on an interval $[a, b]$.

1. *If $f \ge 0$, then $\int_a^b f \ge 0$.*
2. *If $f \le g$, then $\int_a^b f \le \int_a^b g$.*

Proof. To prove the first assertion, suppose that $f \ge 0$. Since the constant zero function 0 is a step function that does not exceed f, it follows from the definition of an integral that

$$0 \le \int_a^b 0 \le \int_a^b f.$$

The second assertion follows automatically from the first one and linearity because if $f \le g$, then $g - f \ge 0$. ∎

11.7.3   Additivity of the Riemann Integral

Suppose that $a \le b \le c$ and that f is a function defined on the interval $[a, c]$.

a	b	c

1. *The function f is integrable on the interval $[a, c]$ if and only if f is integrable on both of the intervals $[a, b]$ and $[b, c]$.*
2. *In the event that f is integrable on these intervals, we have*

$$\int_a^c f = \int_a^b f + \int_b^c f.$$

Proof. Suppose that f is integrable on the interval $[a, c]$ and choose a pair of sequences (s_n) and (S_n) that squeezes f on the interval $[a, c]$. From the inequalities

$$\int_a^b (S_n - s_n) \le \int_a^c (S_n - s_n) \quad \text{and} \quad \int_b^c (S_n - s_n) \le \int_a^c (S_n - s_n)$$

it follows that this pair of sequences squeezes f on the intervals $[a, b]$ and $[b, c]$ and so f is integrable on $[a, b]$ and on $[b, c]$. Furthermore,

$$\int_a^c f = \lim_{n \to \infty} \int_a^c s_n = \lim_{n \to \infty} \left(\int_a^b s_n + \int_b^c s_n \right)$$

$$= \lim_{n \to \infty} \int_a^b s_n + \lim_{n \to \infty} \int_b^c s_n = \int_a^b f + \int_b^c f.$$

Finally, we need to explain why, if f is integrable on each of the intervals $[a, b]$ and $[b, c]$, then f must be integrable on $[a, c]$. We leave the task of writing this explanation as an exercise. ∎

11.7.4   Reversing the Limits of Integration

If a and b are real numbers and $a < b$, and if f is integrable on the interval $[a, b]$, then we define

$$\int_b^a f = - \int_a^b f.$$

A benefit of this definition is that it makes the additivity condition

$$\int_a^c f = \int_a^b f + \int_b^c f$$

independent of the order in which the numbers a, b, and c appear. Suppose, for example, that $b < c < a$ and that f is integrable on the interval $[b, a]$. The additivity condition tells us that

$$\int_b^a f = \int_b^c f + \int_c^a f.$$

Therefore

$$-\int_a^b f = \int_b^c f - \int_a^c f,$$

from which we obtain

$$\int_a^c f = \int_a^b f + \int_b^c f.$$

Another benefit of this definition is the following result, which follows from the nonnegativity property:

Suppose that a and b are real numbers and that f is integrable on the interval that runs from the smaller of these two numbers to the larger. Suppose that k is a positive number and that $|f(x)| \leq k$ for every number x in the interval. Then, applying the nonnegativity property to the inequality

$$-k \leq f \leq k,$$

we obtain

$$\left| \int_a^b f \right| \leq k \, |b - a| .$$

11.8   Upper, Lower, and Oscillation Functions

In this section we shall focus our attention upon step functions that step within a particular given partition. We shall observe that if f is a bounded function defined on an interval $[a, b]$, and if $\mathcal{P}$ is a given partition of $[a, b]$, then, of all the step functions $S \geq f$ that step within this partition $\mathcal{P}$, one of them is least, and of all the step functions $s \leq f$ that step within $\mathcal{P}$, one of them is greatest. These two special step functions will be called the **upper** and **lower** functions of f over the partition $\mathcal{P}$, and the difference between them will be called the **oscillation** function of f over $\mathcal{P}$. With the help of these special step functions we shall present two particularly useful conditions for a given function to be integrable. We call these conditions the **first** and **second criteria** for integrability.

11.8.1　Definition of the Functions $u\left(\mathcal{P}, f\right),$ **and** $l\left(\mathcal{P}, f\right),$ **and** $w\left(\mathcal{P}, f\right)$

Suppose that f is a bounded function on an interval $[a, b]$ and that

$$\mathcal{P} = (x_0, x_1, \cdots, x_n)$$

is a partition of $[a, b]$.

The **upper function** $u\left(\mathcal{P}, f\right)$ of the function f over the partition $\mathcal{P}$ is defined to be the step function whose value at every point x_j of the partition $\mathcal{P}$ is $f(x_j)$ and whose constant value in each open interval (x_{j-1}, x_j) of the partition $\mathcal{P}$ is the number

$$\sup \{f(x) \mid x_{j-1} < x < x_j\}.$$

The **lower function** $l\left(\mathcal{P}, f\right)$ of the function f over the partition $\mathcal{P}$ is defined to be the step function whose value at every point x_j of the partition $\mathcal{P}$ is $f(x_j)$ and whose constant value in each open interval (x_{j-1}, x_j) of the partition $\mathcal{P}$ is the number

$$\inf \{f(x) \mid x_{j-1} < x < x_j\}.$$

The constant values of $u\left(\mathcal{P}, f\right)$ and $l\left(\mathcal{P}, f\right)$ in each interval (x_{j-1}, x_j) are illustrated in Figure 11.18.

We define the **oscillation function** $w\left(\mathcal{P}, f\right)$ of the function f over the partition $\mathcal{P}$ by the equation

$$w\left(\mathcal{P}, f\right) = u\left(\mathcal{P}, f\right) - l\left(\mathcal{P}, f\right).$$

Thus $w\left(\mathcal{P}, f\right)$ takes the value 0 at each point x_j of the partition $\mathcal{P}$, and, in each open interval (x_{j-1}, x_j) of the partition $\mathcal{P}$, the constant value of $w\left(\mathcal{P}, f\right)$ is

$$\sup \{f(x) \mid x_{j-1} < x < x_j\} - \inf \{f(x) \mid x_{j-1} < x < x_j\}.$$

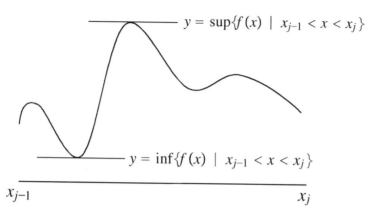

Figure 11.18

11.8.2   The important properties of the step functions $l\,(\mathbf{P}, f)$, and $u\,(\mathbf{P}, f)$, and $w\,(\mathcal{P}, f)$

Suppose that f is a bounded function on an interval $[a, b]$ and that

$$\mathcal{P} = (x_0, x_1, \cdots, x_n)$$

is a partition of $[a, b]$.

1. We have the inequality

$$l\,(\mathbf{P}, f) \le f \le u\,(\mathbf{P}, f).$$

2. The function $u\,(\mathbf{P}, f)$ is the least step function $S \ge f$ that steps within the partition $\mathbf{P}$. and $l\,(\mathbf{P}, f)$ is the greatest step function $s \le f$ that steps within the partition $\mathbf{P}$. Thus, if s and S are any step functions that step within the partition $\mathbf{P}$ and if $s \le f \le S$, then we have

$$s \le l\,(\mathbf{P}, f) \le f \le u\,(\mathbf{P}, f) \le S.$$

3. From the discussion that appears in Subsection 5.8.3 we see that the constant value of $w\,(\mathcal{P}, f)$ in each interval (x_{j-1}, x_j) is the diameter of the set

$$\{f(x) \mid x_{j-1} < x < x_j\}.$$

In other words, the constant value of $w\,(\mathcal{P}, f)$ in each interval (x_{j-1}, x_j) is

$$\sup \left\{ |f(t) - f(x)| \mid t \text{ and } x \text{ belong to } (x_{j-1}, x_j) \right\}.$$

11.8.3   The First Criterion for Integrability

Suppose that f is a bounded function defined on an interval $[a, b]$. Then the following conditions are equivalent:

1. *The function f is integrable on the interval $[a, b]$.*
2. *For every number $\varepsilon > 0$ there exists a partition $\mathcal{P}$ of the interval $[a, b]$ such that*

$$\int_a^b w(\mathcal{P}, f) < \varepsilon.$$

Proof. To prove that condition 1 implies condition 2, we assume that f is integrable on the interval $[a, b]$. Suppose that $\varepsilon > 0$.

Using Theorem 11.5.3, we choose a pair of sequences (s_n) and (S_n) of step functions that squeezes f on $[a, b]$. Choose a positive integer n such that

$$\int_a^b (S_n - s_n) < \varepsilon.$$

Choose a partition $\mathcal{P}$ of $[a, b]$ such that both s_n and S_n step within $\mathcal{P}$. Then, since

$$s_n \leq l(\mathcal{P}, f) \quad \text{and} \quad u(\mathcal{P}, f) \leq S_n,$$

we have

$$\int_a^b w(\mathcal{P}, f) = \int_a^b (u(\mathcal{P}, f) - l(\mathcal{P}, f)) \leq \int_a^b (S_n - s_n) < \varepsilon.$$

Now, to prove that condition 2 implies condition 1 we assume that condition 2 is true. For every positive integer n we use the fact that $1/n > 0$ to choose a partition $\mathcal{P}_n$ of the interval $[a, b]$ such that

$$\int_a^b w(\mathcal{P}_n, f) < \frac{1}{n},$$

and for each n we define

$$s_n = l(\mathcal{P}_n, f) \quad \text{and} \quad S_n = u(\mathcal{P}_n, f).$$

Since the pair of sequences (s_n) and (S_n) squeezes f on the interval $[a, b]$, we conclude that f is integrable on $[a, b]$. ■

11.8.4   The Second Criterion for Integrability

Suppose that f is a bounded function defined on an interval $[a, b]$. Then the following conditions are equivalent:

1. *The function f is integrable on the interval $[a, b]$.*
2. *For every number $\varepsilon > 0$ there exists a partition $\mathcal{P}$ of the interval $[a, b]$ such that, if we define*

$$E = \{x \in [a, b] \mid w\left(\mathcal{P}, f\right)(x) \geq \varepsilon\},$$

then we have $m\left(E\right) < \varepsilon$.

Proof. To prove that condition 1 implies condition 2, we assume that f is integrable on $[a, b]$ and that $\varepsilon > 0$. Using Theorem 11.8.3 and the fact that $\varepsilon^2 > 0$, we choose a partition $\mathcal{P}$ of $[a, b]$ such that

$$\int_a^b w\left(\mathcal{P}, f\right) < \varepsilon^2,$$

and we define

$$E = \{x \in [a, b] \mid w\left(\mathcal{P}, f\right)(x) \geq \varepsilon\}.$$

Since $w\left(\mathcal{P}, f\right)$ is a step function, we see that E is an elementary set and we have

$$\varepsilon^2 > \int_a^b w\left(\mathcal{P}, f\right) \geq \int_E w\left(\mathcal{P}, f\right) \geq \int_E \varepsilon = \varepsilon m\left(E\right),$$

from which we deduce that $m\left(E\right) < \varepsilon$.

Now, to prove that condition 2 implies condition 1, we assume that condition 2 is true. Using the fact that f is bounded, choose a number k such that $|f(x)| < k$ for every $x \in [a, b]$. Since the inequality

$$|f(t) - f(x)| \leq |f(t)| + |f(x)| \leq 2k$$

holds for all t and x in the interval $[a, b]$, we know that if $\mathcal{P}$ is any partition, then $w\left(\mathcal{P}, f\right) \leq 2k$. For each n we use the fact that $1/n > 0$ to choose a partition $\mathcal{P}_n$ of the interval $[a, b]$ such that if

$$E_n = \left\{x \in [a, b] \mid w\left(\mathcal{P}_n, f\right)(x) \geq \frac{1}{n}\right\},$$

then $m\left(E_n\right) < 1/n$. In this way we have separated the interval $[a, b]$ into the "small" set E_n and the set $[a, b] \setminus E_n$, where the function $w\left(\mathcal{P}_n, f\right)$ is "small".

For each n we have

$$\int_a^b w(\mathcal{P}_n, f) = \int_{E_n} w(\mathcal{P}_n, f) + \int_{[a,b]\setminus E_n} w(\mathcal{P}_n, f)$$

$$\leq \int_{E_n} 2k + \int_{[a,b]\setminus E_n} \frac{1}{n} \leq 2km(E_n) + \int_a^b \frac{1}{n}$$

$$< \frac{2k}{n} + \frac{b-a}{n},$$

and since the latter expression approaches 0 as $n \to \infty$, we deduce that condition 2 of Theorem 11.8.3 is satisfied and, consequently, that f is integrable on $[a, b]$. ∎

11.8.5   Some Exercises on Riemann Integrability

1. ⊞ Suppose that

$$f(x) = \begin{cases} 1 & \text{if } x \text{ has the form } \frac{1}{n} \text{ for some positive integer } n \\ 0 & \text{otherwise.} \end{cases}$$

   Prove that f is integrable on the interval $[0, 1]$ and that $\int_0^1 f = 0$.

2. ⊞ Suppose that f is defined on the interval in such a way that, whenever $x \in [0, 1]$ and x has the form $\frac{1}{n}$ for some positive integer n, we have $f(x) = 0$ and whenever x belongs to an interval of the form $\left(\frac{1}{n+1}, \frac{1}{n}\right)$ for some positive integer n we have

$$f(x) = 1 + (-1)^n.$$

   Draw a rough sketch of the graph of this function and explain why it is integrable on the interval $[0, 1]$.

3. Given that f is a bounded function on an interval $[a, b]$, prove that the following conditions are equivalent:

   (a) The function f is integrable on the interval $[a, b]$.

   (b) For every number $\varepsilon > 0$ there exist step functions s and S on the interval $[a, b]$ such that $s \leq f \leq S$ and

$$\int_a^b (S - s) < \varepsilon.$$

   (c) For every number $\varepsilon > 0$ there exist step functions s and S on the interval

$[a, b]$ such that $s \leq f \leq S$ and such that if

$$E = \{x \in [a, b] \mid S(x) - s(x) \geq \varepsilon\},$$

we have $m(E) < \varepsilon$.

4. ▮ Suppose that f is a bounded function on an interval $[a, b]$ and that for every number $\varepsilon > 0$ there exists an elementary subset E of $[a, b]$ such that $m(E) < \varepsilon$ and such that the function $f(1 - \chi_E)$ is Riemann integrable on $[a, b]$. Prove that f must be Riemann integrable on the interval $[a, b]$.

5. (a) Suppose that f is a nonnegative function defined on an interval $[a, b]$ and that for every number $\varepsilon > 0$ the set

$$\{x \in [a, b] \mid f(x) \geq \varepsilon\}$$

is finite. Prove that f must be integrable on $[a, b]$ and that $\int_a^b f = 0$.

   (b) Prove that if f is the ruler function that was introduced in Example 8 of Subsection 8.1.5, then f is an integrable function on the interval $[0, 1]$, even though f is discontinuous at every rational number in the interval.

6. Given that f is a bounded nonnegative function defined on an interval $[a, b]$, prove that the following conditions are equivalent:

   (a) The function f is integrable on the interval $[a, b]$ and $\int_a^b f = 0$.

   (b) For every number $\varepsilon > 0$ there exists an elementary set E such that $m(E) < \varepsilon$ and such that

$$\{x \in [a, b] \mid f(x) \geq \varepsilon\} \subseteq E.$$

If you are reading the on-screen version of this book, you can find a special group of exercises that are designed to be done as a special project. These exercises require you to have read some of the chapter on infinite series and they depend upon the special group of exercises on elementary sets that was mentioned in Subsection 11.4.11. The main purpose of these exercises is to invite you to prove the following interesting fact about integrals:

If f is a nonnegative integrable function on an interval $[a, b]$, where $a < b$, and if $\int_a^b f = 0$, then there must be at least one number $x \in [a, b]$ for which $f(x) = 0$.

To reach this special group of exercises, click on the icon ▮ .

11.9   Riemann Sums and Darboux's Theorem (Optional)

11.9.1   Introduction to Riemann Sums

In the approach to Riemann integration that is taken in most elementary calculus courses, an integral of the form $\int_a^b f$ is defined to be a limit (in some sense that is not always specified clearly) of sums of the type

$$\sum_{j=1}^{n} f(t_j)\,(x_j - x_{j-1})\,,$$

where

$$\mathcal{P} = (x_0, x_1, \cdots, x_n)$$

is a partition of the interval $[a, b]$ and for each j we have

$$x_{j-1} \leq t_j \leq x_j.$$

---

x_{j-1} t_j x_j

Such sums are called **Riemann sums** of the function f over the partition $\mathcal{P}$ and will be introduced precisely in this section. The main theorem of this section is the statement known as **Darboux's theorem**, which tells us that if f is integrable on an interval $[a, b]$, then we can make the Riemann sums of f over a partition lie as close as we like to $\int_a^b f$ by making the mesh (see Subsection 11.2.1) of the partition $\mathcal{P}$ small enough. We shall make use of Darboux's theorem when we prove the theorems on interchange of repeated Riemann integrals in Section 16.6 of Chapter 16.

Another key task of this section is to revisit the criteria for integrability that appear in Section 11.8. As you know, these criteria tell us that if a function f is integrable on an interval $[a, b]$, then it is possible to find a partition $\mathcal{P}$ of $[a, b]$ for which the function $w(\mathcal{P}, f)$ is in some sense small. In this section we shall go a step further and show that if f is integrable on an interval $[a, b]$, then the function $w(\mathcal{P}, f)$ will be small in the sense of the two criteria for *all* partitions $\mathcal{P}$ that have sufficiently small mesh.

Depending on how much time you can devote to the study of this chapter, there are three main ways in which you can choose to approach this optional section:

- Skip the section entirely.

- Read the section as it appears in the hard-copy text and omit the proofs that you would have to access from the on-screen version by clicking on a link.
- Read the section from the on-screen version of the text and use the links that are provided there to jump to the proofs of all of the theorems.

11.9.2   Definition of a Riemann Sum

Suppose that f is a bounded function defined on an interval $[a, b]$ and that

$$\mathcal{P} = (x_0, x_1, \cdots, x_n)$$

is a partition of $[a, b]$. A **Riemann sum** of the function f over the partition $\mathcal{P}$ is defined to be a number that can be expressed in the form

$$\sum_{j=1}^{n} f(t_j)\, (x_j - x_{j-1}),$$

where, for each $j = 1, 2, \cdots, n$, the number t_j is chosen in such a way that

$$x_{j-1} \leq t_j \leq x_j.$$

$$x_{j-1} \hspace{6cm} t_j \hspace{1.5cm} x_j$$

For example, we could choose $t_j = x_{j-1}$ for each j, in which case the sum becomes

$$\sum_{j=1}^{n} f(x_{j-1})\, (x_j - x_{j-1})$$

and is called the **left sum** of the function f over the partition $\mathcal{P}$. If we choose $t_j = x_j$ for each j, then the sum becomes

$$\sum_{j=1}^{n} f(x_j)\, (x_j - x_{j-1})$$

and is called the **right sum** of f over $\mathcal{P}$. The arithmetic mean of the left and right sums is

$$\sum_{j=1}^{n} \left(\frac{f(x_{j-1}) + f(x_j)}{2} \right) (x_j - x_{j-1}),$$

which is called the **trapezoidal sum** of f over $\mathcal{P}$. If we choose t_j to be midway between x_{j-1} and x_j for each j, then the sum becomes

$$\sum_{j=1}^{n} f\left(\frac{x_{j-1} + x_j}{2}\right)(x_j - x_{j-1})$$

and is known as the **midpoint sum** of f over $\mathcal{P}$.

Ⓝ If you would like to experiment with these sums and one or two others and to explore their accuracy with the help of *Scientific Notebook*, click on the icon

▨ . A preview of this interactive process can be seen in the movie 👆 .

11.9.3   Darboux's Theorem

Suppose that f is integrable on an interval $[a, b]$ and that $\varepsilon > 0$. Then there exists a number $\delta > 0$ such that, for every partition

$$\mathcal{P} = (x_0, x_1, \cdots, x_n)$$

of the interval $[a, b]$ satisfying the inequality $\|\mathcal{P}\| < \delta$ and for every possible choice of numbers $t_j \in [x_{j-1}, x_j]$ for each $j = 1, 2, \cdots, n$, we have

$$\left|\int_a^b f - \sum_{j=1}^{n} f(t_j)(x_j - x_{j-1})\right| < \varepsilon.$$

11.9.4   An Extension of the Criteria for Integrability

Suppose that f is integrable on an interval $[a, b]$. Then the following extended versions of the criteria for integrability must hold:

1. *For every number $\varepsilon > 0$ there exists a number $\delta > 0$ such that for every partition $\mathcal{P}$ of $[a, b]$ for which $\|\mathcal{P}\| < \delta$ we have*

$$\int_a^b w(\mathcal{P}, f) < \varepsilon.$$

2. *For every number $\varepsilon > 0$ there exists a number $\delta > 0$ such that for every partition $\mathcal{P}$ of $[a, b]$ for which $\|\mathcal{P}\| < \delta$, if we define*

$$E = \{x \in [a, b] \mid w(\mathcal{P}, f)(x) \geq \varepsilon\},$$

then $m(E) < \varepsilon$.

11.9.5   Another Criterion for Integrability

Suppose that f is a bounded function on an interval $[a, b]$. Suppose that I is a given number and, for each positive integer n, suppose that $\mathcal{P}_n$ is the regular n-partition of the interval $[a, b]$. For each n and for $j = 0, 1, \cdots, n$ we shall write the jth point of the partition $\mathcal{P}_n$ as $x(j, n)$. Thus

$$x(j, n) = a + \frac{j(b - a)}{n}.$$

The following conditions are equivalent:

1. *The function f is Riemann integrable on the interval $[a, b]$ and $\int_a^b f = I$.*
2. *For every possible way of choosing a number $t(j, n)$ in the interval $[x(j - 1, n), x(j, n)]$ for each n and j we have*

$$\lim_{n \to \infty} \frac{b - a}{n} \sum_{j=1}^{n} f(t(j, n)) = I.$$

To see the proofs of the theorems in this section, click on the icon .

11.10   The Role of Continuity in Riemann Integration

In this section we shall observe that continuity plays a major role in Riemann integrability, and we shall introduce some theorems that tell us that a given bounded function will be integrable on a given interval as long as it is continuous at sufficiently many numbers in the interval. We begin by observing that continuous functions are always integrable.

11.10.1   Integrability of Continuous Functions

Suppose that f is a continuous function on an interval $[a, b]$. Then f is integrable on $[a, b]$.

Proof. In order to show that f is integrable, we shall show that f satisfies the second criterion for integrability that appears in Theorem 11.8.4. Suppose that $\varepsilon > 0$. Using the fact that f is uniformly continuous on $[a, b]$, choose a number $\delta > 0$ such that whenever t and x belong to the interval $[a, b]$ and $|t - x| < \delta$ we have $|f(t) - f(x)| < \varepsilon/2$.

Now choose a partition $\mathcal{P}$ of the interval $[a, b]$ whose mesh (see Subsection 11.2.1) is less than δ. (For example, we could choose $\mathcal{P}$ to be the regular n-partition of $[a, b]$ with n sufficiently large.) Now, whenever t and x are numbers that belong to the same interval of this partition, it follows from the fact that

$|t - x| < \delta$ that $|f(t) - f(x)| < \varepsilon/2$. From this fact we deduce that

$$\{x \in [a, b] \mid w(\mathcal{P}, f)(x) \geq \varepsilon\} = \emptyset,$$

and so, of course, the latter set has measure less than ε. ∎

11.10.2   The Lebesgue Criterion for Riemann Integrability

In order to prove that the function f in the preceding theorem is integrable on the interval $[a, b]$, all we needed to show was that if $\varepsilon > 0$, then there is a partition $\mathcal{P}$ of $[a, b]$ for which the measure of the set

$$\{x \in [a, b] \mid w(\mathcal{P}, f)(x) \geq \varepsilon\}$$

is less than ε. In fact, we showed much more than this. We showed that the latter set will actually be empty for every partition that has sufficiently small mesh.

As one might suspect, the preceding theorem can be improved greatly. There is a theorem, known as the **Lebesgue criterion for Riemann integrability,** that tells us that a given bounded function f on an interval $[a, b]$ will be Riemann integrable on $[a, b]$ if and only if the set of numbers $x \in [a, b]$ at which f is discontinuous is, in some sense, small enough. The notion of "small enough" that is required by the Lebesgue criterion is expressed by saying that the set of numbers at which f fails to be continuous has **measure zero**. A study of this concept lies beyond the scope of the present chapter, but, after you have read this chapter, the chapter on infinite series and the chapter on sequences and series of functions, you can reach a special chapter that presents the measure zero concept by clicking on the icon [icon] . That special chapter also presents several other interesting facts about Riemann integration that lie beyond the scope of the present chapter.

In our next theorem we study a simpler, but still very useful, form of the Lebesgue criterion for Riemann integrability that we shall call the **junior Lebesgue criterion**.

11.10.3   The Junior Lebesgue Criterion for Riemann Integrability

Suppose that f is a bounded function defined on an interval $[a, b]$. Then a sufficient condition for the function f to be integrable on $[a, b]$ is that for every number $\varepsilon > 0$ there exists an elementary set E such that $m(E) < \varepsilon$ and such that f is continuous at every number in the set $[a, b] \setminus E$.

Proof. [icon]   Just as in the proof of Theorem 11.10.1, our method of showing that f is integrable will be to show that f satisfies the second criterion for integrability that appears in Theorem 11.8.4.

Suppose that $\varepsilon > 0$. Choose an elementary set E such that $m(E) < \varepsilon$

and such that f is continuous at every number in the set $[a, b] \setminus E$. Now, using Theorem 11.4.8, choose an open elementary set U such that $E \subseteq U$ and $m(U) < \varepsilon$. Since f is continuous on the closed bounded set $[a, b] \setminus U$, we know from Theorem 8.12.6 that f is uniformly continuous on this set. Using this fact, we choose a number $\delta > 0$ such that whenever t and x are numbers in the set $[a, b] \setminus U$ and $|t - x| < \delta$ we have $|f(t) - f(x)| < \varepsilon/2$.

Choose a partition $\mathcal{P}_1$ of $[a, b]$ such that $\|\mathcal{P}_1\| < \delta$ and choose a partition $\mathcal{P}_2$ of $[a, b]$ such that the step function χ_U steps within $\mathcal{P}_2$. We now define $\mathcal{P}$ to be the common refinement of $\mathcal{P}_1$ and $\mathcal{P}_2$ and we write $\mathcal{P}$ in the form

$$\mathcal{P} = (x_0, x_1, \cdots, x_n).$$

What is important about this partition $\mathcal{P}$ is that, for each $j = 1, 2, \cdots, n$, the interval (x_{j-1}, x_j) is either included in the set U or is disjoint from U, depending on whether the constant value of χ_U in (x_{j-1}, x_j) is 1 or 0. Furthermore, if (x_{j-1}, x_j) is disjoint from U, the inequality $x_j - x_{j-1} < \delta$ guarantees that $|f(t) - f(x)| < \varepsilon/2$ for all numbers t and x in (x_{j-1}, x_j). Thus, when (x_{j-1}, x_j) is disjoint from U, the constant value of $w(\mathcal{P}, f)$ in the interval (x_{j-1}, x_j) does not exceed $\varepsilon/2$.

Therefore, if we define

$$A = \{x \in [a, b] \mid w(\mathcal{P}, f)(x) \geq \varepsilon\},$$

we have $A \subseteq U$ and it follows that $m(A) < \varepsilon$. $\blacksquare$

11.10.4   A Remark on the Junior Lebesgue Criterion

Unlike the more sophisticated Lebesgue criterion that can be reached by clicking on the icon 🖼 , the junior Lebesgue criterion is not both necessary and sufficient for Riemann integrability. It is merely sufficient. In other words, it is possible for a function f that fails to satisfy the junior Lebesgue criterion to be integrable. In Exercise 5b of Subsection 11.8.5 you were invited to prove that the ruler function that had been introduced in Example 8 of Subsection 8.1.5 is integrable on $[0, 1]$ even though it is discontinuous at every rational number in the interval. The ruler function does not satisfy the junior Lebesgue criterion for integrability, because any elementary set that contains all of the rational numbers in $[0, 1]$ must contain all but finitely many of the numbers in the interval $[0, 1]$ and must therefore have measure one.

11.10.5   Some Exercises on the Junior Lebesgue Criterion

1. True or false? Every step function satisfies the junior Lebesgue criterion.
2. 🅷 Suppose that (x_n) is a convergent sequence in an interval $[a, b]$ and that

f is a bounded function on $[a, b]$ that is continuous at every member of $[a, b]$ that does not lie in the range of the sequence (x_n). Prove that f is integrable on $[a, b]$.

3. $\blacksquare$ Suppose that (x_n) is a sequence in an interval $[a, b]$, that (x_n) has only finitely many partial limits, and that f is a bounded function on $[a, b]$ that is continuous at every member of $[a, b]$ that does not belong to the range of the sequence (x_n). Prove that f is integrable on $[a, b]$.

4. $\blacksquare$ This exercise does not ask you for a proof. Suppose that (x_n) is a sequence in an interval $[a, b]$ and that f is a bounded function on $[a, b]$ that is continuous at every member of $[a, b]$ that does not belong to the range of the sequence (x_n). Do you think that the function f has to be integrable on $[a, b]$? What does your intuition tell you?

11.11   The Composition Theorem for Riemann Integrability

We begin this section on the composition theorem with two simple special cases that will help to motivate the main result, and which are of importance in their own right.

11.11.1   Integrability of the Absolute Value Function

Suppose that f is integrable on an interval $[a, b]$. Then the function $|f|$ is also integrable and we have

$$\left| \int_a^b f \right| \leq \int_a^b |f| .$$

Proof. Once again, our method of showing at a function is integrable will be to show that it satisfies the second criterion for integrability that appears in Theorem 11.8.4.

Suppose that $\varepsilon > 0$ and, using the fact that f is integrable, choose a partition $\mathcal{P}$ of $[a, b]$ such that if

$$E = \{x \in [a, b] \mid w(\mathcal{P}, f)(x) \geq \varepsilon\} ,$$

then $m(E) < \varepsilon$. Now, since the inequality

$$||f(t)| - |f(x)|| \leq |f(t) - f(x)|$$

holds for all numbers t and x in the interval, we see that if

$$E^* = \{x \in [a, b] \mid w(\mathcal{P}, |f|)(x) \geq \varepsilon\} ,$$

then $E^* \subseteq E$ and so $m(E^*) < \varepsilon$. Now that we have shown that the function $|f|$ is integrable we can use the nonnegativity property and the inequality

$$-|f| \leq f \leq |f|$$

to show that

$$\left| \int_a^b f \right| \leq \int_a^b |f|. \quad \blacksquare$$

11.11.2   Integrability of the Square of a Function

Suppose that f is integrable on an interval $[a, b]$. Then the function f^2 is also integrable on $[a, b]$.

Proof. Using the fact that f is bounded we choose a positive number k such that $|f(x)| \leq k$ for all $x \in [a, b]$. Note that if t and x are any numbers in the interval $[a, b]$, we have

$$\left| f^2(t) - f^2(x) \right| = |f(t) - f(x)| \, |f(t) + f(x)| \leq 2k \, |f(t) - f(x)|.$$

Thus if $\mathcal{P}$ is any partition of $[a, b]$, we have

$$w\left(\mathcal{P}, f^2\right) \leq 2kw\left(\mathcal{P}, f\right).$$

To show that the function f^2 satisfies the second criterion for integrability that appears in Theorem 11.8.4, suppose that $\varepsilon > 0$. We define α to be the smaller of the two numbers ε and $\varepsilon/2k$ and, using the fact that f is integrable, choose a partition $\mathcal{P}$ of $[a, b]$ such that if we define

$$E = \{x \in [a, b] \mid w(\mathcal{P}, f)(x) \geq \alpha\},$$

then $m(E) < \alpha$. We deduce that if

$$E^* = \left\{x \in [a, b] \mid w\left(\mathcal{P}, f^2\right)(x) \geq \varepsilon\right\},$$

then $m(E^*) < \varepsilon$. $\blacksquare$

11.11.3   The Product of Integrable Functions

If f and g are integrable on an interval $[a, b]$, then so is fg.

Proof. Since we already know that sums and differences, squares and constant multiples of integrable functions are integrable, the result follows from the identity

$$fg = \frac{(f+g)^2 - (f-g)^2}{4}. \quad \blacksquare$$

11.11.4   Statement of the Composition Theorem for Integrability

Suppose that f is integrable on an interval $[a, b]$ and that h is continuous and bounded on the range of f. Then the composition $h \circ f$ of the functions f and h is integrable on $[a, b]$.

Unfortunately, this form of the composition theorem seems to be very hard to prove. You can find a proof by clicking on the icon ![icon] , which will take you to the special chapter on sets of measure zero.

In our next theorem we study a simpler, but still very useful, form of the composition theorem for Riemann integrability that we shall call the **junior composition theorem.**

11.11.5   The Junior Composition Theorem for Integrability

Suppose that f is integrable on an interval $[a, b]$ and that h is uniformly continuous on the range of f. Then the composition $h \circ f$ of the functions f and h is integrable on $[a, b]$.

Proof. ![icon]   We write $g = h \circ f$, in other words,

$$g(x) = h\left(f(x)\right)$$

for every $x \in [a, b]$. The function g is bounded. To see why, look at the solution to Exercise 6d of Subsection 8.12.7. Once again, our method of showing that a function is integrable will be to show that it satisfies the second criterion for integrability that appears in Theorem 11.8.4.

Suppose that $\varepsilon > 0$ and, using the fact that h is uniformly continuous on the range of f, choose a number $\delta > 0$ such that whenever y and z are in the range of f and $|y - z| < \delta$ we have

$$|h(y) - h(z)| < \frac{\varepsilon}{2}.$$

We now define α to be the smaller of the two numbers ε and δ and, using the integrability of f, we choose a partition

$$\mathcal{P} = (x_0, x_1, \cdots, x_n)$$

of the interval $[a, b]$ such that, if

$$E = \{x \in [a, b] \mid w\left(\mathcal{P}, f\right)(x) \geq \alpha,\},$$

then $m\left(E\right) < \alpha$. Now for any $j = 1, 2, \cdots, n$, unless the interval (x_{j-1}, x_j) is included in E, we know that for all numbers t and x in (x_{j-1}, x_j) we have

$$|f(t) - f(x)| < \alpha \leq \delta,$$

and for such numbers t and x we have

$$|h\left(f(t)\right) - h\left(f(x)\right)| < \frac{\varepsilon}{2}.$$

Thus, in any interval (x_{j-1}, x_j) that is not included in E, the constant value of $w\left(\mathcal{P}, g\right)$ cannot exceed $\varepsilon/2$. Thus if

$$A = \{x \in [a, b] \mid w\left(\mathcal{P}, g\right)(x) \geq \varepsilon\},$$

then A is an elementary set, $A \subseteq E$ and

$$m\left(A\right) \leq m\left(E\right) < \alpha \leq \varepsilon. \quad \blacksquare$$

11.11.6   Some Exercises on the Composition Theorem

1. Given two functions f and g defined on a set S, we define the functions $f \vee g$ and $f \wedge g$ as follows:

$$f \vee g(x) = \begin{cases} f(x) & \text{if } f(x) \geq g(x) \\ g(x) & \text{if } f(x) < g(x) \end{cases}$$

and

$$f \wedge g(x) = \begin{cases} f(x) & \text{if } f(x) \leq g(x) \\ g(x) & \text{if } f(x) > g(x). \end{cases}$$

Given Riemann integrable functions f and g on an interval $[a, b]$, make the observations

$$f \vee g = \frac{f + g + |f - g|}{2}$$

and

$$f \wedge g = \frac{f + g - |f - g|}{2}$$

and deduce that the functions $f \vee g$ and $f \wedge g$ are also integrable on $[a, b]$.

2. Given that f is a nonnegative integrable function on an interval $[a, b]$, explain why the function $\sqrt{f}$ is integrable.

3. Given that f is integrable on an interval $[a, b]$, that $f(x) \geq 1$ for every $x \in [a, b]$, and that

$$g(x) = \log\left(f(x)\right)$$

for every $x \in [a, b]$, explain why the function g must be integrable on $[a, b]$.

4. Suppose that f is integrable on an interval $[a, b]$ and that $\alpha \leq f(x) \leq \beta$ for every $x \in [a, b]$. Show how the junior version of the composition theorem

for integrability can be used to show that if h is any continuous function on the interval $[\alpha, \beta]$ the function $h \circ f$ is integrable on $[a, b]$.

11.12   The Fundamental Theorem of Calculus

In this section we explore the important link between differential calculus and integral calculus.

11.12.1   Continuity of an Integral

Suppose that f is integrable on an interval $[a, b]$ and suppose that we have defined

$$F(x) = \int_a^x f$$

for every number x in $[a, b]$. Then the function F is uniformly continuous on the interval $[a, b]$.

Proof. Choose a positive number k such that the inequality $|f(x)| \leq k$ holds for every number $x \in [a, b]$. We shall now demonstrate that if x and t are any numbers in the interval $[a, b]$, then

$$|F(x) - F(t)| \leq k \, |x - t| \, .$$

Suppose that x and t belong to $[a, b]$. We may assume, without loss of generality, that $t \leq x$. Now

$$
\begin{aligned}
|F(x) - F(t)| \; &= \; \left| \int_a^x f - \int_a^t f \right| = \left| \int_t^x f \right| \\
&\leq \; \int_t^x |f| \leq \int_t^x k = k \, |x - t| \, . \; \blacksquare
\end{aligned}
$$

11.12.2   Differentiating an Integral

We shall now study the theorem that, in elementary calculus, is stated traditionally in the form

$$\frac{d}{dx} \int_a^x f(t)dt = f(x).$$

In its precise form, the theorem can be stated as follows:

Suppose that f is integrable on an interval $[a, b]$ and suppose that we have *defined*

$$F(x) = \int_a^x f$$

for every number x in $[a, b]$. Suppose that $c \in [a, b]$ and that f is continuous at the number c. Then the function F is differentiable at c and we have $F'(c) = f(c)$.

Proof. We begin by observing that if x is any number in the interval $[a, b]$ and $x \neq c$, we have

$$\frac{F(x) - F(c)}{x - c} - f(c) = \frac{1}{x - c} \int_c^x (f - f(c)).$$

We shall show that the latter expression approaches 0 as $x \to c$. For this purpose, suppose that $\varepsilon > 0$.

Using the fact that f is continuous at the number c, choose a number $\delta > 0$ such that for every number x in the interval $[a, b]$ satisfying the inequality $|x - c| < \delta$ we have

$$|f(x) - f(c)| < \varepsilon.$$

From the version of nonnegativity that appears in Subsection 11.7.4, we see that

$$\left| \frac{F(x) - F(c)}{x - c} - f(c) \right| = \frac{1}{|x - c|} \left| \int_c^x (f - f(c)) \right| \leq \frac{1}{|x - c|} \varepsilon |x - c| = \varepsilon,$$

and so

$$\lim_{x \to c} \left(\frac{F(x) - F(c)}{x - c} - f(c) \right) = 0,$$

as promised. ∎

11.12.3   Integrating a Derivative

Suppose that f is a differentiable function on an interval $[a, b]$ and that its derivative f' is integrable on $[a, b]$. Then

$$\int_a^b f' = f(b) - f(a).$$

Proof. We shall prove the theorem by showing that if s and S are any two step functions on the interval $[a, b]$ satisfying the inequality

$$s \leq f' \leq S,$$

we have

$$\int_a^b s \le f(b) - f(a) \le \int_a^b S.$$

For this purpose, suppose that s and S are step functions on $[a, b]$ and that $s \le f' \le S$. Choose a partition

$$\mathcal{P} = (x_0, x_1, \cdots, x_n)$$

of the interval $[a, b]$ such that both of the step functions s and S step within $\mathcal{P}$. In each interval (x_{j-1}, x_j) of the partition we shall call the constant values of the functions s and S by the names α_j and β_j, respectively. For each $j = 1, 2, \cdots, n$ we now apply the mean value theorem (Theorem 9.4.3) to the function f on the interval $[x_{j-1}, x_j]$ to choose a number $t_j \in (x_{j-1}, x_j)$ such that

$$f(x_j) - f(x_{j-1}) = (x_j - x_{j-1}) f'(t_j).$$

Now, since

$$\alpha_j \le f'(t_j) \le \beta_j$$

for each j, we see that

$$\int_a^b s = \sum_{j=1}^n \alpha_j (x_j - x_{j-1}) \le \sum_{j=1}^n f'(t_j) (x_j - x_{j-1})$$

$$\le \sum_{j=1}^n \beta_j (x_j - x_{j-1}) = \int_a^b S.$$

Thus

$$\int_a^b s \le \sum_{j=1}^n (f(x_j) - f(x_{j-1})) \le \int_a^b S$$

and we conclude that

$$\int_a^b s \le f(b) - f(a) \le \int_a^b S. \quad \blacksquare$$

11.12.4   Integration by Parts

Suppose that f and g are differentiable functions on an interval $[a, b]$ and that their derivatives f' and g' are integrable on $[a, b]$. Then

$$\int_a^b fg' = f(b)g(b) - f(a)g(a) - \int_a^b f'g.$$

Proof. We observe first that since the functions $f'g$ and fg', being products of integrable functions, are integrable by Theorem 11.11.3, the function $f'g + fg'$ is integrable on $[a, b]$. We now define $h = fg$. We deduce from Theorem 11.12.3 that

$$\int_a^b h' = h(b) - h(a),$$

and since the product rule (Theorem 9.3.4) implies that

$$h' = f'g + fg',$$

we have

$$\int_a^b (f'g + fg') = f(b)g(b) - f(a)g(a),$$

which is the conclusion of the theorem. ∎

11.13   The Change of Variable Theorem

11.13.1   Introduction to the Change of Variable Theorem

Just as the integration by parts formula that we have just seen can be thought of as an integral version of the product rule for differentiation, so too can the change of variable theorem be thought of as an integral version of the chain rule. To motivate the theorem, we shall look at an example of the sort of integral that is evaluated with the help of the theorem in elementary calculus.

Suppose that we wanted to evaluate the integral

$$\int_{\sqrt{\pi-1}}^{\sqrt{2\pi-1}} 2t \sin\left(1 + t^2\right) dt.$$

In elementary calculus we evaluate this integral by making the substitution $x = 1 + t^2$. From the equation

$$\frac{dx}{dt} = 2t$$

we deduce that, inside the integral sign, the symbol $2tdt$ can be replaced by the symbol dx provided that we change the limits of integration correctly and replace the expression $1 + t^2$ by x. Thus

$$\int_{\sqrt{\pi-1}}^{\sqrt{2\pi-1}} 2t \sin\left(1 + t^2\right) dt = \int_{1+\left(\sqrt{\pi-1}\right)^2}^{1+\left(\sqrt{2\pi-1}\right)^2} \sin x \, dx = \int_{\pi}^{2\pi} \sin x \, dx = -2.$$

A more precise way of describing this change of variable is to define a function u on the interval

$$\left[\sqrt{\pi - 1}, \sqrt{2\pi - 1}\right]$$

by the equation

$$u(t) = 1 + t^2$$

for each t in the interval and to note that the expression $2tdt$ is the same as $u'(t)dt$. Thus the given integral is

$$\int_{\sqrt{\pi-1}}^{\sqrt{2\pi-1}} \sin\left(u(t)\right) u'(t) dt = \int_{u\left(\sqrt{\pi-1}\right)}^{u\left(\sqrt{2\pi-1}\right)} \sin x \, dx = \int_{\pi}^{2\pi} \sin x \, dx = -2.$$

The change of variable theorem provides the theoretical basis for this technique and may be stated as follows:

11.13.2   The General Change of Variable Theorem

Suppose that u is a differentiable function on an interval $[a, b]$ and that its derivative u' is integrable on $[a, b]$. Then, given any function f that is integrable on the range of u, we have

$$\int_a^b f(u(t))u'(t)dt = \int_{u(a)}^{u(b)} f(x)dx.$$

Although this general form of the theorem has a simple and natural statement, it is unfortunately very difficult to prove, and it will not be proved in this chapter. You can reach a proof by clicking on the icon [skull icon] .

In this section we study two important special cases of the general theorem that are much easier to prove and are sufficient for most common applications of the theorem.

11.13.3   The Continuous Version of the Change of Variable Theorem

Suppose that u is a differentiable function on an interval $[a, b]$ and that its derivative u' is integrable on $[a, b]$. Then, given any function f that is continuous on *the range of u, we have*

$$\int_a^b f(u(t))u'(t)dt = \int_{u(a)}^{u(b)} f(x)dx.$$

Proof. The first observation we need to make is that, since u is continuous on $[a, b]$ and f is continuous on the range of u, the composition $f \circ u$ of these functions is continuous and therefore integrable on $[a, b]$. Since the function u' is also integrable on $[a, b]$ we know from Theorem 11.11.3 that the function $(f \circ u) u'$ is integrable on $[a, b]$. We therefore know that the integral

$$\int_a^b f(u(t))u'(t)dt$$

makes sense. Since u is continuous on $[a, b]$, we know from Theorem 8.10.4 that the range of u is an interval. Call it $[\alpha, \beta]$. For every number x in the interval $[\alpha, \beta]$ we now define

$$F(x) = \int_\alpha^x f.$$

Since f is continuous on the interval $[\alpha, \beta]$, we know from Theorem 11.12.2 that $F'(x) = f(x)$ for every $x \in [\alpha, \beta]$. We now define

$$h(t) = F(u(t))$$

for every number $t \in [a, b]$. From the chain rule we see that if $t \in [a, b]$, then

$$h'(t) = F'(u(t))u'(t) = f(u(t))u'(t),$$

and it follows from Theorem 11.12.3 that

$$\int_a^b f(u(t))u'(t)dt = \int_a^b h'(t)dt = h(b) - h(a) = F(u(b)) - F(u(a))$$

$$= \int_\alpha^{u(b)} f - \int_\alpha^{u(a)} f = \int_{u(a)}^{u(b)} f$$

and so

$$\int_a^b f(u(t))u'(t)dt = \int_{u(a)}^{u(b)} f(x)dx. \blacksquare$$

11.13.4   The Monotone Version of the Change of Variable Theorem

Suppose that u is a monotone differentiable function on an interval $[a, b]$ and that its derivative u' is integrable on $[a, b]$. Then, given any function f that is integrable on the range of u, we have

$$\int_a^b f(u(t))u'(t)dt = \int_{u(a)}^{u(b)} f(x)dx.$$

Proof. We shall assume that the function u is increasing on $[a, b]$ and leave the analogous case u decreasing as an exercise. We begin with the observation that the range of the function u is the interval $[u(a), u(b)]$. In order to prove the theorem we shall first consider the case in which the function f is a step function on the interval $[u(a), u(b)]$. Choose a partition

$$\mathcal{P} = (x_0, x_1, \cdots, x_n)$$

of the interval $[u(a), u(b)]$ within which the function f steps and call the constant value of f in each interval (x_{j-1}, x_j) by the name α_j.

For each $j = 0, 1, 2, \cdots, n$, the set of real numbers t in the interval $[a, b]$ for which

$$u(t) = x_j$$

is a closed subinterval of $[a, b]$ that we shall write as $[s_j, t_j]$. We note that

$$a = s_0 \leq t_0 < s_1 \leq t_1 < s_2 \leq t_2 < \cdots < s_{n-1} \leq t_{n-1} < s_n \leq t_n = b.$$

For each j we observe from the fact that u is constant on the interval $[s_j, t_j]$ that

$$f(u(t))u'(t) = 0$$

for every $t \in (s_j, t_j)$. Furthermore, if t is any number between t_{j-1} and s_j, we have

$$x_{j-1} < u(t) < x_j,$$

and so, for such numbers t we have

$$f(u(t))u'(t) = \alpha_j u'(t).$$

We deduce that the function $(f \circ u) u'$ is integrable on each of the intervals $[s_j, t_j]$ and on each of the intervals $[t_{j-1}, s_j]$, and we have

$$
\begin{aligned}
\int_a^b f(u(t))u'(t)dt &= \sum_{j=0}^n \int_{s_j}^{t_j} f(u(t))u'(t)dt + \sum_{j=1}^n \int_{t_{j-1}}^{s_j} f(u(t))u'(t)dt \\
&= \sum_{j=0}^n 0 + \sum_{j=1}^n \int_{t_{j-1}}^{s_j} \alpha_j u'(t)dt = \sum_{j=1}^n \alpha_j \int_{t_{j-1}}^{s_j} u'(t)dt \\
&= \sum_{j=1}^n \alpha_j \left(u(s_j) - u(t_{j-1}) \right) = \sum_{j=1}^n \alpha_j \left(x_j - x_{j-1} \right) \\
&= \int_{u(a)}^{u(b)} f(x)dx.
\end{aligned}
$$

We can now handle the general case. Using the fact that f is integrable on the interval $[u(a), u(b)]$, choose a pair of sequences of step functions that squeezes f on the interval $[u(a), u(b)]$. In other words,

$$s_n \leq f \leq S_n$$

for each n and

$$\lim_{n \to \infty} \int_{u(a)}^{u(b)} (S_n - s_n) = 0.$$

For each n it follows from the fact that the function u' is nonnegative that

$$(s_n \circ u)\, u' \leq (f \circ u)\, u' \leq (S_n \circ u)\, u'$$

and, by the case that we have already considered, we deduce that

$$\int_a^b (S_n \circ u - s_n \circ u)\, u' = \int_{u(a)}^{u(b)} (S_n - s_n) \to 0$$

as $n \to \infty$.

It follows from Theorem 11.5.3 that the function $(f \circ u)\, u'$ is integrable on the interval $[u(a), u(b)]$ and that

$$\int_a^b f(u(t))u'(t)dt = \lim_{n\to\infty} \int_a^b s_n(u(t))u'(t)dt$$

$$= \lim_{n\to\infty} \int_{u(a)}^{u(b)} s_n(x)dx = \int_{u(a)}^{u(b)} f(x)dx. \ \blacksquare$$

11.13.5   Exercises on the Change of Variable Theorem

1. (a) **H** Given that f is a continuous function on the interval $[-1, 1]$, prove
   that

$$\int_0^{2\pi} f(\sin x)\cos x\,dx = 0.$$

   (b) **H** Given that f is a continuous function on the interval $[0, 1]$, prove that

$$\int_0^{\pi/2} f(\sin x)\,dx = \int_{\pi/2}^{\pi} f(\sin x)\,dx.$$

   (c) **H** Given that $\alpha > 0$, prove that

$$\int_0^{\pi/2} \sin^\alpha x\,dx = 2^\alpha \int_0^{\pi/2} \sin^\alpha x\cos^\alpha x\,dx.$$

2. Given that u is a differentiable function on an interval $[a, b]$ and that its
   derivative u' is integrable on $[a, b]$, and given that $u(a) = u(b)$ and that f is
   integrable on the range of u, prove that

$$\int_a^b f(u(t))u'(t)dt = 0.$$

3. Given that f is integrable on an interval $[a, b]$ and that c is any number, prove
   that

$$\int_a^b f(t)dt = \int_{a+c}^{b+c} f(t-c)dt$$

4. **H** Given that a, b, and c are real numbers; that $ac < bc$; and that f is a
   continuous function on the interval $[ac, bc]$, prove that

$$\int_{ac}^{bc} f(t)dt = c \int_a^b f(ct)dt.$$

5. (a) Suppose that f is a continuous function on an interval $[a, b]$, that g is integrable on $[a, b]$, and that $\int_a^b g > 0$. Prove that the ratio

$$\frac{\int_a^b fg}{\int_a^b g}$$

lies between the minimum and maximum values of f and deduce that there exists a number $c \in [a, b]$ such that

$$\int_a^b fg = f(c) \int_a^b g.$$

This fact is sometimes called the **mean value theorem for integrals**.

(b) Given that f is continuous on an interval $[a, b]$, prove that there exits a number $c \in [a, b]$ such that

$$\int_a^b f = f(c)(b - a).$$

6. (a) Given that f is a nonnegative continuous function on an interval $[a, b]$, where $a < b$, and that $\int_a^b f = 0$, prove that f is the constant zero function.

(b) Given that f is a continuous function on an interval $[a, b]$, where $a < b$, and that $\int_a^x f = 0$ for every $x \in [a, b]$, prove that f is the constant zero function.

7. The on-screen text provides a link to one proof of the "u decreasing" form of the change of variable theorem. In this exercise we consider another proof of this result:

(a) Given that f is an integrable function on an interval $[a, b]$ and that $g(t) = f(-t)$ whenever $-b \leq t \leq -a$, give a direct proof that g is integrable on the interval $[-b, -a]$ and that

$$\int_{-b}^{-a} g(t)dt = \int_a^b f(x)dx.$$

(b) Suppose that u is a decreasing differentiable function on an interval $[a, b]$ and that the derivative u' of u is integrable on $[a, b]$. Apply the form of Theorem 11.13.4 proved above to the function v defined by the equation

$v(t) = -u(t)$ for $-b \leq t \leq -a$ to show that the equation

$$\int_a^b f(u(t))u'(t)dt = \int_{u(a)}^{u(b)} f(x)dx$$

holds for every function f that is integrable on the interval $[u(b), u(a)]$.

To reach some additional exercises that invite you to develop some important inequalities, click on the icon ▨ .

11.14   Integration of Complex Functions (Optional)

To reach this optional section from the on-screen version of the book click on the icon ▨ .

Chapter 12
Infinite Series

This chapter will present the theory of numerical series, the notion of convergence of a series, and the standard tests for convergence, some of which you have probably seen before. Your study of this chapter will pave the way to the notion of expansion of functions in series that appears in Chapter 14.

12.1   Introduction to Infinite Series

12.1.1   Motivating the Concept of a Series

In this chapter we study the idea of an infinite sum

$$a_1 + a_2 + a_3 + \cdots + a_n + \cdots = \sum_{j=1}^{\infty} a_j,$$

where (a_n) is a given sequence of numbers. Before we begin our formal study of this interesting topic we shall toy with some examples that hint at what we can do and what we cannot do with infinite sums.

1. The infinite repeating decimal $0.\overline{258}$ can be thought of as the infinite sum

$$\frac{258}{1000} + \frac{258}{(1000)^2} + \frac{258}{(1000)^3} + \cdots + \frac{258}{(1000)^n} + \cdots.$$

If we call this number s, then, since

$$1000s = 258 + \frac{258}{1000} + \frac{258}{(1000)^2} + \frac{258}{(1000)^3} + \cdots + \frac{258}{(1000)^n} + \cdots = 258 + s,$$

we conclude that

$$0.\overline{258} = \frac{258}{999}.$$

A brief discussion of decimals will be given in Section 12.5 where a link is also provided to a more detailed presentation of this topic.

2. Suppose that c is any number and that s is the infinite sum

$$1 + c + c^2 + c^3 + \cdots + c^n + \cdots.$$

335

Since

$$
\begin{aligned}
cs &= \quad\;\; c + c^2 + \cdots + c^n + \cdots \\
   &= \;\; 1 + c + c^2 + \cdots + c^n + \cdots - 1 = s - 1
\end{aligned}
$$

we deduce that

$$
s = \frac{1}{1 - c}.
$$

If we take $c = 1/2$ in this identity, we obtain the identity

$$
1 + \frac{1}{2} + \left(\frac{1}{2}\right)^2 + \left(\frac{1}{2}\right)^3 + \cdots = 2,
$$

which is a well-known fact. However, if we put $c = 2$, then we obtain the absurd result

$$
1 + 2 + 2^2 + 2^3 + \cdots = -1,
$$

which warns us that we cannot work with an infinite sum until we know that it exists, what it means and how it behaves.

3. Suppose that s is the infinite sum

$$
1 - 1 + 1 - 1 + 1 - \cdots + (-1)^{n-1} + \cdots.
$$

We shall look at this sum in several ways. First we see that

$$
1 - 1 + 1 - 1 + 1 - \cdots = (1 - 1) + (1 - 1) + (1 - 1) + \cdots = 0.
$$

Next we change the bracketing and observe that

$$
1 - 1 + 1 - 1 + 1 - \cdots = 1 - (1 - 1) - (1 - 1) - \cdots = 1.
$$

Finally we substitute $c = -1$ in the identity

$$
1 + c + c^2 + c^3 + \cdots + c^n + \cdots = \frac{1}{1 - c}
$$

and obtain

$$
1 - 1 + 1 - 1 + 1 - \cdots + (-1)^{n-1} + \cdots = \frac{1}{2}.
$$

Which of these three values is correct? Actually, none of them. This infinite sum simply doesn't make sense.

4. In this example we shall attempt to add up all of the numbers in the infinite array

$$
\begin{array}{cccccccccc}
1 & -1 & 0 & 0 & 0 & 0 & 0 & 0 & 0 & 0 & \cdots \\
1 & 1 & -1 & -1 & 0 & 0 & 0 & 0 & 0 & 0 & \cdots \\
1 & 1 & 1 & -1 & -1 & -1 & 0 & 0 & 0 & 0 & \cdots \\
1 & 1 & 1 & 1 & -1 & -1 & -1 & -1 & 0 & 0 & \cdots \\
\vdots & \vdots & \vdots & \vdots & \vdots & \vdots & \vdots & \vdots & \vdots & \vdots & \ddots
\end{array}
$$

We shall perform this summation in two ways. First we add the numbers in each row and place the totals in an extra column to the right of the array. The sum of the numbers in this column is

$$0 + 0 + 0 + \cdots = 0.$$

$$
\begin{array}{cccccccccc|c}
1 & -1 & 0 & 0 & 0 & 0 & 0 & 0 & 0 & 0 & \cdots & 0 \\
1 & 1 & -1 & -1 & 0 & 0 & 0 & 0 & 0 & 0 & \cdots & 0 \\
1 & 1 & 1 & -1 & -1 & -1 & 0 & 0 & 0 & 0 & \cdots & 0 \\
1 & 1 & 1 & 1 & -1 & -1 & -1 & -1 & 0 & 0 & \cdots & 0 \\
\vdots & \vdots & \vdots & \vdots & \vdots & \vdots & \vdots & \vdots & \vdots & \vdots & \ddots & \vdots \\
\hline
\infty & \infty & \infty & \infty & \infty & \infty & \infty & \infty & \infty & \infty & \cdots &
\end{array}
$$

Second, we add the numbers in each column and place the totals in an extra row at the bottom of the array. The sum of the numbers in this row is

$$\infty + \infty + \infty + \cdots = \infty.$$

We have therefore found two different ways of adding the numbers in the array that lead to two dramatically different answers.

5. In this example we shall use the identity

$$\frac{1}{2} + \left(\frac{1}{2}\right)^2 + \left(\frac{1}{2}\right)^3 + \cdots = 1$$

as we attempt to add up all of the numbers in the infinite array

$$
\begin{array}{cccccccccc}
-1 & \frac{1}{2} & \frac{1}{4} & \frac{1}{8} & \frac{1}{16} & \frac{1}{32} & \frac{1}{64} & \frac{1}{128} & \cdots \\
0 & -1 & \frac{1}{2} & \frac{1}{4} & \frac{1}{8} & \frac{1}{16} & \frac{1}{32} & \frac{1}{64} & \cdots \\
0 & 0 & -1 & \frac{1}{2} & \frac{1}{4} & \frac{1}{8} & \frac{1}{16} & \frac{1}{32} & \cdots \\
0 & 0 & 0 & -1 & \frac{1}{2} & \frac{1}{4} & \frac{1}{8} & \frac{1}{16} & \cdots \\
0 & 0 & 0 & 0 & -1 & \frac{1}{2} & \frac{1}{4} & \frac{1}{8} & \cdots \\
\vdots & \vdots & \vdots & \vdots & \vdots & \vdots & \vdots & \vdots & \ddots
\end{array}
$$

We shall perform this summation in two ways. First we add the numbers in each row and place the totals in an extra column to the right of the array. The sum of the numbers in this column is

$$0 + 0 + 0 + \cdots = 0.$$

$$
\begin{array}{cccccccccc|c}
-1 & \frac{1}{2} & \frac{1}{4} & \frac{1}{8} & \frac{1}{16} & \frac{1}{32} & \frac{1}{64} & \frac{1}{128} & \cdots & 0 \\
0 & -1 & \frac{1}{2} & \frac{1}{4} & \frac{1}{8} & \frac{1}{16} & \frac{1}{32} & \frac{1}{64} & \cdots & 0 \\
0 & 0 & -1 & \frac{1}{2} & \frac{1}{4} & \frac{1}{8} & \frac{1}{16} & \frac{1}{32} & \cdots & 0 \\
0 & 0 & 0 & -1 & \frac{1}{2} & \frac{1}{4} & \frac{1}{8} & \frac{1}{16} & \cdots & 0 \\
0 & 0 & 0 & 0 & -1 & \frac{1}{2} & \frac{1}{4} & \frac{1}{8} & \cdots & 0 \\
\vdots & \vdots & \vdots & \vdots & \vdots & \vdots & \vdots & \vdots & \ddots & \vdots \\
\hline
-1 & -\frac{1}{2} & -\frac{1}{4} & -\frac{1}{8} & -\frac{1}{16} & -\frac{1}{32} & -\frac{1}{64} & -\frac{1}{128} & \cdots &
\end{array}
$$

Second, we add the numbers in each column and place the totals in an extra row at the bottom of the array. The sum of the numbers in this row is

$$-1 - \frac{1}{2} - \frac{1}{4} - \frac{1}{8} - \cdots = -2.$$

We have therefore found two different ways of adding the numbers in the array that lead to two different answers.

These examples serve as a warning that infinite sums need to be defined carefully. We need to know when they exist and we need to know exactly what these sums mean when they exist and how they behave. Once this has been done we shall be able to give some genuine examples of infinite sums. The following are just a few of the facts that you will encounter in various parts of this book:

If $-1 < c < 1$, then

$$1 + c + c^2 + \cdots + +c^n + \cdots = \sum_{j=0}^{\infty} c^j = \frac{1}{1-c},$$

$$1 - \frac{1}{2} + \frac{1}{3} - \frac{1}{4} + \cdots + \frac{(-1)^{j-1}}{j} + \cdots = \sum_{j=1}^{\infty} \frac{(-1)^{j-1}}{j} = \log 2,$$

$$1 - \frac{1}{3} + \frac{1}{5} - \frac{1}{7} + \cdots + \frac{(-1)^{j-1}}{2j-1} + \cdots = \sum_{j=1}^{\infty} \frac{(-1)^{j-1}}{2j-1} = \frac{\pi}{4},$$

$$\frac{1}{1^2} + \frac{1}{2^2} + \frac{1}{3^2} + \cdots + \frac{1}{n^2} + \cdots = \sum_{j=1}^{\infty} \frac{1}{j^2} = \frac{\pi^2}{6},$$

$$1 + \frac{1}{2} + \frac{1}{3} + \frac{1}{4} + \cdots + \frac{1}{j} + \cdots = \sum_{j=1}^{\infty} \frac{1}{j} = \infty.$$

If x is any real number, then

$$\sum_{j=0}^{\infty} \frac{x^j}{j!} = e^x.$$

12.1.2   The Basic Definitions

If (a_n) is a given sequence of numbers, then the symbol $\sum a_n$ stands for the sequence whose first term is a_1, whose second term is $a_1 + a_2$, whose third term is $a_1 + a_2 + a_3$, and whose nth term is

$$\sum_{j=1}^{n} a_j$$

for each n. The symbol $\sum a_n$ is known as the **series associated with the sequence** (a_n) and is also called the **series with nth term** a_n. Note that every time we use the word *series* we are actually referring to the series associated with a given sequence. We shall never refer to "a series" as if it were something in its own right.

In the event that the domain of a given sequence (a_n) of numbers starts at an integer k, the series with nth term a_n would have the form

$$a_k, \quad a_k + a_{k+1}, \quad a_k + a_{k+1} + a_{k+2}, \quad \cdots, \quad \sum_{j=k}^{n} a_j, \quad \cdots$$

but, in order to keep the notation as simple as possible, we shall usually take $k = 1$ and regard a series $\sum a_n$ as being

$$a_1, \quad a_1 + a_2, \quad a_1 + a_2 + a_3, \quad \cdots, \quad \sum_{j=1}^{n} a_j, \quad \cdots$$

with the understanding that we can start at any given integer k if it suits us to do so.

Given a sequence (a_n) we shall call each number

$$\sum_{j=1}^{n} a_j$$

the **nth partial sum of** $\sum a_n$. If the limit

$$\lim_{n \to \infty} \sum_{j=1}^{n} a_j$$

exists, this limit is called the **sum** of the series $\sum a_n$ and is written as

$$\sum_{j=1}^{\infty} a_j.$$

Of course, the letter j has no special role here. We could just as well write

$$\lim_{n \to \infty} \sum_{j=1}^{n} a_j = \sum_{j=1}^{\infty} a_j = \sum_{p=1}^{\infty} a_p = \sum_{n=1}^{\infty} a_n.$$

Finally, if the limit

$$\lim_{n \to \infty} \sum_{j=1}^{n} a_j$$

exists and is a finite real number, we say that the series $\sum a_n$ is **convergent**. Otherwise, we say that $\sum a_n$ is **divergent**. Thus there are two ways in which a series $\sum a_n$ can diverge. Either the limit

$$\lim_{n \to \infty} \sum_{j=1}^{n} a_j$$

is $\pm\infty$ or this limit doesn't exist at all.

12.1.3   Some Examples of Series

1. Suppose that $a_n = 1$ for every positive integer n. Since

$$\sum_{j=1}^{\infty} a_j = \lim_{n \to \infty} \sum_{j=1}^{n} a_j = \lim_{n \to \infty} n = \infty,$$

   we conclude that $\sum a_n$ is divergent.
2. Suppose that

$$a_n = \begin{cases} 1 & \text{if } 1 \le n \le 5 \\ 0 & \text{if } n > 5. \end{cases}$$

   Since

$$\sum_{j=1}^{\infty} a_j = \lim_{n \to \infty} \sum_{j=1}^{5} 1 = \lim_{n \to \infty} 5 = 5,$$

   we conclude that $\sum a_n$ is convergent.
3. Suppose that $a_n = (-1)^n$ for each n. Since

$$\sum_{j=1}^{n} a_j = \begin{cases} -1 & \text{if } n \text{ is odd} \\ 0 & \text{if } n \text{ is even}, \end{cases}$$

   we see that the limit

$$\lim_{n \to \infty} \sum_{j=1}^{n} a_j$$

   does not exist and so $\sum a_n$ is divergent. Since $\sum a_n$ has no sum, the symbol $\sum_{j=1}^{\infty} a_j$ has no meaning.

4. Suppose that

$$a_n = \log\left(1 + \frac{1}{n}\right)$$

for every positive integer n. Since

$$\sum_{j=1}^{n} a_j = \sum_{j=1}^{n} \log\left(\frac{j+1}{j}\right) = \sum_{j=1}^{n} \log(j+1) - \sum_{j=1}^{n} \log j = \log(n+1),$$

we conclude that

$$\sum_{j=1}^{\infty} a_j = \lim_{n\to\infty} \sum_{j=1}^{n} a_j = \infty,$$

and therefore $\sum a_n$ is divergent.

5. Suppose that

$$a_n = \frac{1}{n(n+1)}$$

for each n. Since

$$\sum_{j=1}^{\infty} a_j = \lim_{n\to\infty} \sum_{j=1}^{n} a_j = \lim_{n\to\infty} \sum_{j=1}^{n} \left(\frac{1}{j} - \frac{1}{j+1}\right) = \lim_{n\to\infty}\left(1 - \frac{1}{n+1}\right) = 1,$$

we conclude that $\sum a_n$ is convergent.

6. Suppose that $-1 < c < 1$. From the algebraic identity

$$(1 - c)\left(1 + c + c^2 + \cdots + c^{n-1}\right) = 1 - c^n$$

we conclude that

$$\sum_{j=1}^{\infty} c^{j-1} = \lim_{n\to\infty} \sum_{j=1}^{n} c^{j-1} = \lim_{n\to\infty}\left(\frac{1 - c^n}{1 - c}\right) = \frac{1}{1 - c}.$$

To evaluate the latter limit we made use of Subsection 7.7.2. As you may know, the series $\sum c^{n-1}$ is known as the **geometric series with common ratio** c.

7. The nth partial sum of the series $\sum 1/n^2$ is the number

$$\sum_{j=1}^{n} \frac{1}{j^2} = \frac{1}{1^2} + \frac{1}{2^2} + \frac{1}{3^2} + \cdots + \frac{1}{n^2},$$

but, because there is no simple formula for this expression, it is not obvious what happens to this sum as $n \to \infty$. Eventually you will see that the sum of this series is $\pi^2/6$.

12.1.4   Exploring Series Graphically and Numerically with *Scientific Notebook*

Ⓝ This subsection is designed to be read interactively with the on-screen version of the text. To reach this material, click on the icon 🔳 .

12.1.5   Some Elementary Exercises on Series

1. 🔳 Find the nth partial sum of the series

$$\sum \frac{n-3}{n(n+1)(n+3)}.$$

   Deduce that this series is convergent and find its sum.

2. (a) Find the derivative of the nth partial sum of the series $\sum x^n$.
   (b) Find the nth partial sum of the series $\sum nx^{n-1}$. Deduce that if $|x| < 1$, we have

$$\sum_{n=1}^{\infty} nx^{n-1} = \frac{1}{(1-x)^2}.$$

3. Given that $|x| < 1$ and that

$$s_n = \sum_{j=1}^{n} (3j-1) x^{2j}$$

   for every positive integer n, obtain the identity

$$s_n \left(1 - x^2\right) = 2x^2 + \frac{3x^4 - 3x^{2n+4} - (3n-1) x^{2n+2} - x^{2n+4}}{1 - x^2}$$

   and deduce that

$$\sum_{j=1}^{\infty} (3j-1) x^{2j} = 2x^2 + \frac{3x^4}{1 - x^2}.$$

12.2   Elementary Properties of Series

In this section we show that series have the same sort of properties of linearity, nonnegativity, and additivity that we saw for integrals in Chapter 11.

12.2.1   Linearity of Series

Suppose that (a_n) and (b_n) are given sequences, that c is a given number and that both $\sum a_n$ and $\sum b_n$ converge. Then we have:

1. *The series $\sum (a_n + b_n)$ is convergent and*

$$\sum_{n=1}^{\infty} (a_n + b_n) = \sum_{n=1}^{\infty} a_n + \sum_{n=1}^{\infty} b_n.$$

2. *The series $\sum c a_n$ is convergent and*

$$\sum_{n=1}^{\infty} c a_n = c \sum_{n=1}^{\infty} a_n.$$

Proof. To prove the first assertion we observe that

$$\sum_{n=1}^{\infty} (a_n + b_n) = \lim_{n\to\infty} \sum_{j=1}^{n} (a_j + b_j) = \lim_{n\to\infty} \left(\sum_{j=1}^{n} a_j + \sum_{j=1}^{n} b_j \right) = \sum_{n=1}^{\infty} a_n + \sum_{n=1}^{\infty} b_n.$$

The proof of assertion 2 is similar. ∎

12.2.2   Nonnegativity of Series

Suppose that $a_n \leq b_n$ for every n and that both $\sum a_n$ and $\sum b_n$ are convergent. Then we have

$$\sum_{n=1}^{\infty} a_n \leq \sum_{n=1}^{\infty} b_n.$$

Proof. Since $b_n \geq a_n$ for each n, we have

$$\sum_{n=1}^{\infty} b_n - \sum_{n=1}^{\infty} a_n = \sum_{n=1}^{\infty} (b_n - a_n) = \lim_{n\to\infty} \sum_{j=1}^{n} (b_j - a_j) \geq 0. ∎$$

12.2.3   Additivity of Series

Suppose that (a_n) is a sequence and that $N > 1$ is an integer. Then the convergence of $\sum a_n$ is independent of whether we start summing at $n = 1$ or at $n = N$; and if these series are convergent, we have

$$\sum_{n=1}^{\infty} a_n = \sum_{n=1}^{N-1} a_n + \sum_{n=N}^{\infty} a_n.$$

Proof. Whenever $n \geq N$ we have

$$\sum_{j=1}^{n} a_j = \sum_{j=1}^{N-1} a_j + \sum_{j=N}^{n} a_j,$$

and the result follows at once by letting $n \to \infty$. ∎

12.3   Some Elementary Facts About Convergence

12.3.1   The nth Term Criterion for Divergence

If the nth term a_n of the series $\sum a_n$ does not approach zero as $n \to \infty$, then the series $\sum a_n$ must diverge.

Proof. Suppose that $\sum_{i=1}^{n} a_i \to L$ as $n \to \infty$. Then since $\sum_{i=1}^{n-1} a_i \to L$ as $n \to \infty$ we see that

$$\lim_{n \to \infty} a_n = \lim_{n \to \infty} \left[\sum_{i=1}^{n} a_i - \sum_{i=1}^{n-1} a_i \right] = L - L = 0. ∎$$

This criterion lets us see at once that certain series are divergent.

12.3.2   Some Examples to Illustrate the nth Term Criterion

1. Suppose that

$$a_n = \frac{n^3 \cos(n\pi/7)}{1 + n + n^2 + (-1)^n n^3}$$

for each n. Since the sequence (a_n) does not converge to 0 the series $\sum a_n$ must be divergent.

2. Suppose that (a_n) is a given sequence and that for some integer N we have $a_n = 0$ whenever $n \geq N$. Since the equation

$$\sum_{j=1}^{n} a_j = \sum_{j=1}^{N} a_j$$

holds whenever $n \geq N$, we see that

$$\sum_{n=1}^{\infty} a_n = \lim_{n \to \infty} \sum_{j=1}^{n} a_j = \lim_{n \to \infty} \sum_{j=1}^{N} a_j = \sum_{j=1}^{N} a_j,$$

and so $\sum a_n$ is convergent.

3. We saw in Example 4 of Subsection 12.1.3 that the series $\sum \log \left(1 + 1/n\right)$ is divergent. From the fact that

$$\lim_{n \to \infty} \log \left(1 + \frac{1}{n}\right) = \log 1 = 0$$

we see that there is no guarantee that a given series $\sum a_n$ will converge, just because $a_n \to 0$ as $n \to \infty$. What we know is that, unless the sequence (a_n) converges to 0, the series $\sum a_n$ has no chance of converging.

12.4   Convergence of Series with Nonnegative Terms

12.4.1   The Comparison Principle

The theory of convergence of a series $\sum a_n$ is easier when each term a_n is nonnegative than it is when some of the numbers a_n are positive and some are negative. The trouble with a general series is that the sequence of partial sums $\sum_{i=1}^{n} a_i$ increases when we add positive terms and decreases when we add negative terms. Therefore, to show that a sequence of partial sums converges, we may have to work quite hard to show that it doesn't jump around too much. For example, it is quite a difficult job to show that the series

$$\sum \frac{\sin n}{\sqrt[3]{n}}$$

is convergent.

If, however, the condition $a_n \geq 0$ holds for each n, then the sequence of partial sums $\sum_{j=1}^{n} a_j$ is increasing and we know from Theorem 7.7.1 that the limit

$$\lim_{n \to \infty} \sum_{j=1}^{n} a_j$$

must exist and must be the least upper bound of the sequence of partial sums $\sum_{j=1}^{n} a_j$. Thus, if $a_n \geq 0$ for every n, then the series $\sum a_n$ will be convergent if and only if the sequence of partial sums $\sum_{j=1}^{n} a_j$ is bounded above.

One way of showing that a sequence $\left(\sum_{j=1}^{n} a_j\right)$ is bounded above is to find a known bounded sequence (p_n) such that

$$\sum_{j=1}^{n} a_j \leq p_n$$

for each n. Alternatively, we can show that a sequence $\left(\sum_{j=1}^{n} a_j \right)$ is unbounded above if we can find sequence (p_n) that is known to be unbounded above such that

$$\sum_{j=1}^{n} a_j \geq p_n$$

for each n. In each of these cases we are determining the convergence or divergence of the given series $\sum a_n$ by the so-called **comparison principle**.

Take careful note that the comparison principle that we shall be using in this section can be applied only to series with nonnegative terms. It has no analog for series whose terms frequently change sign.

In our first application of the comparison principle we compare a series with an integral to obtain a useful fact known as the **integral test**.

12.4.2   The Integral Test

Those who are reading this chapter before having read the chapter on integration should click on the icon [icon] for a version of the integral test that does not refer explicitly to an integral.

Suppose that f is a nonnegative decreasing function defined on the interval $[1, \infty)$. Then the following conditions are equivalent:[35]

1. *The series $\sum f(n)$ is convergent.*
2. *The sequence of integrals $\int_1^n f(x)dx$ is bounded above.*

Proof. Given any $j = 2, 3, \cdots, n$, the inequality

$$f(j) \leq f(x) \leq f(j-1)$$

holds for all numbers $x \in [j-1, j]$. This inequality is illustrated in Figure 12.1. From this inequality we see that

$$f(j) \leq \int_{j-1}^{j} f(x)dx \leq f(j-1)$$

for each j and, summing, we obtain

$$\sum_{j=2}^{n} f(j) \leq \int_{1}^{n} f(x)dx \leq \sum_{j=1}^{n-1} f(j).$$

---

[35]   As usual, the left endpoint 1 of this interval is unimportant. We could just as well have started at any integer k.

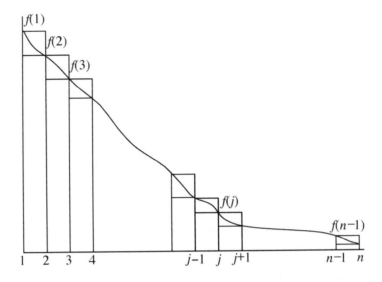

Figure 12.1

Thus the sequence of sums $\sum_{j=1}^{n} f\left(j\right)$ is bounded above if and only if the sequence of integrals $\int_{1}^{n} f(x)dx$ is bounded above. ∎

12.4.3   The p-Series: An Application of the Integral Test

Given any number p, the series

$$\sum \frac{1}{n^p}$$

is called a p-**series.** We shall use the integral test to show that this series converges if $p > 1$ and diverges if $p \leq 1$. For convenience we separate our proof into three cases:

Case 1: Suppose that $p = 1$. From the fact that

$$\lim_{n\to\infty} \int_{1}^{n} \frac{1}{x}dx = \lim_{n\to\infty} \log n = \infty,$$

we see at once that $\sum 1/n$ diverges.

Case 2: Suppose that $p > 1$. From the fact that

$$\lim_{n\to\infty} \int_{1}^{n} \frac{1}{x^p}dx = \lim_{n\to\infty}\left(\frac{1}{1-p}n^{1-p} - \frac{1}{1-p}\right) = \frac{1}{p-1},$$

we see that $\sum 1/n^p$ converges.

Case 3: Suppose that $p < 1$. In the event that $p \leq 0$, the series $\sum 1/n^p$ certainly diverges because $1/n^p$ does not approach 0 as $n \to \infty$. In the event that $0 < p < 1$, we deduce from the fact that

$$\lim_{n \to \infty} \int_1^n \frac{1}{x^p} dx = \lim_{n \to \infty} \left(\frac{1}{1-p} n^{1-p} - \frac{1}{1-p} \right) = \infty$$

that $\sum 1/n^p$ diverges.

We have therefore shown that the p-series $\sum 1/n^p$ converges when $p > 1$ and diverges when $p \leq 1$. ∎

12.4.4   A Refinement of the p-Series

At first sight, the p-series $\sum 1/n^p$ seems to show a boundary between convergence and divergence at $p = 1$. But this isn't so. An easy application of the integral test can be used to show that the series

$$\sum \frac{1}{n (\log n)^p}$$

converges if $p > 1$ and diverges if $p \leq 1$. Note that if p is positive, the terms of the latter series are smaller than the terms of $\sum 1/n$. We can use the integral test again to obtain a further refinement of the p-series: The series

$$\sum \frac{1}{n (\log n) (\log \log n)^p}$$

converges if $p > 1$ and diverges if $p \leq 1$. This refinement process can be taken as far as you like, and it illustrates the fact that it is impossible to put a finger on a series that is a boundary between convergence and divergence.

12.4.5   A Sharper Version of the Integral Test

A sharper version of the integral test tells us that if f is a decreasing positive function, then, regardless of whether $\sum f(n)$ is convergent or divergent, the limit

$$\lim_{n \to \infty} \left(\sum_{j=1}^n f(j) - \int_1^n f(x) dx \right)$$

must exist and be finite.

You can reach a discussion of this form of the integral test and some interesting exercises that make use of it by clicking on the icon .

12.4.6   The Comparison Test

The comparison test tells us that, under certain conditions, if (a_n) and (b_n) are sequences of nonnegative numbers, and if one of the series $\sum a_n$ and $\sum b_n$ converges, then so does the other. The value of this sort of theorem is that it allows us to test a given series by comparing it with another whose behavior is already known to us. For example, we can often determine the convergence or divergence of a given series by comparing it with a p-series $\sum 1/n^p$ or a geometric series $\sum c^n$.

Suppose that (a_n) and (b_n) are sequences of nonnegative numbers.

1. *If it is possible to find a positive number k such that $a_n \leq kb_n$ for all n, and if $\sum b_n$ converges, then $\sum a_n$ must converge.*
2. *If the sequence (a_n/b_n) is bounded and if $\sum b_n$ converges, then $\sum a_n$ must converge.*
3. *If the sequence (a_n/b_n) is convergent and if $\sum b_n$ converges, then $\sum a_n$ must converge.*
4. *If it is possible to find a positive number δ such that $a_n/b_n \geq \delta$ for all n, and if $\sum a_n$ converges, then $\sum b_n$ must converge.*
5. *If the sequence (a_n/b_n) has a positive limit (possibly ∞), and if $\sum a_n$ converges, then $\sum b_n$ must converge.*

Proof. Part 1 follows at once from the comparison principle discussed in Subsection 12.4.1 and the inequality

$$\sum_{j=1}^{n} a_j \leq k \sum_{j=1}^{n} b_j$$

that holds for each n. Parts 2 and 3 follow at once from part 1. Part 4 follows from the fact that if $\delta > 0$ and $a_n/b_n \geq \delta$ for every n, then the sequence (b_n/a_n) is bounded, and part 5 follows at once from part 4. ∎

12.4.7   Some Exercises on The Comparison Test

Test each of the following series for convergence.

1. ▣ $\sum \dfrac{1}{n^{3/2} + n}$.

2. $\sum \dfrac{1}{n^{3/2} - n}$.

3. ▣ $\sum \dfrac{n}{\sqrt{n^4 - n^2 + 2}}$.

4. $\blacksquare$ $\displaystyle\sum \frac{n \log n}{\sqrt{n^5 - n^2 + 2}}$.

5. $\blacksquare$ $\displaystyle\sum \frac{1}{n^{(1+(\log n)/n)}}$.

6. $\blacksquare$ $\displaystyle\sum \frac{1}{n^{\left(1+(\log n)^2/n\right)}}$.

7. $\blacksquare$ $\displaystyle\sum \frac{1}{n^{\left(1+\left((\log n)^{(\log \log n)}\right)/n\right)}}$.

8. $\blacksquare$ $\displaystyle\sum \left(\frac{n}{n+1}\right)^{n \log n}$.

9. $\blacksquare$ $\displaystyle\sum \left(\frac{n}{n+1}\right)^{n(\log n)^2}$.

10. $\blacksquare$ $\displaystyle\sum \left(\frac{1}{\log n}\right)^3$.

11. $\blacksquare$ $\displaystyle\sum \left(\frac{1}{\log n}\right)^n$.

12. $\blacksquare$ $\displaystyle\sum \left(\frac{1}{\log n}\right)^{\log n}$.

13. $\blacksquare$ $\displaystyle\sum \left(\frac{1}{\log \log n}\right)^{\log n}$.

14. $\blacksquare$ $\displaystyle\sum \left(\frac{1}{\log n}\right)^{\log \log n}$.

15. $\displaystyle\sum \left(\frac{1}{\log \log \log n}\right)^{\log n}$.

16. $\displaystyle\sum \left(\frac{1}{\log \log n}\right)^{\log \log n}$.

17. $\blacksquare$ $\displaystyle\sum \left(\left(\frac{1}{\log \log n}\right)^{\log \log n}\right)^{\log \log n}$.

18. $\blacksquare$ $\displaystyle\sum \left(\frac{1}{\log \log n}\right)^{\left((\log \log n)^{(\log \log n)}\right)}$.

19. $\blacksquare$ $\displaystyle\sum \left(\sin \frac{x}{n}\right)^\alpha$, where x and α are given positive numbers.

20. Prove that if (a_n) is a sequence of positive numbers and $\sum a_n$ converges, then so does the series $\sum a_n^2$.

21. In this exercise we encounter a series that diverges very slowly. We begin by defining

$$L(x) = \begin{cases} \log x & \text{if } x \geq e \\ 1 & \text{if } x < e. \end{cases}$$

The graph of this function is illustrated in Figure 12.2. If k is any positive

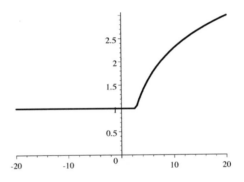

Figure 12.2

number, we shall write L^k for the composition of the function L with itself k times. Thus, if x is a given number, then $L^3(x) = L\left(L\left(L(x)\right)\right)$. We also define $L^0(x) = x$ for every x.

(a) Given any number x, explain why we must have $L^k(x) = 1$ for all sufficiently large values of k. For a given positive integer n, give a simple meaning to the "infinite product"

$$\prod_{k=0}^{\infty} L^k(n) = L^0(n)\, L^1(n)\, L^2(n) \cdots .$$

(b) For each positive integer n we define

$$a_n = \frac{1}{\prod_{k=0}^{\infty} L^k(n)} .$$

Using the integral test (or otherwise), show that the series $\sum a_n$ is divergent.

12.5   Decimals

A **decimal** (with base 10) is a series of the form

$$\sum \frac{a_n}{10^n},$$

where, for each n, the number a_n lies in the set $\{0,1,2,3,4,5,6,7,8,9\}$. The numbers a_n are called the **digits** of the decimal. Since

$$\sum_{n=1}^{\infty} \frac{9}{10^n} = 1,$$

it follows from the comparison test that every decimal converges. It is not hard to show that every number x in the interval $[0,1]$ can be expressed as a decimal

$$x = \sum_{n=1}^{\infty} \frac{a_n}{10^n}$$

and that, if we eliminate those decimals whose digits are all equal to 9 from some position onward, the representation of a number as a decimal is unique. You can reach a more complete discussion of decimals by clicking on the icon 📷 .

12.6   The Ratio Tests

One of the most useful ways of testing a given series $\sum a_n$ for convergence is to look at the size of the ratios

$$\frac{a_{n+1}}{a_n}$$

as n varies. The basic message of this section is that the smaller these ratios are, the more likely it is that the series $\sum a_n$ will converge. We begin by exhibiting a condition on these ratios that guarantees that a given series $\sum a_n$ must diverge.

12.6.1   A Ratio Version of the Criterion for Divergence

Suppose that (a_n) is any sequence of real numbers.

1. *If*

$$\left| \frac{a_{n+1}}{a_n} \right| \geq 1. \tag{12.1}$$

   for all n large enough, then the series $\sum a_n$ must diverge.

2. *If*

$$\lim_{n \to \infty} \left| \frac{a_{n+1}}{a_n} \right| > 1, \tag{12.2}$$

then the series $\sum a_n$ must diverge.

Proof. If the inequality (12.1) holds for all sufficiently large n, then, for all such n, we have $|a_{n+1}| \geq |a_n|$ and therefore we cannot have $a_n \to 0$ as $n \to \infty$.

If the inequality (12.2) holds, then we certainly have

$$\left| \frac{a_{n+1}}{a_n} \right| > 1$$

for all sufficiently large n. ∎

12.6.2   Some Examples to Illustrate the Ratio Criterion for Divergence

1. In this example we test the series $\sum a_n$ where, for each n, we define

$$a_n = \frac{4^n \, (n!)^2}{(2n)!}.$$

From the fact that for each n

$$\frac{a_{n+1}}{a_n} = \frac{2n + 2}{2n + 1} > 1$$

and the ratio version of the criterion for divergence, we deduce that $\sum a_n$ diverges.

2. In this example we test the series $\sum a_n$ where, for each n, we define

$$a_n = \frac{n^n}{e^n n!}.$$

Using L'Hôpital's rule (Theorem 9.6.3) one may see that

$$\lim_{n \to \infty} \frac{a_{n+1}}{a_n} = \frac{e}{2} > 1,$$

and so the series $\sum a_n$ diverges.

12.6.3   Discussion of Series whose Ratios Are Less than One

In view of Theorem 12.6.1, one might reasonably ask whether the inequality

$$\left| \frac{a_{n+1}}{a_n} \right| < 1$$

for all n is enough to guarantee that the series $\sum a_n$ is convergent. The answer is *no!* All this inequality says is that the sequence $(|a_n|)$ is strictly decreasing. Thus, for example, if $a_n = 1/n$ for each n, then, even though the inequality

$$\frac{a_{n+1}}{a_n} < 1 \tag{12.3}$$

holds for all n, the series $\sum a_n$ diverges.

The message of the ratio tests that appear in this section is that, if (a_n) is a decreasing sequence of positive numbers, then the series $\sum a_n$ will converge as long as the ratios a_{n+1}/a_n are not "too close" to 1. Each of the ratio tests provides its own interpretation of how close is "too close". The first ratio test that we shall study is the ratio test of d'Alembert, which tells us that $\sum a_n$ will converge as long as the ratio a_{n+1}/a_n remains below a constant α that is less than 1. This test is the most primitive of the ratio tests.

$$\overline{\quad\underset{a_{n+1}/a_n}{|}\qquad\qquad\qquad\qquad\underset{\alpha}{|}\qquad\qquad\qquad\qquad\underset{1}{|}\quad}$$

After d'Alembert's test we shall study the more powerful ratio test of Raabe, in which the latter requirement is relaxed. Raabe's test tells us that $\sum a_n$ will converge if for some constant $p > 1$ and all sufficiently large n we have

$$\frac{a_{n+1}}{a_n} \leq 1 - \frac{p}{n}.$$

$$\overline{\quad\underset{a_{n+1}/a_n}{|}\qquad\qquad\qquad\qquad\underset{1-p/n}{|}\qquad\qquad\qquad\underset{1}{|}\quad}$$

There are even more powerful ratio tests. For example, it can be shown that $\sum a_n$ will converge if

$$\frac{a_{n+1}}{a_n} \leq 1 - \frac{1}{n} - \frac{p}{n \log n}$$

for some constant $p > 1$ and all sufficiently large n. Going a step further, one may prove that $\sum a_n$ will converge if

$$\frac{a_{n+1}}{a_n} \leq 1 - \frac{1}{n} - \frac{1}{n \log n} - \frac{p}{n (\log n) (\log \log n)}$$

for some constant $p > 1$ and all sufficiently large n, and, as you may guess, $\sum a_n$ will converge if the ratio $\dfrac{a_{n+1}}{a_n}$ does not exceed

$$1 - \frac{1}{n} - \frac{1}{n \log n} - \frac{1}{n (\log n) (\log \log n)} - \frac{p}{n (\log n) (\log \log n) (\log \log \log n)}$$

for some constant $p > 1$ and all sufficiently large n. This sequence of ratio tests can be taken as far as we like, but, unfortunately, the more powerful tests are not as easy to prove as they are to state. Their proofs, while not actually difficult, are rather technical. If, after reading this chapter, you would like to read about the deeper ratio tests, you can do so by clicking on the icon 📓 .

12.6.4   The Ratio Comparison Test

This theorem that we are about to state is the key to the section because it tells us why ratios are so important. Note that, because this theorem and those that follow it are versions of the comparison test, they require the terms of the series to be positive. ✍

Suppose that a_n and b_n are positive for n sufficiently large and that for all sufficiently large n we have

$$\frac{a_{n+1}}{a_n} \leq \frac{b_{n+1}}{b_n}. \tag{12.4}$$

If the series $\sum b_n$ converges, then the series $\sum a_n$ must also converge.

Proof. Choose an integer N such that the inequality (12.4) holds for all $n \geq N$. Then for all $n \geq N$ we have

$$\frac{a_{n+1}}{b_{n+1}} \leq \frac{a_n}{b_n},$$

and so, starting at the integer N, the sequence (a_n/b_n) is decreasing. Therefore

$$\frac{a_n}{b_n} \leq \frac{a_N}{b_N}$$

for all $n \geq N$. Since

$$a_n \leq \left(\frac{a_N}{b_N} \right) b_n$$

whenever $n \geq N$, we deduce from the comparison test (Theorem 12.4.6) that if $\sum b_n$ converges, then so does $\sum a_n$. ∎

12.6.5   d'Alembert's Ratio Test (Sometimes Called *the* Ratio Test)

Suppose that $a_n > 0$ for each n.

1. *If there exists a number $\alpha < 1$ such that*

$$\frac{a_{n+1}}{a_n} \leq \alpha$$

    for each n,

then the series $\sum a_n$ converges.

2. If

$$\lim_{n \to \infty} \frac{a_{n+1}}{a_n} < 1,$$

then the series $\sum a_n$ *must converge.*

Proof of Part 1. Suppose that $\alpha < 1$ and that the inequality

$$\frac{a_{n+1}}{a_n} \le \alpha$$

holds for each n. Now define $b_n = \alpha^n$ for each n. The series $\sum b_n$ is a convergent geometric series. Now we observe that for all n sufficiently large we have

$$\frac{a_{n+1}}{a_n} \le \alpha = \frac{b_{n+1}}{b_n},$$

and so it follows from the ratio comparison test (Theorem 12.6.4) that $\sum a_n$ converges. ∎

Proof of Part 2. Suppose that

$$\lim_{n \to \infty} \frac{a_{n+1}}{a_n} < 1$$

and choose a number α such that

$$\lim_{n \to \infty} \frac{a_{n+1}}{a_n} < \alpha < 1.$$

Since the inequality

$$\frac{a_{n+1}}{a_n} \le \alpha$$

must hold for all sufficiently large, part 2 follows at once from part 1. ∎

Note that d'Alembert's test is really an efficient way of comparing a given series with a geometric series.

12.6.6   Some Examples to Illustrate d'Alembert's Test

1. In order to test the series

$$\sum \frac{(2n)!}{5^n \, (n!)^2}$$

we write

$$a_n = \frac{(2n)!}{5^n \, (n!)^2}$$

for each n. Then, since

$$\lim_{n \to \infty} \frac{a_{n+1}}{a_n} = \lim_{n \to \infty} \frac{2}{5} \frac{2n+1}{n+1} = \frac{4}{5} < 1,$$

the series $\sum a_n$ converges by d'Alembert's test.

2. In this example we demonstrate the fact that d'Alembert's test cannot be used to test the series

$$\sum \frac{(2n)!}{4^n \, (n!)^2}.$$

We define

$$a_n = \frac{(2n)!}{4^n \, (n!)^2}$$

for each n and we observe that, for each n,

$$\frac{a_{n+1}}{a_n} = \frac{2n+1}{2n+2}.$$

Since $a_{n+1}/a_n < 1$ for each n, we do not have an automatic guarantee that $\sum a_n$ diverges. Furthermore, since $a_{n+1}/a_n \to 1$ as $n \to \infty$, neither part of d'Alembert's test can be used to show that $\sum a_n$ converges. We shall analyze this series in a moment using Raabe's form of the ratio test.

12.6.7   Raabe's Ratio Test

Suppose that $a_n > 0$ for each n.

1. *If there exists a number $p > 1$ such that*

$$\frac{a_{n+1}}{a_n} \leq 1 - \frac{p}{n}$$

for each n,

then the series $\sum a_n$ converges.

2. *If there exists a number $p < 1$ such that*

$$\frac{a_{n+1}}{a_n} \geq 1 - \frac{p}{n}$$

for each n,

then the series $\sum a_n$ diverges.

3. *If*

$$\lim_{n \to \infty} n \left(1 - \frac{a_{n+1}}{a_n} \right) = p,$$

then the series $\sum a_n$ converges if $p > 1$ and diverges if $p < 1$.

Proof. The proof of Raabe's test makes use of Example 3 in Subsection 9.6.2, in which we showed that if q is any real number, then

$$\lim_{x \to \infty} x \left(1 - \frac{(x-1)^q}{x^q} \right) = q.$$

To prove part 1 assume that $p > 1$ and that the inequality

$$\frac{a_{n+1}}{a_n} \leq 1 - \frac{p}{n}$$

holds for all n. Choose a number q such that $1 < q < p$ and, for each $n \geq 2$, define

$$b_n = \frac{1}{(n-1)^q}.$$

We note that the series $\sum b_n$ is convergent. Now, since

$$\lim_{n \to \infty} n \left(1 - \frac{b_{n+1}}{b_n} \right) = \lim_{n \to \infty} n \left(1 - \frac{(n-1)^q}{n^q} \right) = q,$$

we see that for all n sufficiently large we have

$$n \left(1 - \frac{b_{n+1}}{b_n} \right) < p.$$

Since the latter inequality can be written as

$$1 - \frac{p}{n} < \frac{b_{n+1}}{b_n},$$

we see that, for all n sufficiently large,

$$\frac{a_{n+1}}{a_n} \leq 1 - \frac{p}{n} < \frac{b_{n+1}}{b_n},$$

and so the convergence of the series $\sum a_n$ follows from the ratio comparison test (Theorem 12.6.4).

The proof of part 2 of the theorem is similar and is left as an exercise.

Finally, suppose that

$$\lim_{n \to \infty} n \left(1 - \frac{a_{n+1}}{a_n} \right) = p.$$

If $p > 1$, we choose a number q such that $1 < q < p$. In this case, since the inequality

$$n \left(1 - \frac{a_{n+1}}{a_n} \right) > q$$

holds for all sufficiently large n, the convergence of $\sum a_n$ follows from part 1 of the theorem. In a similar way we can show that if $p < 1$, the series $\sum a_n$ diverges. ∎

12.6.8   Some Examples to Illustrate Raabe's Test

1. In this example we test the series

$$\sum \frac{(2n)!}{4^n \, (n!)^2}$$

using Raabe's test. We define

$$a_n = \frac{(2n)!}{4^n \, (n!)^2}$$

and we observe that, since

$$\lim_{n \to \infty} n \left(1 - \frac{a_{n+1}}{a_n} \right) = \lim_{n \to \infty} n \left(1 - \frac{1}{2} \frac{2n + 1}{n + 1} \right) = \frac{1}{2} < 1,$$

the series $\sum a_n$ diverges.

2. This example will be useful to us when we study the binomial series in Section 14.6. Suppose that α is a given real number and that for each n we

have

$$a_n = \left| \frac{\alpha \left(\alpha - 1 \right) \left(\alpha - 2 \right) \cdots \left(\alpha - n + 1 \right)}{n!} \right|.$$

In the event that α is a nonnegative integer we have $a_n = 0$ for all $n > \alpha$ and the series $\sum a_n$ clearly converges; so from now on we assume that α is not a nonnegative integer. Whenever $n > \alpha$ we see that

$$\frac{a_{n+1}}{a_n} = \left| \frac{\alpha - n}{n+1} \right| = \frac{n - \alpha}{n+1},$$

and so

$$\lim_{n \to \infty} n \left(1 - \frac{a_{n+1}}{a_n} \right) = 1 + \alpha.$$

We deduce from Raabe's test that the series $\sum a_n$ converges if $\alpha > 0$ and diverges if $\alpha < 0$.

12.6.9   A More Powerful Raabe Test

Suppose that $a_n > 0$ for each n.

1. *If there exists a number $p > 1$ such that*

$$\frac{a_{n+1}}{a_n} \leq 1 - \frac{1}{n} - \frac{p}{n \log n}$$

   for each n, then the series $\sum a_n$ converges.
2. *If there exists a number $p < 1$ such that*

$$\frac{a_{n+1}}{a_n} \geq 1 - \frac{1}{n} - \frac{p}{n \log n}$$

   for each n, then the series $\sum a_n$ diverges.
3. *If*

$$\lim_{n \to \infty} n \left(\log n \right) \left(1 - \frac{1}{n} - \frac{a_{n+1}}{a_n} \right) = p,$$

   then the series $\sum a_n$ converges if $p > 1$ and diverges if $p < 1$.

Proof. You can find the proof by clicking on the icon 📖 .

12.6.10   Exploring the Ratio Tests with *Scientific Notebook*

Ⓝ   To see how the computing features of *Scientific Notebook* can be used to enhance your understanding of the ratio tests, click on the icon ▦ .

12.6.11   A Criterion for the nth Term of a Series to Converge to Zero

As we have seen, the message of the ratio tests is that if (a_n) is a decreasing sequence of positive numbers, then the series $\sum a_n$ will converge as long as the ratio a_{n+1}/a_n is not too close to the number 1. The closer this ratio is to 1, the more slowly the sequence (a_n) will decrease. In this subsection we shall demonstrate that, even though the ratio a_{n+1}/a_n may be too close to 1 to allow convergence of the series $\sum a_n$, the ratio a_{n+1}/a_n may still satisfy a weaker criterion that guarantees that $a_n \to 0$ as $n \to \infty$. This knowledge will be useful to us when we come to Subsection 12.7.9. The precise statement of the weaker criterion is as follows:

Suppose that (a_n) is a sequence of positive numbers and that for each positive integer n we have defined

$$b_n = 1 - \frac{a_{n+1}}{a_n}.$$

1. *If the sequence (a_n) is decreasing, then it will converge to zero if and only if the series $\sum b_n$ diverges.*
    Proof. We assume that (a_n) is decreasing. We observe first that if $n \geq 2$, we have

    $$a_n = a_1 (1 - b_1)(1 - b_2)(1 - b_3) \cdots (1 - b_{n-1}),$$

    and so

    $$\log a_n = \log a_1 + \sum_{j=1}^{n-1} \log(1 - b_j).$$

    Since $a_n \to 0$ as $n \to \infty$ if and only if $\log a_n \to -\infty$ as $n \to \infty$, the condition that $a_n \to 0$ as $n \to \infty$ is equivalent to the condition that

    $$\lim_{n \to \infty} \sum_{j=1}^{n-1} (-\log(1 - b_j)) = \infty.$$

    The latter condition says that the series $\sum(-\log(1 - b_n))$ is divergent. We now consider two cases:
    Case 1: Suppose that the sequence (b_n) does not converge to zero. In this case the series $\sum b_n$ diverges. Furthermore, since the sequence of numbers

$- \log{(1 - b_j)}$ also does not converge to zero, the series $\sum{(- \log{(1 - b_n)})}$ is divergent and $a_n \to 0$ as $n \to \infty$.

Case 2: Suppose that the sequence (b_n) does converge to zero. In this case we deduce from the fact that

$$\lim_{x \to 0} \frac{- \log{(1 - x)}}{x} = 1$$

that

$$\lim_{n \to \infty} \frac{- \log{(1 - b_n)}}{b_n} = 1,$$

and it follows from the comparison test (Theorem 12.4.6) that the series $\sum{(- \log{(1 - b_n)})}$ diverges if and only if $\sum b_n$ diverges. In other words, $a_n \to 0$ as $n \to \infty$ if and only if $\sum b_n$ diverges. ∎

2. *If there exists a positive number p such that the inequality*

$$n \left(1 - \frac{a_{n+1}}{a_n} \right) \geq p$$

holds for every n, then a_n decreases to 0 as $n \to \infty$.
Proof. Suppose that $p > 0$ and that the inequality

$$n \left(1 - \frac{a_{n+1}}{a_n} \right) \geq p$$

holds for all n. Since $1 - a_{n+1}/a_n$ is always positive the sequence (a_n) must be decreasing. Furthermore, since $nb_n \geq p$ for each n the series $\sum b_n$ diverges by the comparison test. ∎

3. *If the limit*

$$\lim_{n \to \infty} n \left(1 - \frac{a_{n+1}}{a_n} \right)$$

exists and is positive, then a_n decreases to 0 as $n \to \infty$.
This part of the theorem follows at once from Part 2. ∎

12.6.12   Some Exercises on the Ratio Tests

Test the following series for convergence. In the event that the series diverges, determine whether or not its nth term approaches 0 as $n \to \infty$.

1. ■ $\sum \dfrac{((2n)!)^3}{((3n)!)^2}$.

2. $\sum \dfrac{3^{\left(n^2\right)}}{n!}$.

3. ⊞ $\sum \dfrac{3^{(n \log n)}}{n!}$.

4. ⊞ $\sum \dfrac{(\log n)^n}{e^n (\log 2)(\log 3) \cdots (\log n)}$.

5. $\sum \left(\dfrac{\alpha (\alpha - 1)(\alpha - 2) \cdots (\alpha - n + 1)}{n!} \right)^2$, where α is a given number

6. ⊞ $\sum \left| \dfrac{\alpha (\alpha - 1)(\alpha - 2) \cdots (\alpha - n + 1)}{n!} \right|^p$, where α and p are given
   numbers and $p(\alpha + 1) \neq 1$.

7. (a) $\sum \dfrac{n^\alpha}{n!}$, where α is a given number.

   (b) $\sum \dfrac{n^{\alpha n}}{n!}$, where α is a given number.

   (c) $\sum \dfrac{n^{n - \log n}}{n!}$.

8. $\sum \dfrac{(2n)!}{4^n (n!)^2} x^n$, where x is a given positive number.

9. $\sum \dfrac{n!}{x(x+1)(x+2) \cdots (x + n - 1)}$, where x is a given positive number.

10. (a) $\sum \dfrac{e^n n!}{n^n}$.

    (b) $\sum \dfrac{n^n}{e^n n!}$.

11. $\sum \left(\dfrac{(2n)!}{4^n (n!)^2} \right)^p$, where p is a given number.

12. ⊞ Ⓝ $\sum \left(\dfrac{\pi^{\frac{n}{2}}}{2^{n(n+1)} e^{\frac{n(n+3)}{2}} n^{11/12}} \right) \prod_{j=1}^n \dfrac{(2j+1)^{2j+\frac{1}{2}}}{(j^j)(j!)}$.

13. $\sum \left| \dfrac{\alpha (\alpha + 1)(\alpha + 2) \cdots (\alpha + n - 1)}{\beta (\beta + 1)(\beta + 2) \cdots (\beta + n - 1)} \right|$, where α and β are given
    numbers.

14. ∎ $\displaystyle\sum \left(\frac{\alpha\,(\alpha+1)\,(\alpha+2)\cdots(\alpha+n-1)}{\beta\,(\beta+1)\,(\beta+2)\cdots(\beta+n-1)}\right)^{2}$, where α and β are given numbers.

15. **Cauchy's root test** says that if (a_n) is a sequence of nonnegative numbers and if $\sqrt[n]{a_n} \to \alpha$ as $n \to \infty$, then the series $\sum a_n$ converges if $\alpha < 1$ and diverges if $\alpha > 1$.

   (a) Prove Cauchy's root test.
   (b) Review Exercise 9 of Subsection 10.6.6 and then prove that if Cauchy's root test can be used to test a given series for convergence, then so can d'Alembert's ratio test.

16. ∎ Prove the following more powerful root test:
   If $a_n \ge 0$ for all n and if

$$\frac{n}{\log n}\left(1 - (a_n)^{1/n}\right) \to p$$

   as $n \to \infty$, then the series $\sum a_n$ converges if $p > 1$ and diverges if $p < 1$.
   This form of the root test is one of the results that are developed in the special document on ratio and root tests that can be reached by clicking on the icon
   ▣ .

12.7   Convergence of Series Whose Terms May Change Sign

12.7.1   Introductory Remarks About General Series

In the introduction to the comparison principle that we saw in Subsection 12.4.1, we observed that a series $\sum a_n$ is considerably easier to test for convergence if all of the numbers a_n are nonnegative. As you know, if $a_n \ge 0$ for each n, then all we have to show to establish the convergence of $\sum a_n$ is that the sequence of partial sums

$$\sum_{j=1}^{n} a_j$$

is bounded above. We relied heavily on this fact in our study of such series. However, the theory of general series has an altogether different flavor because, in the theory of general series, the comparison principle is lost. To show that a general series is convergent we have to do much more than just establish that the sequence of its partial sums is bounded. We have to show why this sequence of partial sums actually converges to some real number. In this sense, the theory of general series is harder than the theory of nonnegative series.

At this point, the one thing that we know about general series is that if a given sequence (a_n) fails to converge to zero, then the series $\sum a_n$ must be divergent. Thus, if we want to test a given series $\sum a_n$, the first thing to do is to determine whether or not $a_n \to 0$ as $n \to \infty$; if not, then $\sum a_n$ is divergent.

In the event that the sequence (a_n) does converge to zero, we have to work much harder. Our initial strategy will be to ignore the given series $\sum a_n$ and to look instead at the series $\sum |a_n|$, which can be tested using the comparison principle. We shall see in a moment that if the series $\sum |a_n|$ happens to converge, then the given series $\sum a_n$ must also be convergent. However, if the series $\sum |a_n|$ is divergent, then we are facing a fundamentally different kind of problem, which we shall confront later in the section.

12.7.2   Absolute Convergence

If (a_n) is a given sequence of real numbers, then we say that the series $\sum a_n$ **converges absolutely** if the series $\sum |a_n|$ is convergent.

12.7.3   Absolute Convergence Implies Convergence

Every absolutely convergence series is convergent.

Proof. Suppose that (a_n) is a given sequence of numbers and that the series $\sum |a_n|$ is convergent. Since the inequality

$$0 \leq a_n + |a_n| \leq 2\,|a_n|$$

holds for every n, it follows from the comparison test (Theorem 12.4.6) that the series $\sum (a_n + |a_n|)$ is convergent. Thus, since

$$a_n = (a_n + |a_n|) - |a_n|$$

for each n, it follows from the linearity property of series (Theorem 12.2.1) that the series $\sum a_n$ is also convergent. ∎

12.7.4   Conditional Convergence

Not every convergent series is absolutely convergent. In the event that a given series is convergent but fails to be absolutely convergent, we say that the series is **conditionally convergent**.

We mention, once again, that the comparison principle cannot be used to deduce conditional convergence. For this purpose we use the following more delicate theorem that is known as *Dirichlet's test*.

12.7.5   Dirichlet's Test

Suppose that (a_n) is a decreasing sequence of positive numbers and that $a_n \to 0$ as $n \to \infty$. Suppose that (b_n) is a sequence of real numbers and that the *set of partial sums of the series $\sum b_n$ is bounded. Then the series $\sum a_n b_n$ is convergent.*

Proof. For each n we define

$$B_n = \sum_{j=1}^{n} b_j.$$

Using the fact that the sequence (B_n) is bounded, we choose a number K such that $|B_n| \leq K$ for every n. We now observe that for each n we have

$$\sum_{j=1}^{n} a_j b_j = a_1 B_1 + a_2 (B_2 - B_1) + a_3 (B_3 - B_2) + \cdots + a_n (B_n - B_{n-1})$$

$$= B_1 (a_1 - a_2) + B_2 (a_2 - a_3) + \cdots + B_{n-1} (a_{n-1} - a_n) + a_n B_n.$$

We therefore need to show that, as $n \to \infty$, both of the expressions

$$\sum_{j=1}^{n-1} B_j (a_j - a_{j+1})$$

and $a_n B_n$ approach finite limits. To establish the existence of the first of these two limits we need to show that the series $\sum B_n (a_n - a_{n+1})$ is convergent. As a matter of fact, we can show that the latter series is absolutely convergent. For this purpose we observe that for each n we have

$$|B_n (a_n - a_{n+1})| \leq K |a_n - a_{n+1}| = K (a_n - a_{n+1}),$$

and so the absolute convergence of $\sum B_n (a_n - a_{n-1})$ will follow from the comparison test (Theorem 12.4.6) if we can show that the series $\sum (a_n - a_{n+1})$ is convergent. But this is easy: For each n,

$$\sum_{j=1}^{n} (a_j - a_{j+1}) = a_1 - a_{n+1},$$

and the latter expression approaches a_1 as $n \to \infty$.

Finally, we need to see that the sequence $(a_n B_n)$ converges. In fact, the limit of this sequence is zero, which we can see at once from the sandwich theorem (Theorem 7.4.6) the inequality

$$|a_n B_n| \leq K a_n,$$

and the fact that $a_n \to 0$ as $n \to \infty$. ■

12.7.6   An Inequality Related to Dirichlet's Test

The theorem we present in this subsection tells us something about the magnitude of the sum of a series whose convergence has been established by Dirichlet's test, and it gives us some idea of the size of the error involved if the sum is replaced by an nth partial sum.

Suppose that (a_n) is a decreasing sequence of positive numbers and that (b_n) is a sequence of real numbers. Suppose that $K \geq 0$ and that, for some positive integer N, we have

$$\left| \sum_{j=N}^{n} b_j \right| \leq K$$

for every integer $n \geq N$. Then for all $n \geq N$ we have

$$\left| \sum_{j=N}^{n} a_j b_j \right| \leq K a_N,$$

and, in the event that the series $\sum a_n b_n$ converges we have

$$\left| \sum_{j=N}^{\infty} a_j b_j \right| \leq K a_N.$$

Proof. For each $n \geq N$ we define

$$B_n = \sum_{j=N}^{n} b_j.$$

(Note that this definition of B_n is not quite the same as the one we used in the preceding theorem.) For every $n \geq N$ we have

$$\sum_{j=N}^{n} a_j b_j = a_N B_N + a_{N+1} \left(B_{N+1} - B_N \right) + \cdots + a_n \left(B_n - B_{n-1} \right)$$

$$= B_N \left(a_N - a_{N+1} \right) + B_{N+1} \left(a_{N+1} - a_{N+2} \right) + \cdots$$
$$+ B_{n-1} \left(a_{n-1} - a_n \right) + a_n B_n,$$

and therefore

$$\left| \sum_{j=N}^{n} a_j b_j \right| \leq \sum_{j=N}^{n-1} |B_j (a_j - a_{j+1})| + |a_n B_n|$$

$$\leq \sum_{j=N}^{n-1} K (a_j - a_{j+1}) + K a_n = K a_N. \quad \blacksquare$$

12.7.7 Abel's Theorem

The theorem of Abel that we present in this subsection is concerned with series of the form

$$\sum a_n x^n.$$

Series of this type are known as **power series**. If you look back to the beginning of this chapter, you will see that we have already encountered several power series.

Suppose that (a_n) is a sequence of real numbers and that the series $\sum a_n$ converges. Then the following assertions are true:

1. *Whenever $0 \leq x < 1$, the series $\sum a_n x^n$ converges absolutely.*
2. *We have*

$$\lim_{x \to 1-} \sum_{n=1}^{\infty} a_n x^n = \sum_{n=1}^{\infty} a_n.$$

Proof. From the convergence of the series $\sum a_n$ we know that $a_n \to 0$ as $n \to \infty$, and so the sequence (a_n) is bounded. Therefore, if $0 \leq x < 1$, the absolute convergence of the series $\sum a_n x^n$ follows from the comparison test.

Now, to prove part 2 of the theorem, suppose that $\varepsilon > 0$ and, using the fact that $\sum a_n$ converges, choose a positive integer N such that whenever $n \geq N$ we have

$$\left| \sum_{j=n}^{\infty} a_j \right| = \left| \sum_{j=1}^{\infty} a_j - \sum_{j=1}^{n-1} a_j \right| < \frac{\varepsilon}{6}.$$

Given $n \geq N$ we see that

$$\left| \sum_{j=N}^{n} a_j \right| = \left| \sum_{j=N}^{\infty} a_j - \sum_{j=n+1}^{\infty} a_j \right| \leq \left| \sum_{j=N}^{\infty} a_j \right| + \left| \sum_{j=n+1}^{\infty} a_j \right| < \frac{\varepsilon}{3},$$

and we deduce from Theorem 12.7.6 that for all $x \in [0, 1]$ we have

$$\left| \sum_{j=N}^{\infty} a_j x^j \right| \leq \frac{\varepsilon x^N}{3} \leq \frac{\varepsilon}{3}.$$

We now use the fact that if p is the polynomial defined by the equation

$$p(x) = \sum_{j=1}^{N-1} a_j x^j$$

for all numbers x, then p is continuous at the number 1. Using this fact we choose a number $\delta > 0$ such that whenever $1 - \delta < x \leq 1$ we have

$$\left| \sum_{j=1}^{N-1} a_j x^j - \sum_{j=1}^{N-1} a_j \right| < \frac{\varepsilon}{3}.$$

Then whenever $1 - \delta < x \leq 1$ we have

$$\left| \sum_{j=1}^{\infty} a_j x^j - \sum_{j=1}^{\infty} a_j \right| \leq \left| \sum_{j=1}^{N-1} a_j x^j - \sum_{j=1}^{N-1} a_j \right| + \left| \sum_{j=N}^{\infty} a_j x^j \right| + \left| \sum_{j=N}^{\infty} a_j \right|$$

$$< \frac{\varepsilon}{3} + \frac{\varepsilon}{3} + \frac{\varepsilon}{3} = \varepsilon. \ \blacksquare$$

12.7.8   Abelian and Tauberian Theorems

A discussion of this interesting topic can be reached by clicking on the icon .

12.7.9   Some Examples of Conditionally Convergent Series

1. In this example we investigate the convergence of the series

$$\sum \frac{(2n)! x^n}{(n!)^2}.$$

Using d'Alembert's test we can see that this series converges absolutely when $|x| < 1/4$. In the event that $|x| > 1/4$, the nth term of the series fails to converge to zero and so the series diverges. When $x = 1/4$ the series diverges, as we saw in Subsection 12.6.8. The main purpose of this example is to discuss the behavior of this series when $x = -1/4$ and the series becomes $\sum (-1)^n a_n$, where, for each n,

$$a_n = \frac{(2n)!}{4^n (n!)^2}.$$

By looking at the ratios a_{n+1}/a_n we can see that the sequence (a_n) is decreasing. Furthermore, if

$$b_n = 1 - \frac{a_{n+1}}{a_n} = \frac{1}{2(n+1)}$$

for each n, then, since $\sum b_n$ is divergent, it follows from Theorem 12.6.11 that $a_n \to 0$ as $n \to \infty$. Thus the given series is conditionally convergent when $x = -1/4$.

2. In this example we investigate the convergence of the series

$$\sum \frac{\sin nx}{n},$$ (12.5)

where x is any real number. We begin with the observation that if $\sin x/2 = 0$, then x is an integer multiple of 2π and every term of the given series is zero. From now on we suppose that $\sin x/2 \neq 0$. From the trigonometric identity

$$\cos (\alpha - \beta) - \cos (\alpha + \beta) = 2 \sin \alpha \sin \beta,$$

we see that if n is any positive integer, then

$$\left| \sum_{j=1}^{n} \sin jx \right| = \left| \sum_{j=1}^{n} \frac{2 \sin \frac{x}{2} \sin jx}{2 \sin \frac{x}{2}} \right|$$

$$= \left| \frac{1}{2 \sin \frac{x}{2}} \sum_{j=1}^{n} \left(\cos (2j-1) \frac{x}{2} - \cos (2j+1) \frac{x}{2} \right) \right|$$

$$= \left| \frac{\cos \frac{x}{2} - \cos (2n+1) \frac{x}{2}}{2 \sin \frac{x}{2}} \right| \leq \frac{1}{\left| \sin \frac{x}{2} \right|},$$

and we can therefore deduce from Dirichlet's test that the series (12.5) converges for every real number x. By a similar argument we can observe that the series

$$\sum \frac{\cos nx}{n}$$

converges if x is not an integer multiple of 2π and diverges if x is an integer multiple of 2π. We shall now show that these series converge conditionally. Given any positive integer n we see from the identity

$$\frac{1}{n} = \frac{\cos 2nx}{n} + \frac{2\sin^2 nx}{n}$$

that, unless x is an integer multiple of π, the convergence of $\sum [\cos 2nx]/n$ and the divergence of $\sum 1/n$ guarantee that the series

$$\sum \frac{\sin^2 nx}{n}$$

is divergent. In view of the inequality

$$0 \le \sin^2 nx \le |\sin nx| ,$$

we can deduce from the comparison test that the series $\sum |\sin nx|/n$ diverges whenever x is not an integer multiple of π. Therefore the series (12.5) is conditionally convergent whenever x is not an integer multiple of π.

3. Given any real number α and a positive integer n, the **binomial coefficient** $\binom{\alpha}{n}$ is defined by the equation

$$\binom{\alpha}{n} = \frac{\alpha(\alpha-1)(\alpha-2)\cdots(\alpha-n+1)}{n!}.$$

We also define $\binom{\alpha}{0} = 1$. We saw in Subsection 12.6.8 that the series $\sum \binom{\alpha}{n}$ is absolutely convergent whenever $\alpha > 0$ and fails to be absolutely convergent when $\alpha < 0$. In the event that $\alpha \le -1$, it is easy to see that $\left|\binom{\alpha}{n}\right| \ge 1$ for each n, and so in this case the series $\sum \binom{\alpha}{n}$ diverges. Suppose finally that $-1 < \alpha < 0$. For each n we have

$$\binom{\alpha}{n} = (-1)^n \left|\binom{\alpha}{n}\right|.$$

Now, since

$$\frac{\left|\binom{\alpha}{n+1}\right|}{\left|\binom{\alpha}{n}\right|} = \frac{n-\alpha}{n+1} < 1,$$

we see that the sequence of numbers $\left|\binom{\alpha}{n}\right|$ is decreasing. Furthermore, if

$$b_n = 1 - \frac{\left|\binom{\alpha}{n+1}\right|}{\left|\binom{\alpha}{n}\right|} = \frac{\alpha+1}{n+1}$$

for each n, then since $\sum b_n$ is divergent, it follows from Theorem 12.6.11 that $\left| \binom{\alpha}{n} \right| \to 0$ as $n \to \infty$. We deduce from Dirichlet's test that the series $\sum \binom{\alpha}{n}$ is conditionally convergent when $-1 < \alpha < 0$.

12.7.10 Exercises on Conditionally Convergent Series

1. A common test for convergence that one encounters in an elementary calculus course is the **alternating series test**, sometimes known as the **Leibniz test**, which says that if (a_n) is a decreasing sequence of positive numbers and if $a_n \to 0$ as $n \to \infty$, then the series $\sum (-1)^n a_n$ is convergent. Prove that the alternating series test follows at once from Dirichlet's test.

2. Given that (a_n) is a decreasing sequence of positive numbers and that for each n we have

$$b_n = 1 - \frac{a_{n+1}}{a_n},$$

prove that the series $\sum (-1)^n a_n$ is convergent if and only if the series $\sum b_n$ is divergent.

3. Test the following series for convergence and for absolute convergence:

   (a) $\sum \dfrac{(-1)^n \log n}{n}$.

   (b) $\sum \dfrac{\sin (n\pi/4)}{n}$.

   (c) ■ $\sum \left(\dfrac{1}{2} - 1 \right) \left(\dfrac{1}{3} - 1 \right) \cdots \left(\dfrac{1}{n} - 1 \right)$.

   (d) ■ $\sum \left(\dfrac{1}{2^\delta} - 1 \right) \left(\dfrac{1}{3^\delta} - 1 \right) \cdots \left(\dfrac{1}{n^\delta} - 1 \right)$, where $\delta > 0$.

   (e) $\sum \dfrac{(2 \log 2 - 1)(3 \log 3 - 1) \cdots (n \log n - 1)}{(n!)(\log 2)(\log 3) \cdots (\log n)}$.

4. Determine for what values of x the following series converge and for what values of x the series converge absolutely.

   (a) $\sum \dfrac{(3x - 2)^n}{n}$.

   (b) $\sum \dfrac{(\log x)^n}{n}$.

   (c) $\sum \dfrac{(-1)^n x^n}{(\log n)^x}$.

(d) $\displaystyle\sum \frac{((3n)!)\, x^n}{((2n)!)\,(n!)}$.

(e) $\displaystyle\sum \frac{n^n x^n}{n!}$.

5. Find the values of x for which the following series converge and when they converge absolutely:

(a) $\displaystyle\sum \frac{\sin nx \cos nx}{n}$.

(b) $\displaystyle\sum \frac{(-1)^n \cos nx}{n}$.

(c) ∎ $\displaystyle\sum \frac{\cos^2 nx}{n}$.

(d) $\displaystyle\sum \frac{|\cos nx|}{n}$.

(e) ∎ $\displaystyle\sum \frac{\cos^3 nx}{n}$.

(f) $\displaystyle\sum \frac{\cos^4 nx}{n}$.

6. With an eye on Exercise 5, give an example of a convergent series $\sum a_n$ such that the series $\sum a_n^3$ is divergent.

7. Find the values of x and α for which the **binomial series**

$$\sum \binom{\alpha}{n} x^n$$

is convergent.

8. ∎ Prove that if x is not an integer multiple of 2π, then

$$\left| \sum_{j=1}^{\infty} \frac{\sin jx}{j} \right| \le \frac{1}{\left|\sin \frac{x}{2}\right|}.$$

9. Prove **Abel's test** for convergence of a series which states that if (a_n) is a decreasing sequence of positive numbers and if $\sum b_n$ is a convergent series, then the series $\sum a_n b_n$ is convergent. This theorem may be proved by the method of proof of Dirichlet's test, but it also follows very simply from the statement of Dirichlet's test. Which proof do you prefer?

10. Give an example of a sequence (a_n) of positive numbers and a sequence (b_n) of real numbers such that each of the following conditions holds:

(a) We have $a_n \to 0$ as $n \to \infty$.

(b) The sequence of number $\sum_{j=1}^{n} b_j$ is bounded.

(c) The series $\sum a_n b_n$ is divergent.

11. Give an example of sequences (a_n) and (b_n) such that the following conditions hold:

(a) The sequence (a_n) is a decreasing sequence of positive numbers.

(b) The sequence of number $\sum_{j=1}^{n} b_j$ is bounded.

(c) The series $\sum a_n b_n$ is divergent.

12.8   Rearrangements of Series

The material of this section is optional and can be reached in the on-screen version of the text by clicking on the icon .

12.9   Iterated Series

12.9.1   Introduction and Definition of Iterated Series

Suppose that f is a function from $\mathbf{Z}^+ \times \mathbf{Z}^+$ into $\mathbf{R}$. If, for each positive integer n, the series

$$\sum_{m} f(m, n)$$

converges, and if we define

$$\phi(n) = \sum_{m=1}^{\infty} f(m, n)$$

for each n, then the **iterated series**

$$\sum_{n} \sum_{m} f(m, n)$$

is defined to be the series $\sum \phi(n)$. In the event that this iterated series converges, its sum, which we write as

$$\sum_{n=1}^{\infty} \sum_{m=1}^{\infty} f(m, n),$$

is defined by the equation

$$\sum_{n=1}^{\infty}\sum_{m=1}^{\infty} f(m,n) = \sum_{n=1}^{\infty} \phi(n) = \sum_{n=1}^{\infty}\left(\sum_{m=1}^{\infty} f(m,n)\right).$$

A natural question to ask about iterated series is whether the convergence of one of the iterated series $\sum_n \sum_m f(m,n)$ and $\sum_m \sum_n f(m,n)$ implies that the other is also convergent. In the event that these two series are convergent, a natural question to ask is whether

$$\sum_{n=1}^{\infty}\sum_{m=1}^{\infty} f(m,n) = \sum_{m=1}^{\infty}\sum_{n=1}^{\infty} f(m,n).$$

The answer to both of these questions is *no*, as we saw in Examples 4 and 5 of Subsection 12.1.1.

12.9.2   Iterated Series with Nonnegative Terms

In this subsection we shall observe that if $f(m,n) \geq 0$ for all m and n, then the order of summation can be interchanged.

Suppose that f is a nonnegative function defined on the set $Z^+ \times Z^+$. Then

$$\sum_{n=1}^{\infty}\sum_{m=1}^{\infty} f(m,n) = \sum_{m=1}^{\infty}\sum_{n=1}^{\infty} f(m,n).$$

Proof. We begin with the observation that if M is any positive integer, then, by the algebraic rules for limits we saw in Section 7.5, we have

$$\sum_{n=1}^{\infty}\sum_{m=1}^{M} f(m,n) = \lim_{N\to\infty} \sum_{n=1}^{N}\sum_{m=1}^{M} f(m,n)$$

$$= \lim_{N\to\infty} \sum_{m=1}^{M}\sum_{n=1}^{N} f(m,n) = \sum_{m=1}^{M}\sum_{n=1}^{\infty} f(m,n).$$

Now, given any positive integer M, since the inequality

$$\sum_{m=1}^{M} f(m,n) \leq \sum_{m=1}^{\infty} f(m,n)$$

holds for every n, we have

$$\sum_{m=1}^{M}\sum_{n=1}^{\infty} f(m,n) = \sum_{n=1}^{\infty}\sum_{m=1}^{M} f(m,n) \le \sum_{n=1}^{\infty}\sum_{m=1}^{\infty} f(m,n).$$

Thus

$$\sum_{m=1}^{\infty}\sum_{n=1}^{\infty} f(m,n) = \lim_{M\to\infty}\sum_{m=1}^{M}\sum_{n=1}^{\infty} f(m,n) \le \sum_{n=1}^{\infty}\sum_{m=1}^{\infty} f(m,n).$$

Thus the left side of the desired equation cannot exceed the right side, and, in the same way, one may show that the right side cannot exceed the left side. ∎

12.9.3   Absolutely Convergent Iterated Series

Suppose that f is a function defined on the set $\mathbf{Z}^+ \times \mathbf{Z}^+$ and that the iterated series $\sum_m \sum_n f(m,n)$ is absolutely convergent. Then so is the series $\sum_n \sum_m f(m,n)$ and we have

$$\sum_{n=1}^{\infty}\sum_{m=1}^{\infty} f(m,n) = \sum_{m=1}^{\infty}\sum_{n=1}^{\infty} f(m,n).$$

Proof. We can prove this theorem very easily by applying the nonnegative case to each of the functions $|f|+f$ and $|f|$ and then subtracting. We leave the details as an exercise. ∎

12.10   Multiplication of Series

12.10.1   Introduction to the Idea of a Cauchy Product

For convenience we shall start our series at $n = 0$ in this section. The question we have to consider is how a product

$$\left(\sum_{n=0}^{\infty} a_n\right)\left(\sum_{n=0}^{\infty} b_n\right)$$

can be expanded if $\sum a_n$ and $\sum b_n$ are two given convergent series. There are a number of different approaches to this problem, leading to concepts known as the *Cauchy product*, the *Laurent product*, the *Fourier product*, and the *Dirichlet product*; and, if you wish, you can read about these approaches in Hardy [12].[36]

---

[36]   Hardy's book contains difficult and advanced material, but you may enjoy skimming through it.

We shall be concerned only with the first of these approaches, the Cauchy product, which we shall motivate by considering two series of the form $\sum a_n x^n$ and $\sum b_n x^n$. Such series are known as power series and will be studied in Chapter 14. Under reasonable circumstances, we might expect the expansion of

$$\left(a_0 + a_1 x + a_2 x^2 + a_3 x^3 + \cdots\right)\left(b_0 + b_1 x + b_2 x^2 + b_3 x^3 + \cdots\right)$$

to be

$$a_0 b_0 + \left(a_1 b_0 + a_0 b_1\right) x + \left(a_2 b_0 + a_1 b_1 + a_0 b_2\right) x^2$$

$$+ \left(a_3 b_0 + a_2 b_1 + a_1 b_2 + a_0 b_3\right) x^3 + \cdots$$

$$+ \left(a_n b_0 + a_{n-1} b_1 + \cdots + a_1 b_{n-1} + a_0 b_n\right) x^n + \cdots.$$

Putting $x = 1$ in this identity, we arrive at the idea of a Cauchy product:

We define the **Cauchy product** of two given series $\sum a_n$ and $\sum b_n$ to be the series $\sum c_n$ where, for each n, we have defined

$$c_n = a_n b_0 + a_{n-1} b_1 + \cdots + a_1 b_{n-1} + a_0 b_n = \sum_{j=0}^{n} a_{n-j} b_j.$$

Ideally we would like to be able to say that if $\sum c_n$ is the Cauchy product of two given convergent series $\sum a_n$ and $\sum b_n$, then $\sum c_n$ is also convergent and

$$\sum_{n=0}^{\infty} c_n = \left(\sum_{n=0}^{\infty} a_n\right)\left(\sum_{n=0}^{\infty} b_n\right).$$

However, the series $\sum c_n$ can diverge, as we see in the following two examples:

12.10.2   Some Examples of Divergent Cauchy Products

1. For each integer $n \geq 0$ we define

$$a_n = b_n = \frac{(-1)^n}{\log{(n+2)}}.$$

The convergence of each of the series $\sum a_n$ and $\sum b_n$ follows from Dirichlet's test. We now define $\sum c_n$ to be the Cauchy product of $\sum a_n$ and

$\sum b_n$, and we observe that

$$|c_n| = \left| \frac{(-1)^n}{\log (n + 2) \log 2} + \frac{(-1)^n}{\log (n + 1) \log 3} + \cdots + \frac{(-1)^n}{\log 2 \log (n + 2)} \right|$$

$$\geq \frac{n}{(\log (n + 2))^2}.$$

Since $|c_n| \to \infty$ as $n \to \infty$, we conclude that the series $\sum c_n$ is (very) divergent.

2. For each integer $n \geq 0$ we define

$$a_n = b_n = \frac{(-1)^n}{\sqrt{n + 1}}.$$

Once again, each of the series $\sum a_n$ and $\sum b_n$ converges, but, if $\sum c_n$ is the Cauchy product of $\sum a_n$ and $\sum b_n$, then it is not hard to show that $|c_n| \geq 1$ for each n. So, once again, $\sum c_n$ diverges. We leave the details as an exercise.

12.10.3   The Theorems on Cauchy Products

Although the Cauchy product of two convergent series can be divergent, this is the only pathology that can occur. We shall see soon in the theorem of Abel that if $\sum c_n$ is the Cauchy product of two convergent series $\sum a_n$ and $\sum b_n$, and if $\sum c_n$ happens to converge, then the identity

$$\sum_{n=0}^{\infty} c_n = \left(\sum_{n=0}^{\infty} a_n \right) \left(\sum_{n=0}^{\infty} b_n \right)$$

is assured. We shall also prove Cauchy's theorem that says that if the two series $\sum a_n$ and $\sum b_n$ happen to be absolutely convergent, then so is their Cauchy product.

In the on-screen version of this text you can find some further interesting theorems about Cauchy products.

12.10.4   Cauchy's Theorem for Products of Series

Suppose that $\sum a_n$ and $\sum b_n$ are absolutely convergent series and that $\sum c_n$ is their Cauchy product. Then the series $\sum c_n$ is absolutely convergent and we *have*

$$\sum_{n=0}^{\infty} c_n = \left(\sum_{n=0}^{\infty} a_n \right) \left(\sum_{n=0}^{\infty} b_n \right).$$

Proof. Given nonnegative integers j and n, we define

$$f(n,j) = \begin{cases} a_{n-j} b_j & \text{if } j \le n \\ 0 & \text{if } j > n. \end{cases}$$

Since

$$\sum_{j=0}^{\infty} \sum_{n=0}^{\infty} |f(n,j)| = \sum_{j=0}^{\infty} \sum_{n=j}^{\infty} |a_{n-j} b_j| = \sum_{j=0}^{\infty} |b_j| \left(\sum_{n=j}^{\infty} |a_{n-j}| \right)$$

$$= \left(\sum_{j=0}^{\infty} |b_j| \right) \left(\sum_{n=0}^{\infty} |a_n| \right) < \infty,$$

and, similarly,

$$\sum_{j=0}^{\infty} \sum_{n=0}^{\infty} f(n,j) = \sum_{j=0}^{\infty} \sum_{n=j}^{\infty} a_{n-j} b_j$$

$$= \sum_{j=0}^{\infty} b_j \left(\sum_{n=j}^{\infty} a_{n-j} \right) = \left(\sum_{j=0}^{\infty} b_j \right) \left(\sum_{n=0}^{\infty} a_n \right),$$

we deduce from Theorem 12.9.3 that

$$\left(\sum_{n=0}^{\infty} b_n \right) \left(\sum_{n=0}^{\infty} a_n \right) = \sum_{j=0}^{\infty} \sum_{n=0}^{\infty} f(n,j)$$

$$= \sum_{n=0}^{\infty} \sum_{j=0}^{\infty} f(n,j) = \sum_{n=0}^{\infty} \sum_{j=0}^{n} a_{n-j} b_j. \blacksquare$$

12.10.5   Abel's Theorem on Products of Series

If $\sum c_n$ is the Cauchy product of two convergent series $\sum a_n$ and $\sum b_n$, and if the series $\sum c_n$ happens to converge, then

$$\sum_{n=0}^{\infty} c_n = \left(\sum_{n=0}^{\infty} a_n \right) \left(\sum_{n=0}^{\infty} b_n \right).$$

Proof. Whenever $0 \leq x < 1$, it follows from Abel's theorem (Theorem 12.7.7) that the series $\sum a_n x^n$ and $\sum b_n x^n$ are absolutely convergent. Furthermore, it is easy to see that $\sum c_n x^n$ is the Cauchy product of $\sum a_n x^n$ and $\sum b_n x^n$, and so we can deduce from Cauchy's theorem that whenever $0 \leq x < 1$ we have

$$\sum_{n=0}^{\infty} c_n x^n = \left(\sum_{n=0}^{\infty} a_n x^n \right) \left(\sum_{n=0}^{\infty} b_n x^n \right).$$

We now obtain the desired result by letting $x \to 1$ from the left and using Abel's theorem again. ∎

12.10.6   Some Exercises on Products of Series

1. Calculate the Cauchy product of the series $\sum (-1)^n x^n$ and $\sum x^n$. By looking at the sums of these three series, verify that Cauchy's theorem (Theorem 12.10.4) is true for these series when $|x| < 1$.
2. This exercises requires a knowledge of the binomial theorem (see Exercise 4 of Subsection 9.5.4). Show that the Cauchy product of the two series $\sum x^n/n!$ and $\sum y^n/n!$ is $\sum (x+y)^n/n!$. As you may know, the sums of these series are e^x and e^y and e^{x+y}, respectively, and you will see this fact officially in Subsection 14.5.2. What does Cauchy's theorem say for these three series?

12.10.7   The More Powerful Theorems on Cauchy Products

As we promised in Subsection 12.10.3, the on-screen version of this text contains some more powerful theorems on Cauchy products than we have seen here. It contains the theorem of Franz Mertens, which says that if at least one of two convergent series $\sum a_n$ and $\sum b_n$ happens to be absolutely convergent, then their Cauchy product converges. You will also find an interesting theorem of Sheila Edmonds, which says that if $\sum a_n$ and $\sum b_n$ are given convergent series, and if the terms of these series are small enough to make the sequences (na_n) and (nb_n) bounded, then the Cauchy product of the two series will converge. The latter theorem is deduced with the help of an interesting theorem about Cauchy products that was discovered by Ludwig Neder. To reach these more powerful theorems, click on the icon ▩ .

12.11   The Cantor Set

The Cantor set is defined to be the set of all those real numbers that can be expressed in the form

$$\sum_{n=1}^{\infty} \frac{a_n}{3^n},$$

where each of the numbers a_n is either 0 or 2. You can find a discussion of the Cantor set and an associated function called the **Cantor function** by clicking on the icon .

Chapter 13
Improper Integrals

13.1　Introduction to Improper Integrals

Some of the integrals that play a major role in elementary calculus do not fit into Chapter 11 because they run into difficulty at one or the other of the two endpoints of the interval of integration. To remind ourselves how this can happen, we shall look at some examples.

13.1.1　Some Examples to Motivate Improper Integrals

1. We would naturally expect that

$$\int_0^1 \frac{1}{\sqrt{x}} dx = 2\sqrt{x}\Big|_0^1 = 2,$$

but the trouble with this integral is that the integrand (the function that we are integrating) is not defined at 0, and, even worse, this integrand is not bounded. Since the theory of Riemann integration as presented in Chapter 11 was confined to bounded functions, we have to ask what the latter integral really means.

In elementary calculus this question is answered with the observation that whenever $0 < w \le 1$ we have

$$\int_w^1 \frac{1}{\sqrt{x}} dx = 2 - 2\sqrt{w},$$

and consequently that

$$\lim_{w \to 0+} \int_w^1 \frac{1}{\sqrt{x}} dx = \lim_{w \to 0+} \left(2 - 2\sqrt{w}\right) = 2.$$

This is the sense in which we could say that

$$\int_0^1 \frac{1}{\sqrt{x}} dx = 2.$$

2. The integral

$$\int_0^1 \frac{1}{\sqrt{1 - x^2}} dx$$

fails to exist in the sense of Chapter 11 because its integrand is unbounded, but the fact that

$$\lim_{w \to 1-} \int_0^w \frac{1}{\sqrt{1-x^2}} dx = \lim_{w \to 1-} \left(\arcsin w - \arcsin 0\right) = \frac{\pi}{2}$$

provides us with a way of saying that

$$\int_0^1 \frac{1}{\sqrt{1-x^2}} dx = \frac{\pi}{2}.$$

3. The integral

$$\int_1^\infty \frac{1}{x^2} dx$$

fails to exist in the sense of Chapter 11 because the interval of integration is unbounded, but the fact that

$$\lim_{w \to \infty} \int_1^w \frac{1}{x^2} dx = \lim_{w \to \infty} \left(1 - \frac{1}{w}\right) = 1$$

provides us with a way of saying that

$$\int_1^\infty \frac{1}{x^2} dx = 1.$$

These examples suggest the definition of an improper integral that follows:

13.1.2   Definition of an Improper Integral

We define an **improper integral** of the type

$$\int_a^{\to b} f(x) dx$$

as follows: Suppose that $-\infty < a < b \le \infty$. Suppose that f is a function defined on the interval $[a, b)$ and that, whenever $a < w < b$, the function f is Riemann integrable on the interval $[a, w]$. If the limit

$$\lim_{w \to b-} \int_a^w f(x) dx$$

exists, then we define

$$\int_a^{\to b} f(x) dx = \lim_{w \to b-} \int_a^w f(x) dx.$$

In the event that the improper integral $\int_a^{\to b} f(x)dx$ exists and is finite, we say that f is **improper Riemann integrable** on the interval $[a, b)$. Somewhat less precisely, we also say that the integral $\int_a^{\to b} f(x)dx$ is **convergent**.

An improper integral of the type

$$\int_{a\leftarrow}^b f(x)dx$$

is defined similarly: Suppose that $-\infty \leq a < b < \infty$, that f is a function defined on the interval $(a, b]$, and that f is Riemann integrable on the interval $[w, b]$ whenever $a < w < b$. If the limit

$$\lim_{w\to a+} \int_w^b f(x)dx$$

exists, then we define

$$\int_{a\leftarrow}^b f(x)dx = \lim_{w\to a+} \int_w^b f(x)dx.$$

An improper integral that fails to converge is said to **diverge**. This means that there are two ways in which a given improper integral $\int_{a\leftarrow}^b f(x)dx$ or $\int_a^{\to b} f(x)dx$ may diverge. Either the appropriate limit

$$\lim_{w\to a+} \int_w^b f(x)dx \quad \text{or} \quad \lim_{w\to b-} \int_a^w f(x)dx$$

is infinite or the limit does not exist at all. Our use of the words *convergent* and *divergent* here is therefore analogous to the way in which we used these words for infinite series in Chapter 12.

13.1.3 Some Examples of Improper Integrals

1. In our first example we observe that if an integral $\int_a^b f$ exists in the ordinary Riemann sense, then each of the improper integrals $\int_a^{\to b} f$ and $\int_{a\leftarrow}^b f$ exists and

$$\int_a^{\to b} f = \int_{a\leftarrow}^b f = \int_a^b f.$$

This assertion follows at once from Theorem 11.12.1.

2. Returning to the examples that appear in Subsection 13.1, we see that

$$\int_{0\leftarrow}^{1}\frac{1}{\sqrt{x}}dx = 2 \quad \text{and} \quad \int_{0}^{\rightarrow 1}\frac{1}{\sqrt{1-x^2}}dx = \frac{\pi}{2} \quad \text{and} \quad \int_{1}^{\rightarrow \infty}\frac{1}{x^2}dx = 1.$$

3. Since

$$\lim_{w\to\infty}\int_{1}^{w}\frac{1}{\sqrt{x}}dx = \lim_{w\to\infty}\left(2\sqrt{w}-2\right) = \infty,$$

we have

$$\int_{1}^{\infty}\frac{1}{\sqrt{x}}dx = \infty,$$

and so this integral is divergent.

4. Whenever $w \geq 0$ we have

$$\int_{0}^{w}\cos x\,dx = \sin w,$$

and since $\sin w$ does not approach a limit as $w \to \infty$, we see that the integral $\int_{0}^{\rightarrow\infty}\cos x\,dx$ diverges.

13.1.4   An Extension of our Notation

We shall often write improper integrals of the form $\int_{a}^{\rightarrow b} f$ or $\int_{a\leftarrow}^{b} f$ more simply as $\int_{a}^{b} f$ and an improper integral of the form $\int_{a}^{\rightarrow\infty} f$ will be written more simply as $\int_{a}^{\infty} f$. In adopting this convention, however, we need to bear in mind that it is not really precise because it leaves us with the task of determining that an ordinary looking Riemann integral is actually improper. If there are several numbers in an interval $[a,b]$ near which a given function f is unbounded, we interpret the improper integral $\int_{a}^{b} f$ to mean the sum of several improper integrals of the type discussed above. Thus, for example, the integral

$$\int_{0}^{\infty}\frac{1}{\sqrt[3]{x}\,(x-5)}dx$$

means the sum

$$\int_{0\leftarrow}^{1}\frac{1}{\sqrt[3]{x}\,(x-5)}dx + \int_{1}^{\rightarrow 5}\frac{1}{\sqrt[3]{x}\,(x-5)}dx$$

$$+ \int_{5\leftarrow}^{6}\frac{1}{\sqrt[3]{x}\,(x-5)}dx + \int_{6}^{\rightarrow\infty}\frac{1}{\sqrt[3]{x}\,(x-5)}dx.$$

The choice of the intermediate numbers 1 and 6 in this interpretation of the integral is unimportant.

13.2 Elementary Properties of Improper Integrals

13.2.1 Linearity of Improper Integrals

Suppose that f and g are improper Riemann integrable functions on an interval $[a, b)$ and that α is a given real number.

1. *The function $f + g$ is improper Riemann integrable on $[a, b)$ and we have*

$$\int_a^{\to b} (f + g) = \int_a^{\to b} f + \int_a^{\to b} g.$$

2. *The function αf is improper Riemann integrable on $[a, b)$ and we have*

$$\int_a^{\to b} \alpha f = \alpha \int_a^{\to b} f.$$

Proof. To prove the first part we observe that

$$\lim_{w \to b-} \int_a^w (f + g) = \lim_{w \to b-} \left(\int_a^w f + \int_a^w g \right) = \int_a^{\to b} f + \int_a^{\to b} g,$$

and the proof of the second part is similar. ∎

13.2.2 Additivity of Improper Riemann Integrals

Suppose that f is an improper Riemann integrable function on an interval $[a, b)$ and that $a \leq c < b$. Then we have

$$\int_a^{\to b} f = \int_a^c f + \int_c^{\to b} f.$$

We leave the proof of this assertion as an exercise.

13.2.3 Some Exercises on Improper Integrals

1. Ⓝ Evaluate each of the following improper integrals, when possible, and specify those that diverge. If you can't see how to evaluate the integral exactly yourself, ask *Scientific Notebook* to evaluate it for you. (Before asking *Scientific Notebook* to evaluate one of these integrals, remove the arrow sign from the limits of integration.)

   (a) $\int_0^{\to \infty} \frac{1}{(1+x^2)^{3/2}} dx$.

(b) $\int_2^{\to\infty} \frac{1}{x\sqrt{x^2-1}} dx.$

(c) $\int_{1\leftarrow}^2 \frac{1}{x\sqrt{x^2-1}} dx.$

(d) $\int_{1\leftarrow}^{\to\infty} \frac{1}{x\sqrt{x^2-1}} dx.$

(e) $\int_0^{\to\pi/2} \tan x\, dx.$

(f) $\int_0^{\to\pi/2} \sqrt{\tan x \sin x}\, dx.$

(g) $\int_{0\leftarrow}^{\pi/2} \frac{x\cos x - \sin x}{x^2} dx.$

(h) $\int_1^{\to\infty} e^{-x} \sin x\, dx.$

2. (a) Prove that the integral

$$\int_{0\leftarrow}^1 \frac{1}{x^p} dx$$

converges when $p < 1$ and diverges when $p \geq 1$.

(b) Prove that the integral

$$\int_1^{\to\infty} \frac{1}{x^p} dx$$

converges when $p > 1$ and diverges when $p \leq 1$.

3. Prove that the integral

$$\int_2^{\to\infty} \frac{1}{x\,(\log x)^p} dx$$

is convergent when $p > 1$ and divergent when $p \leq 1$.

4. Interpret the integral

$$\int_0^2 \frac{1}{(x-1)^{1/3}} dx$$

as the sum of two improper integrals and evaluate it.

5. Prove that if f is bounded on an interval $[a,b]$ and is improper Riemann integrable on $[a,b)$, then f is Riemann integrable on $[a,b]$ and

$$\int_a^{\to b} f = \int_a^b f.$$

6. In the discussion of improper integrals that appears in Subsection 13.1.2 we used the words:

... Somewhat less precisely, we also say that the integral $\int_a^{\to b} f(x)dx$ is convergent. ...

Why is the statement that the integral $\int_a^{\to b} f(x)dx$ is convergent less precise than the statement that f is improper Riemann integrable on $[a, b)$? Hint: In our study of infinite series we made a careful distinction between the symbols $\sum a_n$ and $\sum_{n=1}^{\infty} a_n$.

13.3   Convergence of Integrals of Nonnegative Functions

This section contains the analog for integrals of the theory of convergence of series with nonnegative terms that we presented In Section 12.4. We shall present the main theorems for integrals of the type $\int_a^{\to b}$. The analogous results for other kinds of improper integral can be obtained in the same way.

13.3.1   The Comparison Principle for Integrals

Suppose that $-\infty < a < b \leq \infty$, suppose that f is a nonnegative function defined on the interval $[a, b)$, and suppose that whenever $a < w < b$, the function f is Riemann integrable on the interval $[a, w]$. For each number w in the interval $[a, b)$ suppose that

$$F(w) = \int_a^w f(x)dx.$$

The function F is increasing because if w_1 and w_2 belong to the interval $[a, b)$ and $w_1 \leq w_2$, then we have

$$F(w_2) - F(w_1) = \int_{w_1}^{w_2} f(x)dx \geq 0.$$

Therefore, either $F(w) \to \infty$ as $w \to b-$ or $F(w)$ approaches a finite limit as $w \to b-$.

13.3.2   The Comparison Test for Integrals

By analogy with the comparison test for series that we saw in Subsection 12.4.6, we have a comparison test for improper integrals that tells us that if f and g are nonnegative functions defined on an interval $[a, b)$, and if the integrals $\int_a^w f$ and $\int_a^w g$ exist whenever $a \leq w < b$, then, under certain conditions, if one of the integrals $\int_a^{\to b} f$ and $\int_a^{\to b} g$ converges, then so does the other. The value of this sort of theorem is that it allows us to test a given integral by comparing it with another whose behavior is already known to us.

Statement of the Comparison Test

   Suppose that f and g are nonnegative functions defined on an interval $[a, b)$ and that the integrals $\int_a^w f$ and $\int_a^w g$ exist whenever $a \leq w < b$.

1. *If it is possible to find a positive number k such that $f(x) \leq kg(x)$ for all $x \in [a, b)$, and if $\int_a^{\to b} g$ converges, then $\int_a^{\to b} f$ must converge.*
2. *If the function f/g is bounded and if $\int_a^{\to b} g$ converges, then $\int_a^{\to b} f$ must converge.*
3. *If $f(x)/g(x)$ approaches a finite limit as $x \to b$ (from the left), and if $\int_a^{\to b} g$ converges, then $\int_a^{\to b} f$ must converge.*
4. *If it is possible to find a positive number δ such that $f(x)/g(x) \geq \delta$ for all $x \in [a, b)$, and if $\int_a^{\to b} f$ converges, then $\int_a^{\to b} g$ must converge.*
5. *If $f(x)/g(x)$ approaches a positive limit (possibly ∞) as $x \to b$, and if $\int_a^{\to b} f$ converges, then $\int_a^{\to b} g$ must converge.*

Proof. Assertion 1 follows at once from the comparison principle discussed in Subsection 13.3.1 and the inequality

$$\int_a^w f \leq k \int_a^w g$$

that holds for each n.

   Assertion 2 follows at once from assertion 1.

   To prove assertion 3, suppose that $f(x)/g(x)$ approaches a finite limit q as $x \to b-$ and choose a number c in the interval (a, b) such that $f(x)/g(x) < q+1$ whenever $c < x < b$. Whenever $c < w < b$ we have

$$\int_a^w f(x)dx = \int_a^c f(x)dx + \int_c^w f(x)dx \leq \int_a^c f(x)dx + (q+1)\int_c^w g(x)dx,$$

and therefore $\int_a^w f(x)dx$ cannot approach ∞ as $w \to b-$.

   Assertion 4 follows at once from assertion 2 and the fact that if $\delta > 0$ and if $f(x)/g(x) \geq \delta$ for every $x \in [a, b)$, then the function g/f is bounded and assertion 5 follows at once from assertion 3. ■

13.3.3   Some Exercises on the Comparison Test for Integrals

1. Determine the convergence or divergence of the following integrals:

   (a) $\displaystyle\int_1^{\to\infty} \frac{\sqrt{x}}{x^2 - x + 1}dx.$

(b) $\displaystyle\int_{0\leftarrow}^{1} \frac{1}{x+x^2}dx.$

(c) $\displaystyle\int_{1}^{\rightarrow\infty} \frac{\sin^2 x}{x^2}dx.$

(d) $\displaystyle\int_{0\leftarrow}^{\rightarrow\infty} \frac{\sin^2 x}{x^2\sqrt{x}}dx.$

(e) $\displaystyle\int_{0}^{\rightarrow\pi/2} \sqrt{\tan x}dx.$

(f) $\displaystyle\int_{1\leftarrow}^{2} \frac{1}{\log x}dx.$

(g) $\displaystyle\int_{0\leftarrow}^{\pi/2} \log\left(\sin x\right)dx.$

(h) ■ $\displaystyle\int_{2}^{\rightarrow\infty} \frac{1}{(\log x)^{\log x}}dx.$

(i) $\displaystyle\int_{3}^{\rightarrow\infty} \frac{1}{(\log\log x)^{\log x}}dx.$

(j) $\displaystyle\int_{30}^{\rightarrow\infty} \frac{1}{(\log\log\log x)^{\log x}}dx.$

(k) $\displaystyle\int_{3}^{\rightarrow\infty} \frac{1}{(\log x)^{\log\log x}}dx.$

(l) $\displaystyle\int_{1}^{\rightarrow\infty} \frac{1}{\exp\left(\sqrt{\log x}\right)}dx.$

2. (a) Prove that the integral

$$\int_{1}^{\rightarrow\infty} x^{\alpha-1}e^{-x}dx$$

converges for every number α.

(b) Prove that the integral

$$\int_{0\leftarrow}^{1} x^{\alpha-1}e^{-x}dx$$

converges if and only if $\alpha > 0$ and deduce that the integral

$$\int_{0\leftarrow}^{\to\infty} x^{\alpha-1}e^{-x}dx$$

converges if and only if $\alpha > 0$. The latter integral defines the value at α of the **gamma function** and is denoted as $\Gamma(\alpha)$. You will see more about this important function in Subsection 16.6.7.

3. Prove that the integral

$$\int_{0\leftarrow}^{\to 1} (1-t)^{\alpha-1}\, t^{\beta-1}dt$$

converges if and only if both α and β are positive. This integral defines the value at the point (α, β) of the **beta function** and is denoted as $B(\alpha, \beta)$.

13.4   Absolute and Conditional Convergence

In this section we make a more careful study of the convergence of improper integrals of a function f that is no longer required to be nonnegative. Without the assumption that f is nonnegative, we have no analog of the comparison tests that we studied in the preceding section. Just as we did for series in Section 12.7, we shall discuss the notions of absolute and conditional convergence of an improper integral, and we shall show that every absolutely convergent improper integral is convergent. Then we shall focus our attention on conditionally convergent integrals, and we shall obtain an integral analog of Dirichlet's test, Theorem 12.7.5.

13.4.1   Definition of Absolute Convergence

Suppose that $-\infty < a < b \leq \infty$ and that f is a function defined on the interval $[a, b)$. Suppose that f is Riemann integrable on the interval $[a, w]$ whenever $a < w < b$. Note that for each such w the function $|f|$ must also be Riemann integrable on $[a, w]$. If the improper integral $\int_a^{\to b} |f|$ converges, then we say that the integral $\int_a^{\to b} f$ **converges absolutely**.

13.4.2   Convergence of Absolutely Convergent Integrals

Every absolutely convergent improper integral is convergent. Furthermore, if $\int_a^{\to b} f$ is absolutely convergent, then

$$\left| \int_a^{\to b} f \right| \leq \int_a^{\to b} |f|.$$

Proof. Suppose that $\int_a^{\to b} f$ is absolutely convergent. From the inequality

$$0 \le (f + |f|) \le 2|f|$$

and the comparison test (Theorem 13.3.2) it follows that the integral

$$\int_a^{\to b} (f + |f|)$$

is convergent. The convergence of the integral $\int_a^{\to b} f$ therefore follows from the linearity property (Theorem 13.2.1) and the identity

$$f = (f + |f|) - |f|.$$

The final assertion of the theorem now follows from the fact that whenever $a < w < b$ we have

$$\left| \int_a^w f \right| \le \int_a^w |f|. \quad \blacksquare$$

13.4.3   Definition of Conditional Convergence

If an improper integral $\int_a^{\to b} f$ is convergent but is not absolutely convergent, then we say that the integral is **conditionally convergent**.

13.4.4   Example of a Conditionally Convergent Integral

We define a function f on the interval $[1, \infty)$ by defining

$$f(x) = \frac{(-1)^{n-1}}{n}$$

whenever n is a positive integer and $n \le x < n + 1$. The graph of this function is illustrated in Figure 13.1.

Since

$$\lim_{w \to \infty} \int_1^w f = \lim_{n \to \infty} \sum_{j=1}^n \frac{(-1)^{j-1}}{j} = \sum_{j=1}^\infty \frac{(-1)^{j-1}}{j},$$

which converges by Dirichlet's test for series (Theorem 12.7.5), and since

$$\lim_{w \to \infty} \int_1^w |f| = \lim_{n \to \infty} \sum_{j=1}^n \frac{1}{j} = \infty,$$

the integral $\int_1^{\to \infty} f$ is conditionally convergent.

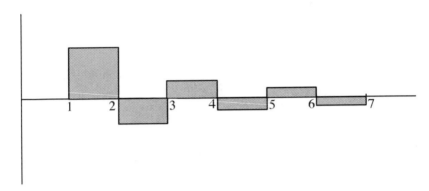

Figure 13.1

13.4.5   Dirichlet's Test for Improper Integrals

Suppose that $-\infty < a < b \leq \infty$ and that f is a positive decreasing differentiable function defined on the interval $[a, b)$. Suppose that the derivative of f is continuous and that $f(x) \to 0$ as $x \to b$ (from the left).

Suppose that g is a continuous function on the interval $[a, b)$, that $K \geq 0$, and that the inequality

$$\left| \int_a^x g \right| \leq K$$

holds whenever $a < x < b$. Then the integral $\int_a^{\to b} fg$ converges and

$$\left| \int_a^{\to b} fg \right| \leq Kf(a).$$

Proof. For each number x in the interval $[a, b)$ we define

$$G(x) = \int_a^x g(t)dt.$$

We observe that $G(a) = 0$ and that for every $x \in [a, b)$ we have $G'(x) = g(x)$. Integrating by parts, we obtain

$$\int_a^x fg = \int_a^x fG' = f(x)G(x) - f(a)G(a) - \int_a^x f'G = f(x)G(x) - \int_a^x f'G.$$

Since $|f(x)G(x)| \leq Kf(x)$ for each x and since $f(x) \to 0$ as $x \to b$, we have

$$\lim_{x \to b-} f(x)G(x) = 0.$$

Therefore, to prove that the integral $\int_a^{\to b} fg$ is convergent we need only show that the integral $\int_a^{\to b} f'G$ is convergent. For this purpose we observe that whenever $a \le x < b$ we have

$$0 \le |f'(x)G(x)| \le K\left(-f'(x)\right).$$

Now, since

$$\lim_{x \to b-} \int_a^x K\left(-f'(t)\right) dt = \lim_{x \to b} \left(Kf(a) - Kf(x)\right) = Kf(a)$$

we know that the integral $\int_a^{\to b} K\left(-f'\right)$ is convergent, and so it follows from the comparison test that the integral $\int_a^{\to b} f'G$ converges absolutely.

Finally, since

$$\int_a^x fg = f(x)G(x) + \int_a^x \left(-f'\right)G$$

for each $x \in [a, b)$, we see that for each such x

$$\left| \int_a^x fg \right| \le |f(x)G(x)| + \int_a^x \left(-f'\right)G \le Kf(x) + K\int_a^x \left(-f'\right)$$

$$\le Kf(x) - Kf(x) + Kf(a) = Kf(a)$$

and therefore

$$\left| \int_a^{\to b} fg \right| \le Kf(a). \ \blacksquare$$

13.4.6 Abel's Theorem for Integrals

Abel's theorem for integrals is an analog of Abel's theorem for series seen in Theorem 12.7.7.

Suppose that g is continuous and improper Riemann integrable on an interval $[a, \infty)$. Then we have

$$\lim_{p \to 0+} \int_a^{\to \infty} e^{-px} g(x) dx = \int_a^{\to \infty} g(x) dx.$$

Proof. We begin with the observation that the integral

$$\int_a^{\to \infty} e^{-px} g(x) dx$$

converges by Dirichlet's test whenever $p > 0$. Suppose that $\varepsilon > 0$ and choose a number w such that whenever $x \geq w$ we have

$$\left| \int_x^{\to \infty} g \right| = \left| \int_a^{\to \infty} g - \int_a^x g \right| < \frac{\varepsilon}{8}.$$

Thus whenever $x \geq w$ we have

$$\left| \int_w^x g \right| = \left| \int_w^{\to \infty} g - \int_x^{\to \infty} g \right| \leq \left| \int_w^{\to \infty} g \right| + \left| \int_x^{\to \infty} g \right| < \frac{\varepsilon}{4}.$$

Therefore whenever $p > 0$ we can apply Theorem 13.4.5 on the interval $[w, \infty)$ to obtain

$$\left| \int_w^{\to \infty} e^{-px} g(x) dx \right| \leq e^{-pw} \frac{\varepsilon}{4} < \frac{\varepsilon}{4}.$$

Therefore whenever $p > 0$ we have

$$\left| \int_a^{\to \infty} e^{-px} g(x) dx - \int_a^{\to \infty} g(x) dx \right|$$

$$\leq \left| \int_a^w e^{-px} g(x) dx - \int_a^w g(x) dx \right| + \left| \int_w^{\to \infty} e^{-px} g(x) dx \right|$$

$$+ \left| \int_w^{\to \infty} g(x) dx \right|$$

$$< \left| e^{-pa} - 1 \right| \int_a^w |g| + \frac{\varepsilon}{4} + \frac{\varepsilon}{4},$$

and we can make the latter expression less than ε by making p small enough. ∎

13.4.7   Another Conditionally Convergent Integral

Suppose that w is any positive number. Using the inequality

$$\left| \int_w^x \sin t\, dt \right| = |\cos w - \cos x| \leq 2$$

that holds whenever $x \geq w$ and Dirichlet's test, we deduce that the integral

$$\int_w^{\to \infty} \frac{\sin x}{x} dx$$

is convergent and that

$$\left| \int_w^{\to\infty} \frac{\sin x}{x} dx \right| \le \frac{2}{w},$$

and we see similarly that if c is any nonzero number, then the integral

$$\int_w^{\to\infty} \frac{\cos cx}{x} dx$$

converges. It therefore follows from the identity

$$\frac{\sin^2 x}{x} = \frac{1}{2x} - \frac{\cos 2x}{2x}$$

that the integral

$$\int_w^{\to\infty} \frac{\sin^2 x}{x} dx$$

diverges. Since

$$0 \le \frac{\sin^2 x}{x} \le \left| \frac{\sin x}{x} \right|$$

for every $x \ge w$, we deduce from the comparison test that the integral

$$\int_w^{\to\infty} \left| \frac{\sin x}{x} \right| dx$$

diverges and so the integral

$$\int_w^{\to\infty} \frac{\sin x}{x} dx$$

is conditionally convergent. We mention finally that, since

$$\lim_{x\to 0} \frac{\sin x}{x} = 1$$

the integral

$$\int_0^1 \frac{\sin x}{x} dx$$

is not improper and so the integral

$$\int_0^{\to\infty} \frac{\sin x}{x} dx$$

is convergent. We shall show in Section 16.5 that the value of the latter integral is $\pi/2$.

13.4.8   Some Further Exercises on Improper Integrals

1. Determine the convergence or divergence of the following integrals:

   (a) $\displaystyle\int_0^{\to\infty} \frac{\sin x}{\sqrt{x}}\,dx.$

   (b) $\displaystyle\int_2^{\to\infty} \frac{\sin^3 x}{x}\,dx.$

   (c) $\displaystyle\int_1^{\to\infty} \frac{e^x \sin (e^x)}{x}\,dx.$

2. Integrate by parts to obtain the identity

$$\int_0^w \frac{\sin^2 x}{x^2}\,dx = -\frac{\sin^2 w}{w} + \int_0^w \frac{2\sin x \cos x}{x}\,dx$$

   and deduce that each of the following four improper integrals equals

$$\int_0^{\to\infty} \frac{\sin x}{x}\,dx:$$

   (a) $\displaystyle\int_0^{\to\infty} \frac{2\sin x \cos x}{x}\,dx.$

   (b) $\displaystyle\int_0^{\to\infty} \frac{\sin^2 x}{x^2}\,dx.$

   (c) $\displaystyle\int_0^{\to\infty} \frac{2\sin^2 x \cos^2 x}{x^2}\,dx.$

   (d) $\displaystyle\int_0^{\to\infty} \frac{2\sin^4 x}{x^2}\,dx.$

Chapter 14
Sequences and Series of Functions

In addition to the theory of limits of numerical sequences that we studied in Chapter 7, there is also a very important theory of limits of sequences of the form (f_n) where, for each n, the symbol f_n stands for a function. In this chapter, we shall introduce three important ways in which the convergence of such sequences can be described. These types of convergence of a sequence of functions are known as **pointwise convergence**, **bounded convergence**, and **uniform convergence**, and, for each of these three types, we shall ask ourselves questions like the following:

1. If the sequence (f_n) converges to a function f and each of the functions f_n is continuous, to what extent is it true that f must also be continuous?
2. If the sequence (f_n) converges to a function f and each of the functions f_n is integrable on a given interval $[a, b]$, to what extent is it true that f must also be integrable on $[a, b]$? In the event that f is also integrable, to what extent is it true that

$$\lim_{n \to \infty} \int_a^b f_n = \int_a^b f?$$

3. If the sequence (f_n) converges to a function f and each of the functions f_n is differentiable, to what extent is it true that f must also be differentiable? In the event that f is also differentiable, to what extent is it true that the sequence (f_n') must converge to the function f'?
4. If (f_n) is a sequence of integrable functions on an interval $[a, b]$, when do we have

$$\int_a^b \sum_{n=1}^{\infty} f_n = \sum_{n=1}^{\infty} \int_a^b f_n?$$

5. If (f_n) is a sequence of differentiable functions, when do we have

$$\left(\sum_{n=1}^{\infty} f_n \right)' = \sum_{n=1}^{\infty} f_n'?$$

The theorems that assert positive answers to questions of this type are among some of the most useful theorems in mathematical analysis.

14.1   The Three Types of Convergence

14.1.1   Pointwise Convergence

Pointwise convergence is the simplest type of convergence of a sequence of functions, but, as we shall see, it is not related simply to other types of limit concepts such as continuity, derivatives, and integrals.

Suppose that (f_n) is a sequence of real valued functions defined on a set S and that $f : S \to \mathbf{R}$. We say that the sequence (f_n) **converges pointwise** to the function f if we have

$$\lim_{n \to \infty} f_n(x) = f(x)$$

for every number x in the set S.

14.1.2   Bounded Convergence

A sequence (f_n) of real valued functions defined on a set S is said to converge **boundedly on** S to a given function f if the following two conditions are satisfied:

1. The sequence (f_n) converges pointwise to f on the set S.
2. There exists a number α such that $\sup |f_n| \leq \alpha$ for every n.

14.1.3   Uniform Convergence

A sequence (f_n) of real valued functions defined on a set S is said to converge **uniformly on** S to a given function f if we have

$$\lim_{n \to \infty} \sup |f_n - f| = 0.$$

In other words, the sequence (f_n) converges uniformly to f if

$$\sup \{|f_n(x) - f(x)| \, x \in S\} \to 0$$

as $n \to \infty$.

14.1.4   Some Examples of Sequences of Functions

1. Ⓝ We begin by defining

$$f_n(x) = x^n$$

for each positive integer n and each number x in the interval $[0, 1]$. For each n, the function f_n is a continuous function from the interval $[0, 1]$ onto $[0, 1]$. Figure 14.1 illustrates the graphs of the functions f_n for the first few values of n. Since

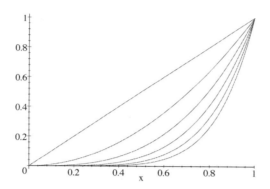

Figure 14.1

$$\lim_{n \to \infty} x^n = \begin{cases} 0 & \text{if } 0 \le x < 1 \\ 1 & \text{if } x = 1, \end{cases}$$

the sequence (f_n) converges pointwise to the function f defined by the equation

$$f(x) = \begin{cases} 0 & \text{if } 0 \le x < 1 \\ 1 & \text{if } x = 1. \end{cases}$$

Since $\sup |f_n| = 1$ for each n, we see that the sequence (f_n) converges boundedly to the function f. However, since

$$\sup |f_n - f| = \sup \{x^n - 0 \mid 0 \le x < 1\} = 1$$

for each n, the sequence (f_n) does not converge uniformly to the function f. Since the limit function f fails to be continuous at the number 1, we conclude that a sequence of continuous functions can converge boundedly to a function that is not continuous. To view this sequence of functions interactively in the on-screen version of the text, click on the icon 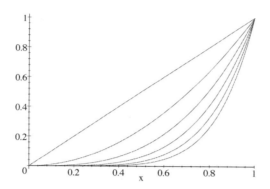 .

2. In this example we provide another sequence of continuous functions with a discontinuous limit. We define

$$f_n(x) = \frac{x^{2n}}{1 + x^{2n}}$$

for each positive integer n and each number x in the interval $[-2, 2]$. Figure 14.2 illustrates the graphs of the functions f_n for the first three values of n. This sequence of continuous functions converges boundedly to the function

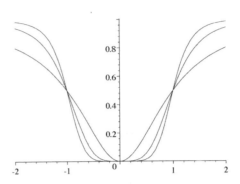

Figure 14.2

f defined by the equation

$$f(x) = \begin{cases} 0 & \text{if} \quad |x| < 1 \\ \frac{1}{2} & \text{if} \quad |x| = 1 \\ 1 & \text{if} \quad |x| > 1. \end{cases}$$

However, since

$$\sup |f_n - f| \geq \sup \left\{ \frac{x^{2n}}{1 + x^{2n}} \mid -1 < x < 1 \right\} = \frac{1}{2}$$

for each n, the sequence (f_n) does not converge uniformly to the function f. Observe that f is discontinuous at each of the numbers -1 and 1. To view this sequence of functions interactively in the on-screen version of the text, click on the icon 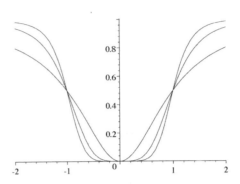 .

3. In this example we define

$$f_n(x) = \frac{\sin nx}{1 + nx}$$

for each positive integer n and each number x in the interval $[0, \pi]$. Figure 14.3 illustrates the graphs of the functions f_n for $n = 1, n = 5,$ and $n = 20$. It is easy to see that this sequence (f_n) converges boundedly to the constant function zero. Since

$$\sup |f_n| = \sup \left\{ |f_n(x)| \mid 0 \leq x \leq \pi \right\} \geq f_n \left(\frac{1}{n} \right) = \frac{\sin 1}{2}$$

for each n, the sequence (f_n) does not converge uniformly to the function 0.

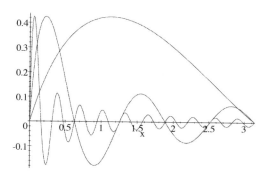

Figure 14.3

To view this sequence of functions interactively in the on-screen version of
the text, click on the icon 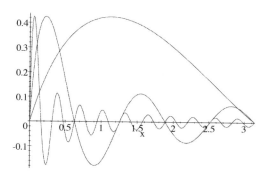 .

4. In this example we exhibit a sequence (f_n) of functions that are integrable
    on the interval $[0, 1]$ such that (f_n) converges boundedly to a function f that
    is not integrable. We begin by defining a set E_n for each positive integer n.
    For each n we define E_n to be the set of all those rational numbers in the
    interval $[0, 1]$ that can be expressed as ratios p/q of positive integers p and q
    in which $q \leq n$. For each n we see at once that the set E_n is finite, and so
    the function χ_{E_n} is Riemann integrable on $[0, 1]$ and

    $$\int_0^1 \chi_{E_n} = 0.$$

    The sequence $\left(\chi_{E_n}\right)$ converges boundedly to the function f defined by the
    equation

    $$f(x) = \begin{cases} 0 & \text{if} \quad x \text{ is irrational} \\ 1 & \text{if} \quad x \text{ is rational,} \end{cases}$$

    and, as we know from Example 11.6.4, this function f is not integrable.
    Since $\sup |f_n - f| = 1$ for each n, the sequence (f_n) does not converge
    uniformly to f.

5. If (f_n) is a sequence of integrable functions on an interval $[a, b]$ and
    (f_n) converges pointwise to an integrable function f, then it may seem

natural to expect that

$$\lim_{n \to \infty} \int_a^b f_n = \int_a^b f.$$

However, this apparently "obvious fact" is false. In this example we exhibit a sequence (f_n) of functions that are integrable on the interval $[0, 1]$ such that (f_n) converges pointwise to the constant function zero even though

$$\lim_{n \to \infty} \int_0^1 f_n = 1.$$

For each positive integer n and each number x in the interval $[0, 1]$ we define

$$f_n(x) = \frac{2n^2 x}{(1 + n^2 x^2)^2}.$$

We see easily that the sequence (f_n) converges pointwise to the constant function zero and that

$$\lim_{n \to \infty} \int_0^1 f_n = \lim_{n \to \infty} \int_0^1 \frac{2n^2 x}{(1 + n^2 x^2)^2} dx = \lim_{n \to \infty} \left(1 - \frac{1}{1 + n^2}\right) = 1.$$

By looking at the graphs of these functions we can gain some idea of what prevented the integral $\int_0^1 f_n$ from approaching zero. For small values of n, the graph of f_n hovers quite low above the x-axis, but then, as n increases, the graph begins to grow a peak that becomes ever higher and narrower as it moves toward the left. Any individual positive number x will lie to the right of these peaks as long as n is sufficiently large but the area under the nth peak is approximately 1. Figure 14.4 illustrates the graphs of the functions f_n for the first few values of n. To see how high the peaks are, we observe that each function f_n takes its maximum value when $f_n'(x) = 0$, in other words, when

$$\frac{2n^2 \left(1 - 3n^2 x^2\right)}{(1 + n^2 x^2)^3} = 0.$$

Thus the maximum value of each function f_n is the number

$$f_n\left(\frac{1}{\sqrt{3n}}\right) = \frac{2n^2 \left(\frac{1}{\sqrt{3n}}\right)}{\left(1 + n^2 \left(\frac{1}{\sqrt{3n}}\right)^2\right)^2} = \frac{3\sqrt{3n}}{8}.$$

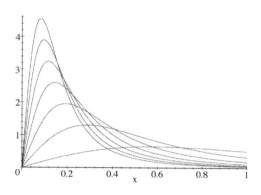

Figure 14.4

Thus, although (f_n) converges pointwise to the function 0, it does not converge boundedly and it does not converge uniformly. To view this sequence of functions interactively in the on-screen version of the text, click on the icon ▮ .

6. In this example we consider a variation of the preceding example. We define

$$f_n(x) = \frac{2nx}{\left(1 + n^2 x^2\right)^2}$$

for all $x \in [0, 1]$ and each positive integer n. Once again, the sequence converges pointwise to the constant function 0. And once again, the graph of the function f_n has a peak that becomes narrower and moves toward the left as n increases. This time, however, the height of each peak is $3\sqrt{3}/8$, and so the sequence (f_n) converges boundedly, but not uniformly, to the function 0. Figure 14.5 illustrates the graphs of the functions f_n for the first few values of n. As you would expect, since the peaks become narrower as n increases without rising any higher, the area under the graph must approach zero. In fact

$$\lim_{n \to \infty} \int_0^1 f_n = \lim_{n \to \infty} \int_0^1 \frac{2nx}{\left(1 + n^2 x^2\right)^2} dx = \lim_{n \to \infty} \left(\frac{1}{n} - \frac{1}{n\left(1 + n^2\right)} \right) = 0.$$

7. In this example we take a second look at the phenomenon that occurred in Example 5, but this time we make each function f_n a step function. For each

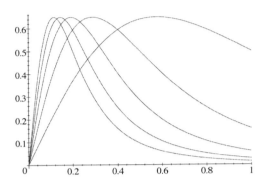

Figure 14.5

positive integer n we define

$$f_n(x) = \begin{cases} n & \text{if } 0 < x \le \frac{1}{n} \\ 0 & \text{if } \frac{1}{n} < x \le 1 \text{ or } x = 0. \end{cases}$$

Although this example isn't as satisfying as Example 5, it is worth looking at because it is so simple. For each n, the integral $\int_0^1 f_n$ is the area of the shaded region in Figure 14.6, and so $\int_0^1 f_n = 1$.

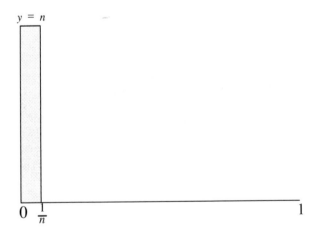

Figure 14.6

8. In this example we exhibit a sequence (f_n) of differentiable functions that converges uniformly to zero even though the sequence $(f_n'(x))$ is

divergent at every number x. We define

$$f_n(x) = \frac{\sin nx}{\sqrt{n}}$$

for each positive integer n and each real number x. Since

$$0 \leq \sup |f_n| \leq \frac{1}{\sqrt{n}}$$

for each n, it follows from the sandwich theorem that (f_n) converges uniformly to the function 0. The graphs of the functions f_1, f_4, f_{16}, and f_{64} are illustrated in Figure 14.7. As we see in this figure, the graph of f_n

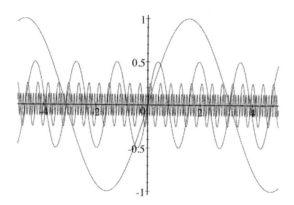

Figure 14.7

oscillates with greater and greater frequency as n increases and the derivative of f_n becomes more and more chaotic. In fact, for all n and x we have

$$f_n'(x) = \sqrt{n} \cos nx.$$

In the event that x is an integer multiple of 2π, it is clear that

$$\lim_{n \to \infty} f_n'(x) = \infty.$$

If x is not an integer multiple of 2π, then the behavior of $\sqrt{n} \cos nx$ as $n \to \infty$ needs to be considered more carefully. From the examples on conditional convergence that appeared in Subsection 12.7.9, we know that if x is not an integer multiple of 2π, then the series $\sum \frac{\cos nx}{n}$ is conditionally convergent. From the fact that

$$\left|\frac{\cos nx}{n}\right| = \left|\frac{1}{n^{3/2}}\right|\left|\sqrt{n}\cos nx\right|$$

and the fact that the series $\sum\left|\frac{\cos nx}{n}\right|$ is divergent, we deduce from the comparison test that the sequence $(\sqrt{n}\cos nx)$ cannot be bounded. Furthermore, since the series $\sum\frac{\cos nx}{n}$ is conditionally convergent, the numbers $\cos nx$ must frequently change sign. Therefore the sequence $(\sqrt{n}\cos nx)$ cannot have any limit, finite or infinite.

This example shows that even if a sequence (f_n) of differentiable functions converges uniformly to a differentiable function f, there is no guarantee that $f_n' \to f'$ as $n \to \infty$.

To view this sequence of functions interactively in the on-screen version of the text, click on the icon .

14.1.5   The Comparison Tests for Series of Functions

Suppose that (f_n) and (g_n) are sequences of functions defined on a set S and that $|f_n| \leq g_n$ for every n.

1. *If the series $\sum g_n$ converges pointwise, then so does the series $\sum f_n$.*
2. *If the series $\sum g_n$ converges boundedly, then so does the series $\sum f_n$.*
3. *If the series $\sum g_n$ converges uniformly, then so does the series $\sum f_n$.*

Proof. If $\sum g_n$ converges pointwise, then, given any number $x \in S$ we deduce from the comparison test (Theorem 12.4.6) and the fact that $|f_n(x)| \leq g_n(x)$ for every n that the series $\sum f_n(x)$ converges (absolutely). This completes the proof of part 1.

Now suppose that $\sum g_n$ converges boundedly and choose a number K such that

$$\sum_{j=1}^{n} g_j(x) \leq K$$

for every $x \in S$ and every positive integer n. Given given any such x and n we have

$$\left|\sum_{j=1}^{n} f_j(x)\right| \leq \sum_{j=1}^{n}|f_j(x)| \leq \sum_{j=1}^{n} g_j(x) \leq K.$$

This completes the proof of part 2.

Finally, suppose that $\sum g_n$ converges uniformly. To prove that $\sum f_n$ converges uniformly to its sum, suppose that $\varepsilon > 0$. Choose N such that the inequality

$$\sup \left(\sum_{j=1}^{\infty} g_j - \sum_{j=1}^{n} g_j \right) < \varepsilon$$

holds whenever $n \geq N$. Then, given any number $x \in S$ and any $n \geq N$, we have

$$\left| \sum_{j=1}^{\infty} f_j(x) - \sum_{j=1}^{n} f_j(x) \right| = \left| \sum_{j=n+1}^{\infty} f_j(x) \right| \leq \sum_{j=n+1}^{\infty} |f_j(x)| \leq \sum_{j=n+1}^{\infty} g_j(x) < \varepsilon.$$

This completes the proof of part 3. ∎

14.1.6   Exercises on Convergence of Sequences of Functions

1. (N) For each of the following definitions of the function f_n on the interval $[0, 1]$ prove that the sequence (f_n) converges pointwise to the function 0 on $[0, 1]$ and determine whether the sequence converges boundedly and whether it converges uniformly. In each case, determine whether or not we have

$$\lim_{n \to \infty} \int_0^1 f_n = 0.$$

   In each case, use *Scientific Notebook* to sketch some graphs of the given function and ask yourself whether your conclusion is compatible with what you see in the graphs.

   (a) ▣ $f_n(x) = nx \exp(-nx)$ for each $x \in [0, 1]$ and each positive integer n.

   (b) $f_n(x) = n^2 x \exp(-nx)$ for each $x \in [0, 1]$ and each positive integer n.

   (c) $f_n(x) = nx \exp(-n^2 x^2)$ for each $x \in [0, 1]$ and each positive integer n.

   (d) $f_n(x) = nx \exp(-nx^2)$ for each $x \in [0, 1]$ and each positive integer n.

   (e) $f_n(x) = nx \exp(-n^2 x)$ for each $x \in [0, 1]$ and each positive integer n.

2. Given that $f_n(x) = x^n$ for all x and n, prove that the series $\sum f_n$ converges pointwise, but not uniformly, on the interval $[0, 1)$ and that $\sum f_n$ converges uniformly on the interval $[0, \delta]$ whenever $0 \leq \delta < 1$.

3. (N) Given that $f_n(x) = (\sin nx)/n^2$ for all n and x, prove that the series $\sum f_n$ converges uniformly on $\mathbf{R}$. Use *Scientific Notebook* to sketch some of the graphs of these functions to motivate your conclusions.

4. ▣ Prove that the series $\sum x^n/n!$ converges uniformly in x on every bounded interval but does not converge uniformly in x on the entire line $\mathbf{R}$.

5. ■ Prove that the series

$$\sum \frac{(2n)!}{4^n \, (n!)^2} x^n$$

does not converge uniformly in x on the interval $(-1, 1)$ but that it does converge uniformly on the interval $[-\delta, \delta]$ whenever $0 \le \delta < 1$.

6. Prove that the series $\sum (x \log x)^n$ converges uniformly in x on the interval $(0, 1]$.

7. Given that (f_n) and (g_n) are sequences of real valued functions defined on a set S, that f and g are functions defined on S, and that $f_n \to f$ and $g_n \to g$ pointwise as $n \to \infty$, prove that

    (a) $f_n + g_n \to f + g$ pointwise as $n \to \infty$.
    (b) $f_n - g_n \to f - g$ pointwise as $n \to \infty$.
    (c) $f_n g_n \to f g$ pointwise as $n \to \infty$.
    (d) In the event that $g(x) \ne 0$ for every number x in the set S, we have $f_n/g_n \to f/g$ pointwise as $n \to \infty$.

8. Given that (f_n) and (g_n) are sequences of real valued functions defined on a set S, that f and g are functions defined on S, and that $f_n \to f$ and $g_n \to g$ boundedly as $n \to \infty$, prove that

    (a) $f_n + g_n \to f + g$ boundedly as $n \to \infty$.
    (b) $f_n - g_n \to f - g$ boundedly as $n \to \infty$.
    (c) $f_n g_n \to f g$ boundedly as $n \to \infty$.
    (d) In the event that there exists a number $\delta > 0$ such that $|g_n(x)| \ge \delta$ for each n and every number x in the set S, we have $f_n/g_n \to f/g$ boundedly as $n \to \infty$.

9. Suppose that (f_n) is a sequence of real valued functions defined on a set S and that f is a given function defined on S. Prove that the following conditions are equivalent:

    (a) The sequence (f_n) converges uniformly to the function f on the set S.
    (b) For every number $\varepsilon > 0$ there exists an integer N such that the inequality

    $$\sup |f_n - f| < \varepsilon$$

    holds for all $n \ge N$.
    (c) For every number $\varepsilon > 0$ there exists an integer N such that the inequality

    $$|f_n(x) - f(x)| < \varepsilon$$

    holds for all $n \ge N$ and all $x \in S$.

10. Suppose that (f_n) is a sequence of real valued functions defined on a set S and that f is a given function defined on S. Examine the following two conditions:

    (a) For every number $\varepsilon > 0$ there exists an integer N such that the inequality

    $$|f_n(x) - f(x)| < \varepsilon$$

    holds for all $n \geq N$ and all $x \in S$.

    (b) For every number $\varepsilon > 0$ and every number $x \in S$ there exists an integer N such that the inequality

    $$|f_n(x) - f(x)| < \varepsilon$$

    holds for all $n \geq N$.

    The first of these conditions asserts that the sequence (f_n) converges uniformly to the function f, while the second one asserts that (f_n) converges pointwise to f. Make sure that you can distinguish between the two conditions and see that they are not saying the same thing.

11. Given that a sequence (f_n) converges uniformly to a function f on a set S and that the function f is bounded, prove that (if we start the sequence at a sufficiently large value of n) the sequence (f_n) converges boundedly to f.

12. Prove that if the sequences (f_n) and (g_n) converge uniformly on a set S to functions f and g, respectively, then $f_n + g_n \to f + g$ uniformly on S.

13. ▦ Give an example of sequences (f_n) and (g_n) that converge uniformly on a set S to functions f and g, respectively, such that the sequence $(f_n g_n)$ fails to converge uniformly to the function fg.

14. Prove that if the sequences (f_n) and (g_n) converge uniformly and boundedly on a set S to functions f and g, respectively, then $f_n g_n \to fg$ uniformly on S.

15. ▦ Given that (f_n) is a decreasing sequence of nonnegative continuous functions on a closed bounded set S and that (f_n) converges pointwise to the function 0, prove that (f_n) converges uniformly to the function 0.

14.1.7   Some Special Tests for Uniform Convergence

This optional material presents a more detailed study of uniform convergence and includes the Cauchy criterion for uniform convergence, Abel's test for uniform convergence, and Dirichlet's test for uniform convergence. To reach this material from the on-screen version of the book, click on the icon ▦ .

14.2   The Important Properties of Uniform Convergence

Looking back at the examples that appear in Subsection 14.1.4, we see that some of the pathology that they reveal disappears when the sequence converges uniformly. In this section we shall focus our attention on this phenomenon by presenting a few theorems that guarantee good behavior whenever a sequence converges uniformly.

14.2.1   Uniform Convergence and Continuity at a Number

Suppose that a sequence (f_n) converges uniformly to a function f on a set S, suppose that $x \in S$, and suppose that each function f_n is continuous at the number x. Then the function f must also be continuous at the number x.

Proof. Suppose that $\varepsilon > 0$. Using the fact that $f_n \to f$ uniformly on S, we choose an integer N such that the inequality

$$\sup |f_n - f| < \frac{\varepsilon}{3}$$

holds whenever $n \geq N$. Now, using the fact that the function f_N is continuous at the number x, we choose a neighborhood U of x such that whenever $t \in S \cap U$ we have

$$|f_N(t) - f_N(x)| < \frac{\varepsilon}{3}.$$

For every number t in the set $U \cap S$ we have

$$
\begin{aligned}
|f(t) - f(x)| &= |f(t) - f_N(t) + f_N(t) - f_N(x) + f_N(x) - f(x)| \\
&\leq |f(t) - f_N(t)| + |f_N(t) - f_N(x)| + |f_N(x) - f(x)| \\
&< \frac{\varepsilon}{3} + \frac{\varepsilon}{3} + \frac{\varepsilon}{3} = \varepsilon. \ \blacksquare
\end{aligned}
$$

14.2.2   Uniform Convergence and Continuity on a Set

Suppose that a sequence (f_n) converges uniformly to a function f on a set S and suppose that each function f_n is continuous an the set S. Then the function f must also be continuous on the set S

This theorem follows at once from the corresponding result for continuity at a number.

14.2.3   Some Special Facts About the Continuity of a Limit Function

As we saw in the examples of Subsection 14.1.4, the pointwise limit of a sequence of continuous functions does not have to be continuous. However, the news is not totally bad. If (f_n) is a sequence of continuous functions on an interval $[a, b]$, where $a < b$, and if (f_n) converges pointwise on $[a, b]$ to a function f, then, even though f does not have to be continuous at every number $x \in [a, b]$, it will be continuous at many of those numbers. The on-screen version of this book contains some special exercises that will enable you to explore this remarkable fact. To access those exercises, click on the icon 🖼 .

14.2.4   Uniform Convergence and Riemann Integrability

Suppose that a sequence (f_n) converges uniformly to a function f on an interval $[a, b]$ and that each function f_n is Riemann integrable on $[a, b]$. Then the function f is also Riemann integrable on $[a, b]$.

Proof. To prove that the function f is Riemann integrable on $[a, b]$, we shall show that f satisfies the criterion for integrability that appears in Theorem 11.8.4. Suppose that $\varepsilon > 0$ and, using the fact that the $f_n \to f$ uniformly on $[a, b]$, choose an integer N such that

$$\sup |f_N - f| < \frac{\varepsilon}{4}.$$

Now, using the fact that the function f_N is Riemann integrable on $[a, b]$, we choose a partition $\mathcal{P}$ of $[a, b]$ such that if

$$E = \left\{ x \in [a, b] \mid w\left(\mathcal{P}, f_N\right)(x) \geq \frac{\varepsilon}{4} \right\},$$

then $m(E) < \varepsilon/4$. Now, given any two successive points x_{j-1} and x_j of $\mathcal{P}$, unless the interval (x_{j-1}, x_j) is included in E, we know that for all numbers s and t in this interval

$$|f(t) - f(s)| \leq |f(t) - f_N(t)| + |f_N(t) - f_N(s)| + |f_N(s) - f(s)| < \varepsilon.$$

We deduce that

$$w(\mathcal{P}, f)(x) \leq \frac{3\varepsilon}{4}$$

whenever $x \in [a, b] \setminus E$, and so f satisfies the criterion for integrability in Theorem 11.8.4. ∎

14.2.5   Uniform Convergence and Riemann Integration

Suppose that a sequence (f_n) converges uniformly to a function f on an interval $[a, b]$ and that each function f_n is Riemann integrable on $[a, b]$. Then

$$\lim_{n \to \infty} \int_a^b f_n = \int_a^b f.$$

Proof. As we know, the function f is integrable. Now for each n we have

$$0 \leq \left| \int_a^b f_n - \int_a^b f \right| \leq \int_a^b |f_n - f| \leq (b-a) \sup |f_n - f| ,$$

and so the result follows at once from the sandwich theorem. ∎

14.2.6   Some Exercises on the Properties of Uniform Convergence

1. Determine whether the following statement is true or false:
    If f_n is uniformly continuous on a set S for every positive integer n, and if the sequence (f_n) converges uniformly on S to a function f, then f must be uniformly continuous on S.
2. Determine whether the following statement is true or false:
    If f_n is uniformly continuous on a set S for infinitely many positive integers n, and if the sequence (f_n) converges uniformly on S to a function f, then f must be uniformly continuous on S.
3. A family $\mathcal{F}$ of functions is said to be **equicontinuous** on a set S if for every $\varepsilon > 0$ and every number $x \in S$ there exists a number $\delta > 0$ such that whenever $f \in \mathcal{F}$ and whenever t lies in the set $S \cap (x - \delta, x + \delta)$ we have

    $$|f(t) - f(x)| < \varepsilon.$$

    Prove that if a sequence (f_n) converges uniformly on S and if each function f_n is continuous on S, then the family $\{f_n \mid n = 1, 2, 3, \cdots\}$ is equicontinuous on S.
4. Invent a meaning for *equi-uniform continuity* of a family $\mathcal{F}$ on a set S and decide whether or not your definition provides an analog of Exercise 3.

14.3   The Important Property of Bounded Convergence

14.3.1   Introduction to the Bounded Convergence Theorem

The purpose of this section is to present a very useful theorem known as the **Arzela bounded convergence theorem** that is a dramatic improvement of Theorem 14.2.5. We know from Example 4 of Subsection 14.1.4, that, in the absence of uniform convergence, a sequence of Riemann integrable functions can converge to a function that is not integrable. Furthermore, even if a sequence (f_n) of integrable functions on an interval $[a, b]$ converges pointwise to a function f that happens to be integrable on $[a, b]$, we have seen that the condition

$$\lim_{n \to \infty} \int_a^b f_n = \int_a^b f$$

can fail to hold. We saw this failure in Example 5 of Subsection 14.1.4, which we illustrate again in Figure 14.8. In this example, the graphs of the functions

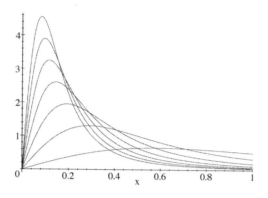

Figure 14.8

f_n have peaks that become narrower and higher as n increases and the area under these peaks does not approach zero. On the other hand, in Example 6, which we illustrate in Figure 14.9, and in which we restrict the height of the peaks, the

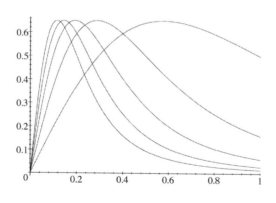

Figure 14.9

condition

$$\lim_{n \to \infty} \int_0^1 f_n = \int_0^1 0$$

does hold true. It holds true in spite of the fact that (f_n) fails to converge uniformly. The important feature of Example 6 is that (f_n) converges to 0 *boundedly.*

With these examples in mind we might guess that if a sequence (f_n) of functions integrable on an interval $[a, b]$ converges boundedly to an integrable function f, then the condition

$$\lim_{n \to \infty} \int_a^b f_n = \int_a^b f$$

should always hold. This, in a nutshell, is the statement of the bounded convergence theorem. As simple and intuitive as this statement may appear, it is actually very powerful. The proof of the bounded convergence theorem, while not actually difficult, is a far cry from the almost trivial proof of Theorem 14.2.5. But it is well worth studying. The bounded convergence theorem opens the door to several immensely important topics that are easy to study with it but are almost impossible to study without it.

The heart of the bounded convergence theorem lies in the following important theorem that we shall prove first.

14.3.2   The Contracting Sequences Theorem

Suppose that (S_n) is a contracting sequence of bounded sets and that

$$\bigcap_{n=1}^{\infty} S_n = \emptyset.$$

For each n, suppose that

$$\alpha_n = \sup \left\{ m\left(E \right) \mid E \text{ is an elementary subset of } S_n \right\}.$$

Then $\alpha_n \to 0$ as $n \to \infty$.

Proof. The sequence (α_n) is clearly decreasing. To obtain a contradiction, assume that this sequence does not converge to 0, and choose a number $\delta > 0$ such that $\alpha_n > \delta$ for every n. For each n we now choose an elementary subset E_n of the set S_n such that

$$m\left(E_n \right) > \alpha_n - \frac{\delta}{2^n}.$$

Now for each n we use Theorem 11.4.8 to choose a closed elementary subset F_n of the set E_n such that

$$m\left(F_n \right) > \alpha_n - \frac{\delta}{2^n},$$

and then we define

$$H_n = \bigcap_{j=1}^{n} F_j.$$

Observe that the sequence (H_n) is a contracting sequence of closed bounded elementary sets. The plan of the proof is to show that every one of the sets H_n is nonempty. It will then follow from the Cantor intersection theorem (Theorem 7.8) that the intersection of all of the sets H_n is nonempty, in spite of the fact that the larger sets S_n have an empty intersection. This will be the desired contradiction. In order to show that $H_n \neq \emptyset$ for each n, we make two observations.

First Observation: Suppose that n is any positive integer. Given any elementary subset E of the set $S_n \setminus F_n$, we have

$$m(E) + m(F_n) = m(E \cup F_n) \leq \alpha_n,$$

and it therefore follows from the inequality $m(F_n) > \alpha_n - \delta/2^n$ that $m(E) < \delta/2^n$. This observation can be thought of as saying that, because F_n nearly fills the set S_n, it leaves very little room for an elementary set E that is included in $S_n \setminus F_n$. This notion is illustrated in Figure 14.10.

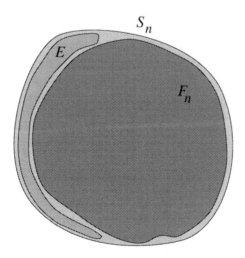

Figure 14.10

Second Observation: Suppose that n is any positive integer. Given any elementary subset E of the set $S_n \setminus H_n$, since

$$E = (E \setminus F_1) \cup (E \setminus F_2) \cup \cdots \cup (E \setminus F_n)$$

and since each set $E \setminus F_j$ is a subset of the corresponding set $S_j \setminus F_j$, we have

$$m\left(E\right) \leq \sum_{j=1}^{n} m\left(E \setminus F_j\right) < \sum_{j=1}^{n} \frac{\delta}{2^j} = \delta \left(1 - \frac{1}{2^n}\right) < \delta.$$

Look now at the two sets S_n and $S_n \setminus H_n$. The set S_n *must* have an elementary subset whose measure is greater than δ while the $S_n \setminus H_n$ *cannot*. Thus for each n we have

$$S_n \neq S_n \setminus H_n,$$

and it follows that the set H_n must be nonempty. ∎

14.3.3   The Bounded Convergence Theorem

Suppose that (f_n) is a sequence of functions that are integrable on a given interval $[a, b]$ and that (f_n) converges boundedly to a function f that is integrable on $[a, b]$. Then

$$\lim_{n \to \infty} \int_a^b f_n = \int_a^b f.$$

Proof.   From the inequality

$$\left| \int_a^b f_n - \int_a^b f \right| = \left| \int_a^b (f_n - f) \right| \leq \int_a^b |f_n - f|$$

we observe that the theorem will be proved when we have shown that

$$\lim_{n \to \infty} \int_a^b |f_n - f| = 0.$$

Using the fact that the sequence (f_n) converges to f boundedly, we choose a number K such that $|f_n(x)| \leq K$ for every $x \in [a, b]$ and every n. Given any $x \in [a, b]$, it follows from the fact that

$$\lim_{n \to \infty} f_n(x) = f(x)$$

that $|f(x)| \leq K$. Therefore, for every $x \in [a, b]$ and every n we have

$$0 \leq |f_n(x) - f(x)| \leq |f_n(x)| + |f(x)| \leq 2K.$$

Now, to prove that $\int_a^b |f_n - f| \to 0$ as $n \to \infty$, suppose that $\varepsilon > 0$. For each n we define S_n to be the set of all those numbers x in the interval $[a, b]$ for which the inequality

$$|f_j(x) - f(x)| \geq \frac{\varepsilon}{2(b-a)}$$

holds for at least one integer $j \geq n$. We see at once that the sequence (S_n) is contracting. Furthermore, if x is any number in the interval $[a, b]$, then it follows from the fact that $f_j(x) \to f(x)$ as $j \to \infty$ that if n is sufficiently large, then we have

$$|f_j(x) - f(x)| < \frac{\varepsilon}{2(b-a)}$$

for all $j \geq n$. In other words, if n is sufficiently large, we have $x \notin S_n$. Therefore

$$\bigcap_{n=1}^{\infty} S_n = \emptyset.$$

We now use the contracting sequences theorem to choose an integer N such that whenever $n \geq N$ and whenever E is an elementary subset of S_n we have

$$m(E) < \frac{\varepsilon}{4K}.$$

We shall complete the proof by showing that whenever $n \geq N$ we have

$$\int_a^b |f_n - f| \leq \varepsilon.$$

From the definition of the integral that we saw in Subsection 11.5.2, we need only show that if $n \geq N$ and g is a step function satisfying the inequality

$$0 \leq g \leq |f_n - f|,$$

we have $\int_a^b g \leq \varepsilon$. Suppose then that $n \geq N$, that g is a step function, and that

$$0 \leq g \leq |f_n - f|.$$

We define

$$E = \left\{ x \in [a, b] \mid g(x) \geq \frac{\varepsilon}{2(b-a)} \right\}$$

and $F = [a, b] \setminus E$. The sets E and F are elementary subsets of $[a, b]$ and, since

$E \subseteq S_n$, we have $m(E) < \varepsilon/4K$. Therefore

$$\int_a^b g \;=\; \int_E g + \int_F g \le \int_E 2K + \int_F \frac{\varepsilon}{2(b-a)} = 2Km(E) + \frac{\varepsilon}{2(b-a)}m(F)$$

$$< \; 2K\frac{\varepsilon}{4K} + \frac{\varepsilon}{2(b-a)}(b-a) = \varepsilon. \quad \blacksquare$$

14.3.4   A Bounded Convergence Theorem for Stieltjes Integrals

If you chose to read the integration chapter in its optional Riemann-Stieltjes form, you can reach the Stieltjes form of the bounded convergence theorem by clicking on the icon 🖼 .

14.3.5   A Bounded Convergence Theorem for Derivatives

Suppose that (f_n) is a sequence of differentiable functions on an interval $[a,b]$ and that each function f_n' is continuous on $[a,b]$. Suppose that the sequence (f_n') converges boundedly on $[a,b]$ and has a continuous limit, and suppose that the sequence $(f_n(c))$ converges for at least one number c in the interval $[a,b]$.

Then the sequence (f_n) converges pointwise on $[a,b]$ to a differentiable function f and for each x we have

$$f'(x) = \lim_{n\to\infty} f_n'(x).$$

Proof. We write the limit function of the sequence (f_n') as g and we choose a number $c \in [a,b]$ such that the sequence $(f_n(c))$ is convergent. For every $x \in [a,b]$ and every n it follows from the fundamental theorem of calculus (Theorem 11.12.3) that

$$f_n(x) = f_n(c) + \int_c^x f_n'(t)dt.$$

From the bounded convergence theorem we deduce that the right side of the latter identity converges as $n \to \infty$ and

$$\lim_{n\to\infty} f_n(x) = \lim_{n\to\infty} f_n(c) + \int_c^x g(t)dt.$$

We have therefore shown that the sequence (f_n) converges pointwise on the interval $[a,b]$. Furthermore, if we define

$$f(x) = \lim_{n\to\infty} f_n(x)$$

for each $x \in [a, b]$, then for each such x we have

$$f(x) = \lim_{n \to \infty} f_n(c) + \int_c^x g(t)dt.$$

Using the fundamental theorem (Theorem 11.12.2) once again we conclude that f is differentiable and that if $x \in [a, b]$, then

$$f'(x) = g(x) = \lim_{n \to \infty} f_n'(x). \quad \blacksquare$$

14.3.6   Term-by-Term Integration of the Sum of a Series

Suppose that (f_n) is a sequence of functions that are Riemann integrable on an interval $[a, b]$ and that the series $\sum f_n$ converges boundedly on $[a, b]$ to an integrable function f. This means, of course, that the sequence of functions

$$\sum_{j=1}^{n} f_j$$

converges boundedly to f. Then we have

$$\int_a^b f(x)dx = \sum_{n=1}^{\infty} \int_a^b f_n(x)dx.$$

Since the latter statement can also be written as

$$\int_a^b \sum_{n=1}^{\infty} f_n(x)dx = \sum_{n=1}^{\infty} \int_a^b f_n(x)dx,$$

it is often described as **term-by-term integration** of the sum of the series $\sum f_n$.

Proof. The theorem follows at once from the bounded convergence theorem:

$$\int_a^b f = \int_a^b \lim_{n \to \infty} \sum_{j=1}^{n} f_j = \lim_{n \to \infty} \int_a^b \sum_{j=1}^{n} f_j = \lim_{n \to \infty} \sum_{j=1}^{n} \int_a^b f_j = \sum_{j=1}^{\infty} \int_a^b f_j. \quad \blacksquare$$

14.3.7   Term-by-Term Differentiation of the Sum of a Series

Suppose that (f_n) is a sequence of differentiable functions on an interval $[a, b]$ and that each function f_n' is continuous on $[a, b]$. Suppose that the series $\sum f_n'$ converges boundedly on $[a, b]$ and has a continuous sum, and suppose that the series $\sum f_n(c)$ converges for at least one number c in the interval $[a, b]$.

Then the series $\sum f_n$ converges pointwise on $[a, b]$ to a differentiable function and we have

$$\left(\sum_{n=1}^{\infty} f_n\right)' = \sum_{n=1}^{\infty} f_n'.$$

We leave the proof of this theorem as an exercise. 🖳

14.3.8   A Convergence Theorem for Improper Integrals

In this subsection we obtain an analog of the bounded convergence theorem for improper integrals. The theorem that we shall state here is a form of a result known as the **dominated convergence theorem.**

Suppose that $-\infty < a < b \leq \infty$ and that (f_n) is a sequence of functions that are improper Riemann integrable on the interval $[a, b)$. Suppose that the sequence (f_n) converges pointwise to an improper Riemann integrable function f and that there exists an improper Riemann integrable function g such that $|f_n| \leq g$ for each n. Then

$$\lim_{n \to \infty} \int_a^{\to b} f_n = \int_a^{\to b} f.$$

Proof. From the comparison test for improper integrals (Theorem 13.3.2) we see that the integrals $\int_a^{\to b} f_n$ and $\int_a^{\to b} f$ converge absolutely. The key to the proof now lies in the fact that whenever $w \in [a, b)$ we have

$$\left| \int_a^{\to b} f_n - \int_a^{\to b} f \right| \leq \int_a^{\to b} |f_n - f| = \int_a^w |f_n - f| + \int_w^{\to b} |f_n - f|$$

$$\leq \int_a^w |f_n - f| + \int_w^{\to b} 2g.$$

Suppose that $\varepsilon > 0$. Using the fact that the integral $\int_a^{\to b} g$ is convergent, we choose a number $w \in [a, b)$ such that

$$\int_a^{\to b} g - \int_a^w g < \frac{\varepsilon}{4},$$

in other words, $\int_w^{\to b} g < \varepsilon/4$. Since g is bounded on the interval $[a, w]$, and since $|f_n - f| \leq 2g$ for every n, and since $|f_n - f| \to 0$ pointwise as $n \to \infty$, we may apply the bounded convergence theorem to choose a positive integer N

such that

$$\int_a^{\to w} |f_n - f| < \frac{\varepsilon}{2}$$

whenever $n \geq N$. Whenever $n \geq N$ we have

$$\left| \int_a^{\to b} f_n - \int_a^{\to b} f \right| \leq \int_a^{\to w} |f_n - f| + \int_w^{\to b} 2g < \frac{\varepsilon}{2} + \frac{\varepsilon}{2} = \varepsilon. \blacksquare$$

14.3.9   A Convergence Theorem for Integrals of Nonnegative Functions

As you continue your studies, you will find that in many theories of integration the theorems are stated once for absolutely convergent integrals and once again for possibly infinite integrals of nonnegative functions. The present theorem is an analog of the preceding one, but it requires the functions to be nonnegative and the integrals are allowed to be infinite.

Suppose that (f_n) is a sequence of nonnegative functions defined on an interval $[a, b)$ and that (f_n) converges pointwise on $[a, b)$ to a function f. Suppose that $f_n \leq f$ for every n and suppose that the improper integrals $\int_a^{\to b} f$ and $\int_a^{\to b} f_n$ are all defined. Then, even if the integrals are infinite, we have

$$\lim_{n \to \infty} \int_a^{\to b} f_n = \int_a^{\to b} f.$$

Proof. In the event that the integral $\int_a^{\to b} f$ is finite, the theorem follows at once from Theorem 14.3.8. Suppose now that $\int_a^{\to b} f = \infty$. To show that

$$\lim_{n \to \infty} \int_a^{\to b} f_n = \infty,$$

we suppose that K is any real number and we choose a number $w \in [a, b)$ such that

$$\int_a^{\to w} f > K.$$

From Theorem 14.3.8 we see that

$$\lim_{n \to \infty} \int_a^{\to w} f_n = \int_a^{\to w} f,$$

and therefore, if n is sufficiently large, we have

$$\int_a^{\to b} f_n \geq \int_a^w f_n > K. \quad \blacksquare$$

14.3.10   Some Exercises on Bounded Convergence

1. Prove that the sequence (f_n) in Theorem 14.3.5 actually converges boundedly to the function f.
2. Suppose that (a_n) is a strictly increasing sequence of positive integers and that

$$f_n(x) = (\sin a_n x - \sin a_{n+1} x)^2$$

   for every positive integer n and every number x.

   (a) Work out the integral

$$\int_0^{2\pi} f_n(x) dx$$

   and deduce that there must be at least one number $x \in [0, 2\pi]$ such that the sequence $(f_n(x))$ does not converge to zero.
   (b) Prove that there must be at least one number $x \in [0, 2\pi]$ for which the sequence $(\sin a_n x)$ diverges.
3. Suppose that for each positive integer n we have

$$f_n(x) = \begin{cases} \sum_{j=0}^n (-x)^j & \text{if } -1 < x < 1 \\ \frac{1}{2} & \text{if } x = 1. \end{cases}$$

   (a) Prove that if

$$f(x) = \frac{1}{1+x}$$

   for $-1 < x \leq 1$, then the sequence (f_n) converges boundedly to the function f on the interval $[0, 1]$.
   (b) Explain why each function f_n is integrable on the interval $[0, 1]$ and why

$$\int_0^1 f_n = \sum_{j=0}^n \frac{(-1)^j}{j+1},$$

and deduce that

$$\sum_{j=0}^{\infty} \frac{(-1)^j}{j+1} = \log 2.$$

(c) Given any number x satisfying $-1 < x \leq 1$, prove that the sequence (f_n) converges boundedly to f on the closed interval running between 0 and x and deduce that the equation

$$\sum_{j=0}^{\infty} \frac{(-1)^j x^{j+1}}{j+1} = \log(1+x)$$

holds whenever $-1 < x \leq 1$.

4. Suppose that for each positive integer n we have

$$f_n(x) = \begin{cases} \sum_{j=0}^{n} (-x^2)^j & \text{if } 0 \leq x < 1 \\ \frac{1}{2} & \text{if } x = 1. \end{cases}$$

Repeat the steps of Exercise 3 for this sequence of functions and deduce that

$$\sum_{j=0}^{\infty} \frac{(-1)^j}{2j+1} = \frac{\pi}{4}.$$

5. Prove that if

$$f_n(x) = \begin{cases} \frac{1}{n} & \text{if } 0 \leq x \leq n \\ 0 & \text{if } x > n \end{cases}$$

whenever n is a positive integer, then the sequence (f_n) converges uniformly on the interval $[0, \infty)$ to the constant function 0 even though

$$\lim_{n \to \infty} \int_0^{\to \infty} f_n \neq \int_0^{\to \infty} 0.$$

6. Suppose that $\alpha > 0$ and that

$$f_n(x) = \begin{cases} \left(1 - \frac{x}{n}\right)^n x^{\alpha-1} & \text{if } \frac{1}{n} \leq x \leq n \\ 0 & \text{if } x < \frac{1}{n} \text{ or } x > n \end{cases}$$

whenever n is a positive integer.

(a) **H** Prove that if x is any positive number, then

$$\lim_{n \to \infty} f_n(x) = e^{-x} x^{\alpha-1},$$

and that for each n we have

$$|f_n(x)| \leq e^{-x}x^{\alpha-1}.$$

(b) Use Theorem 14.3.8 to show that

$$\lim_{n\to\infty} \int_{1/n}^{n} \left(1 - \frac{x}{n}\right)^{n} x^{\alpha-1} dx = \int_{0}^{\to\infty} e^{-x}x^{\alpha-1} dx = \Gamma(\alpha).$$

For the definition of the gamma function Γ, see Exercise 2b of Subsection 13.3.3.

14.3.11   Some Applications of the Bounded Convergence Theorem

From the on-screen version of this book you can reach some optional exercises that will lead you to the sums of some interesting series. For example, you will show that

$$\sum_{n=1}^{\infty} \frac{(-1)^{n-1}\cos n\theta}{n} = \frac{1}{2}\log(2 + 2\cos\theta)$$

and

$$\sum_{n=1}^{\infty} \frac{\cos n\theta}{n^2} = \frac{\pi^2}{6} - \frac{\pi\theta}{2} + \frac{\theta^2}{4}$$

for certain appropriate values of θ. To reach these exercises, click on the icon ⬛.

14.4   Power Series

The concept of a power series that we discuss in this section first appeared when we discussed Abel's theorem (Theorem 12.7.7).

14.4.1   Definition of a Power Series

A **power series** is a series of the form $\sum a_n x^n$ or, more generally, of the form $\sum a_n(x-c)^n$, where c is a given real number and (a_n) is a sequence of numbers. By analogy with the way we described Taylor polynomials in Section 9.5, we call a power series of the form $\sum a_n(x-c)^n$ a power series in x, centered at c and with (a_n) as its sequence of coefficients.

In this section we shall show that a power series of the form $\sum a_n(x-c)^n$ converges absolutely in an interval of the form $(c-r, c+r)$ for some nonnegative number r and that, within this interval, the series may be differentiated and integrated term by term. We shall use this information to determine the sums of some important power series.

14.4.2   Some Examples of Power Series

1. The simplest power series is the geometric series $\sum x^n$, which, as you know, converges whenever $|x| < 1$ and diverges otherwise. If $|x| < 1$, then

$$\sum_{n=0}^{\infty} x^n = \frac{1}{1-x}.$$

2. By using d'Alembert's test (Theorem 12.6.5) we can see that the series $\sum x^n/n!$ converges absolutely for every real number x. We shall see soon that if we start summing at $n = 0$, then the sum of this series is e^x.
3. By using d'Alembert's test we can see that the series $\sum (-1)^{n-1} x^n/n$ converges absolutely for every number $x \in (-1, 1)$ and an easy application of Dirichlet's test (Theorem 12.7.5) shows that the series converges when $x = 1$. For all other values of x, the series diverges. In Exercise 3 of Subsection 14.3.10 we saw that, at each number x at which the series converges, its sum is $\log (1 + x)$. We shall encounter this fact again in the present section.
4. In Example 3 of Subsection 12.7.9 we defined the binomial coefficients

$$\binom{\alpha}{n} = \frac{\alpha (\alpha - 1) (\alpha - 2) \cdots (\alpha - n + 1)}{n!},$$

and we defined $\binom{\alpha}{0} = 1$. The series $\sum \binom{\alpha}{n} x^n$ is known as the **binomial series**, which we shall discuss in Section 14.6.

14.4.3   Convergence Behavior of Power Series

Given any sequence (a_n) of real numbers, there exists an extended real number $r \geq 0$ such that the power series $\sum a_n x^n$ converges absolutely whenever $|x| < r$ and diverges whenever $|x| > r$.

Proof. We define

$$r = \sup \{|t| \mid \text{the sequence } (a_n t^n) \text{ is bounded}\}.$$

Whenever $|x| > r$, it follows from the fact that the sequence $(a_n x^n)$ is unbounded that the series $\sum a_n x^n$ is divergent. Now suppose that $|x| < r$. Using the fact that $|x|$ is not an upper bound of the set

$$\{|t| \mid \text{the sequence } (a_n t^n) \text{ is bounded}\},$$

we choose a number δ such that $|x| < \delta$ and such that the sequence $(a_n \delta^n)$ is bounded. Choose a number K such that $|a_n \delta^n| \le K$ for each n. Now, given any n, we see that

$$|a_n x^n| = |a_n \delta^n| \left|\frac{x}{\delta}\right|^n \le K \left|\frac{x}{\delta}\right|^n ,$$

and it follows from the convergence of the geometric series $\sum |x/\delta|^n$ and the comparison test that the series $\sum a_n x^n$ converges absolutely. ■

14.4.4   Radius of Convergence of a Power Series

The **radius of convergence** of a given power series $\sum a_n (x - c)^n$ is defined to be the extended real number $r \ge 0$ for which the series converges absolutely whenever $|x - c| < r$ and diverges whenever $|x - c| > r$. Thus the set of numbers x for which the series converges is one of the four intervals

$$(c - r, c + r) \qquad [c - r, c + r) \qquad (c - r, c + r] \qquad [c - r, c + r]$$

and is called the **interval of convergence** of the series. By using the tests for convergence that appear in Chapter 12, one may check that the radii of convergence of the series

$$\sum x^n, \quad \sum \frac{x^n}{n!}, \quad \sum \frac{(-1)^{n-1} x^n}{n}, \quad \sum \frac{x^n}{2^n n^2}, \quad \sum (n!)\, x^n, \quad \sum \frac{(n!)\, x^n}{n^n}$$

are $1, \infty, 1, 2, 0,$ and e, respectively. The intervals of convergence of these series are $(-1, 1), \mathbf{R}, (-1, 1], [-2, 2], \{0\},$ and $(-e, e)$.

14.4.5   Uniform Convergence of Power Series

Suppose that r is the radius of convergence of the series $\sum a_n (x - c)^n$. Then, given any number δ satisfying $0 \le \delta < r$, the series will converge uniformly in x in the closed interval $[c - \delta, c + \delta]$.

$$\overline{\begin{array}{ccccc} c - r & c - \delta & c & c + \delta & c + r \end{array}}$$

Proof. Since

$$|a_n (x - c)^n| \le |a_n \delta^n|$$

for every $x \in [c - \delta, c + \delta]$ and since the series $\sum |a_n \delta^n|$ converges, the result follows from the comparison test for uniform convergence (Theorem 14.1.5). ■

14.4.6   Continuity of the Sum of a Power Series

Suppose that r is the radius of convergence of the series $\sum a_n (x - c)^n$ and suppose that we define a function f on the open interval $(c - r, c + r)$ by the *equation*

$$f(x) = \sum_{n=0}^{\infty} a_n (x - c)^n$$

for every $x \in (c - r, c + r)$. *Then* f *is continuous on the interval* $(c - r, c + r)$.
Proof. Suppose that $x \in (c - r, c + r)$. Choose a number δ such that

$$|x - c| < \delta < r.$$

$c - r \qquad c - \delta \qquad\qquad c \qquad x \qquad c + \delta \qquad c + r$

Since the series converges uniformly on the interval $[c - \delta, c + \delta]$, we deduce
from Theorem 14.2.2 that f is continuous at the number x. ∎

14.4.7   Term-by-Term Integration of a Power Series

Suppose that r is the radius of convergence of the power series $\sum a_n (x - c)^n$.

1. Given any number $x \in (c - r, c + r)$ we have

$$\int_c^x \sum_{n=0}^{\infty} a_n (t - c)^n \, dt = \sum_{n=0}^{\infty} a_n \int_c^x (t - c)^n \, dt = \sum_{n=0}^{\infty} a_n \frac{(x - c)^{n+1}}{n + 1}.$$

2. Even if the series $\sum a_n r^n$ diverges, as long as the sequence of partial sums
of this series is bounded, we have

$$\int_c^{c+r} \sum_{n=0}^{\infty} a_n (t - c)^n \, dt = \sum_{n=0}^{\infty} a_n \int_c^{c+r} (t - c)^n \, dt = \sum_{n=0}^{\infty} a_n \frac{r^{n+1}}{n + 1}.$$

Proof. Part 1 of the theorem follows at once from the fact that the series converges uniformly on the interval running between the numbers c and x, and
from the bounded convergence theorem. As a matter of fact, we don't even need
the bounded convergence theorem. The simpler theorem, Theorem 14.2.5, that
appeared earlier is all we need.

To prove 2 we define

$$f_n(x) = \begin{cases} \displaystyle\sum_{j=0}^{n} a_j (x - c)^j & \text{if } |x - c| < r \\[2mm] 0 & \text{if } x = c + r \end{cases}$$

whenever n is a positive integer, and we define

$$f(x) = \begin{cases} \displaystyle\sum_{j=0}^{\infty} a_j \, (x - c)^j & \text{if} \quad |x - c| < r \\[2em] 0 & \text{if} \quad x = c + r. \end{cases}$$

From the fact that the sequence of partial sums of the series $\sum a_n r^n$ is bounded and the fact that if $c \le x < c + r$ we have

$$a_n \, (x - c)^n = (a_n r^n) \left(\frac{x - c}{r} \right)^n,$$

we can use Theorem 12.7.6 to deduce that the sequence (f_n) converges boundedly to the function f on the interval $[c, c + r]$. Part 2 therefore follows from the bounded convergence theorem. ∎

14.4.8   Term-by-Term Differentiation of a Power Series

Suppose that r is the radius of convergence of the series $\sum a_n \, (x - c)^n$.

1. The number r is also the radius of convergence of the "derived" series $\sum n a_n \, (x - c)^{n-1}$.
2. If we define

$$f(x) = \sum_{n=0}^{\infty} a_n \, (x - c)^n$$

whenever $|x - c| < r$, then for every such number x we have

$$f'(x) = \sum_{n=1}^{\infty} n a_n \, (x - c)^{n-1} .$$

Proof of Part 1.   Whenever $|x - c| > r$ we know from the reasoning given in the proof of Theorem 14.4.3 that the sequence $(a_n \, (x - c)^n)$ is unbounded. Therefore, since

$$n a_n \, (x - c)^{n-1} = \left(\frac{n}{x - c} \right) (a_n \, (x - c)^n),$$

we deduce that the sequence of numbers $n a_n \, (x - c)^{n-1}$ is also unbounded and consequently that the series $\sum n a_n \, (x - c)^{n-1}$ diverges.

Now we want to show that whenever $|x - c| < r$, the series $\sum n a_n \, (x - c)^{n-1}$

converges absolutely. Suppose that $|x - c| < r$. Choose a number δ such that

$$|x - c| < \delta < r.$$

Since

$$\left|na_n (x - c)^{n-1}\right| = |a_n\delta^n| \left(\frac{n}{|x - c|}\right)\left(\frac{|x - c|}{\delta}\right)^n,$$

and since the fact that $|x - c| < \delta$ guarantees that

$$\lim_{n\to\infty}\left(\frac{n}{|x - c|}\right)\left(\frac{|x - c|}{\delta}\right)^n = 0,$$

we deduce from the comparison test (Theorem 12.4.6) that the series $\sum na_n (x - c)^{n-1}$ converges absolutely. ∎

Proof of Part 2. For every number x in the interval $(c - r, c + r)$ we define

$$g(x) = \sum_{n=1}^{\infty} na_n (x - c)^{n-1}.$$

From Theorem 14.4.7 we deduce that if $x \in (c - r, c + r)$, we have

$$\int_c^x g(t)dt = \int_c^x \sum_{n=1}^{\infty} na_n (t - c)^{n-1}\,dt = \sum_{n=1}^{\infty} na_n \int_c^x (t - c)^{n-1}\,dt = \sum_{n=1}^{\infty} a_nx^n,$$

and so

$$f(x) = a_0 + \int_c^x g(t)dt.$$

Since the function g is continuous, we deduce from the fundamental theorem of calculus (Theorem 11.12.2) that

$$f'(x) = g(x) = \sum_{n=1}^{\infty} na_n (x - c)^{n-1}$$

for each $x \in (c - r, c + r)$. ∎

14.4.9   Calculating the Coefficients in a Power Series

The theorem on term-by-term differentiation of a power series has some far-reaching consequences. Suppose that r is the radius of convergence of a given power series $\sum a_n \left(x - c\right)^n$, and suppose that we have defined

$$f(x) = \sum_{n=0}^{\infty} a_n \left(x - c\right)^n$$

whenever $|x - c| < r$. As we know from the preceding theorem, if $x \in (c - r, c + r)$, then

$$f'(x) = \sum_{n=1}^{\infty} n a_n \left(x - c\right)^{n-1}.$$

Differentiating again we see that for each such x we have

$$f''(x) = \sum_{n=2}^{\infty} n \left(n - 1\right) a_n \left(x - c\right)^{n-2},$$

and, continuing this process, we see that if k is any nonnegative integer, then

$$f^{(k)}(x) = \sum_{n=k}^{\infty} n \left(n - 1\right) \cdots \left(n - k + 1\right) a_n (x - c)^{n-k}.$$

Putting $x = c$ in the latter equation yields

$$f^{(k)}(c) = (k!)\, a_k,$$

and we conclude that

$$a_n = \frac{f^{(n)}(c)}{n!}$$

for every integer $n \geq 0$, and so

$$f(x) = \sum_{n=0}^{\infty} \frac{f^{(n)}(c)}{n!} \left(x - c\right)^n.$$

We can also look at this question in another way. Suppose that f is a given function that has derivatives of all orders in a neighborhood of a given number c. There is no guarantee that f can be represented as the sum of a power series with center c at every number x in some neighborhood of c, but, if there is such a series, it can only be the series

$$\sum \frac{f^{(n)}(c)}{n!} \left(x - c\right)^n.$$

14.4.10   Taylor Series of a Function

Suppose that f is a given function that has derivatives of all orders in a neighborhood of a given number c. The series

$$\sum \frac{f^{(n)}(c)}{n!} (x - c)^n$$

is called the **Taylor series** with center c of the function f. Observe that for each n the nth partial sum of this series is just the nth Taylor polynomial of the function f that we defined in Subsection 9.5.2. When $c = 0$ we usually refer to this Taylor series as the **Maclaurin series** of the function f. What we have seen is that the only power series with center c that can converge to a given function f in a neighborhood of c is the Taylor series, center c, of the function f. In other words, either the Taylor series center c of a given function f converges to f at every number x in a neighborhood of c or there is *no* power series center c that converge to f at each number x in a neighborhood of c.

An interesting question now arises. Suppose that a given function f has derivatives of all orders in an interval $(c - r, c + r)$, where c is a given real number and $r > 0$. Can we conclude that the equation

$$f(x) = \sum_{n=0}^{\infty} \frac{f^{(n)}(c)}{n!} (x - c)^n$$

must hold for every number x in the interval $(c - r, c + r)$? The answer is *no!* As a matter of fact, this question is one of the important partings of the ways between **real analysis** – the analysis of functions defined on subsets of $\mathbf{R}$ – and **complex analysis** – the analysis of functions defined on subsets of the set $\mathbf{C}$ of complex numbers. At the heart of complex analysis lies a theorem known as **Taylor's theorem**, which guarantees that every function that is differentiable near a given complex number c will be the sum of its Taylor series near c. When you study complex analysis you will state this theorem precisely and prove it.

One of the most striking features of Taylor's theorem for functions of a complex variable is that it requires only that the function be differentiable once. The fact that the function will have derivatives of all orders follows automatically. This is in sharp contrast to real analysis, in which a function can quite easily be differentiable in an interval even though its second derivative doesn't exist everywhere.

14.4.11   Some Examples Illustrating the Behavior of Taylor Series

1. Ⓝ In this example we use *Scientific Notebook* to explore the Maclaurin

series of the function f defined by the equation

$$f(x) = x \sin\left(1 + x^2\right) + 2x \cos\left(1 + x - x^2\right)$$

for every number x, and we play a sound movie to illustrate how well the partial sums of the series approximate f. You can reach the details by clicking on a link in the on-screen book.

2. Suppose that

$$f(x) = x^2 \sin\frac{1}{x}$$

for every number x. The behavior of this function near 0 is illustrated in

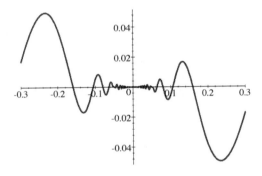

Figure 14.11

Figure 14.11. Even though this function is differentiable at every number, it fails to have a second derivative at 0, and so it cannot have a Taylor series center 0.

3. Suppose that

$$f(x) = \frac{1}{1 + x^2}$$

for every number x. The graph of this function is illustrated in Figure 14.12. By differentiating this function repeatedly we see easily that it has derivatives of all orders at every number. Furthermore, since the formula for summing geometric series gives us

$$\frac{1}{1 + x^2} = \sum_{n=0}^{\infty} (-1)^n x^{2n}$$

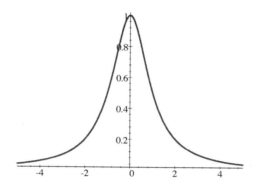

Figure 14.12

whenever $-1 < x < 1$, we see that the Maclaurin series of f is $\sum (-1)^n x^{2n}$. However, in spite of the fact that f is well behaved everywhere, this Maclaurin series converges only in the interval $(-1, 1)$.

4. In view of the theorem on integration of power series term by term and the preceding example we see that

$$\arctan x = \sum_{n=0}^{\infty} \frac{(-1)^n x^{2n+1}}{2n + 1}$$

whenever $-1 \le x \le 1$. Thus the Maclaurin series of the function $\arctan$ is $\sum (-1)^n x^{2n+1} / (2n + 1)$, which converges only in the interval $[-1, 1]$ in spite of the good behavior of $\arctan$ on the entire line. The graph of the function $\arctan$ is illustrated in Figure 14.13.

5. Suppose that α is a given real number and that $\sin \alpha \ne 0$. Suppose that

$$f(x) = \frac{1}{x^2 - 2 (\cos \alpha) x + 1}$$

for every real number x. Note that since $|\cos \alpha| < 1$, this definition is possible at every real number. By expanding the expression

$$(x^2 - 2 (\cos \alpha) x + 1) \sum_{n=0}^{\infty} (\sin (n + 1) \alpha) x^n$$

and collecting the terms, it is not hard to show that whenever $|x| < 1$ we

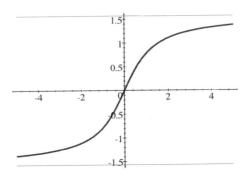

Figure 14.13

have

$$\left(x^2 - 2\left(\cos \alpha\right) x + 1\right) \sum_{n=0}^{\infty} \left(\sin \left(n + 1\right) \alpha\right) x^n = \sin \alpha,$$

and we conclude that if $|x| < 1$, we have

$$f(x) = \sum_{n=0}^{\infty} \frac{\sin \left(n + 1\right) \alpha}{\sin \alpha} x^n.$$

Therefore the Maclaurin series of this function f is

$$\sum \frac{\sin \left(n + 1\right) \alpha}{\sin \alpha} x^n,$$

and the series converges to $f(x)$ whenever $|x| < 1$. To see that the radius of convergence of this series is 1 we shall observe that the series diverges when $x = 1$ and, for this purpose, we shall show that $\sin n\alpha$ does not approach 0 as $n \to \infty$. To obtain a contradiction, assume that

$$\lim_{n\to\infty} \sin n\alpha = 0.$$

Then we have

$$\lim_{n\to\infty} 2\sin \alpha \cos n\alpha = \lim_{n\to\infty} \left(\sin \left(n + 1\right) \alpha - \sin \left(n - 1\right) \alpha\right) = 0$$

and since $\sin \alpha \neq 0$, we deduce that $\cos n\alpha \to 0$ as $n \to \infty$. However, the identity $\cos^2 n\alpha + \sin^2 n\alpha = 1$ makes it impossible for both $\sin n\alpha$ and $\cos n\alpha$ to approach 0 as $n \to \infty$. This is the desired contradiction.

We conclude that, although the function f has derivatives of all orders at every real number, its Maclaurin series converges only in the interval $(-1, 1)$.

6. In this example we define

$$f(x) = \begin{cases} \exp\left(-\frac{1}{x^2}\right) & \text{if } x \neq 0 \\ 0 & \text{if } x = 0. \end{cases}$$

The graph of this function is illustrated in Figure 14.14.Our purpose in this

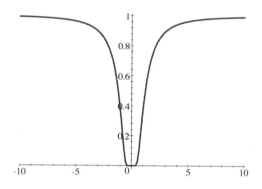

Figure 14.14

example is to demonstrate that this function f has the interesting property that $f^{(n)}(0) = 0$ for every n, and so the Maclaurin series of f is the series $\sum 0x^n$. Thus, although f has derivatives of all orders at every real number, and although the Maclaurin series of f converges at every real number, the series $\sum 0x^n$ fails to converge to the value $f(x)$, except at the number 0. To help us find the derivatives of f at 0, we shall begin by showing that if g is any polynomial, then

$$\lim_{x \to 0} g\left(\frac{1}{x}\right) \exp\left(-\frac{1}{x^2}\right) = 0.$$

If this limit is taken from the right, then, by substituting $u = 1/x$, we obtain it as

$$\lim_{u \to \infty} \frac{g(u)}{\exp(u^2)},$$

which can easily be shown to be zero using L'Hôpital's rule. If the limit is taken from the left, then, by substituting $u = -1/x$, we obtain it as

$$\lim_{u \to \infty} \frac{g(-u)}{\exp\left(u^2\right)}$$

which is also zero.

We now observe that if $x \neq 0$, we have

$$f'(x) = \frac{2}{x^3} \exp\left(-\frac{1}{x^2}\right).$$

Therefore

$$f''(0) = \lim_{x \to 0} \frac{f'(x) - 0}{x} = \lim_{x \to 0} \frac{2}{x^4} \exp\left(-\frac{1}{x^2}\right) = 0.$$

Furthermore, since

$$f''(x) = \left(-\frac{6}{x^4} + \frac{4}{x^6}\right) \exp\left(-\frac{1}{x^2}\right)$$

whenever $x \neq 0$, we have

$$f^{(3)}(0) = \lim_{x \to 0} \frac{f''(x) - 0}{x} = \lim_{x \to 0} \left(-\frac{6}{x^5} + \frac{4}{x^7}\right) \exp\left(-\frac{1}{x^2}\right) = 0;$$

and, in general, if n is any nonnegative integer, then the expression $f^{(n)}(x)$ has the form

$$h_n\left(\frac{1}{x}\right) \exp\left(-\frac{1}{x^2}\right),$$

where h_n is a polynomial, and the fact that $f^{(n)}(0) = 0$ for every n follows from the fact that for each n we have

$$\lim_{x \to 0} \frac{f^{(n)}(x) - 0}{x} = \lim_{x \to 0} \frac{h_n\left(\frac{1}{x}\right)\exp\left(-\frac{1}{x^2}\right) - 0}{x}$$

$$= \lim_{x \to 0} \left(\frac{1}{x} h_n\left(\frac{1}{x}\right)\right) \exp\left(-\frac{1}{x^2}\right) = 0.$$

14.4.12   Some Exercises on Power Series

1. Find the radius of convergence and the interval of convergence of each of the following series:

(a) $\displaystyle\sum \frac{(n!)^2 x^n}{(2n)!}$.

(b) $\displaystyle\sum \frac{((2n)!) x^n}{(n!)^2}$.

(c) $\displaystyle\sum \frac{(n!) x^n}{n^n}$.

(d) $\displaystyle\sum \frac{n^n x^n}{n!}$.

2. [H] Given that $c \neq 1$ and that

$$f(x) = \frac{1}{1-x}$$

whenever $x \neq 1$, expand the function f in a power series with center c and find the interval of convergence of this series.

3. Does a power series have to have the same interval of convergence as its derived series?

4. Suppose that f and g are given functions, that c is a given number, and that $r > 0$. Suppose that both of the functions f and g are the sums of their Taylor series at every number x in the interval $(c - r, c + r)$. Suppose finally that there exists a number $\delta > 0$ such that $f(x) = g(x)$ whenever x belongs to the interval $(c - \delta, c + \delta)$.

Prove that $f(x) = g(x)$ for every number x in the interval $(c - r, c + r)$.

5. True or false? If f and g have derivatives or all orders in a neighborhood of a number c, and if $f^{(n)}(c) = g^{(n)}(c)$ for every nonnegative integer n, then we have $f(x) = g(x)$ for every number x sufficiently close to c.

6. (N) Use *Scientific Notebook* to calculate a variety of nth partial sums of the Maclaurin series of the function f defined by the equation

$$f(x) = \frac{e^x \sin x}{1 + x^6}$$

for every real number x. Then use *Scientific Notebook* to plot some of these nth partial sums with the graph of f on the interval $[-2, 2]$ and explore the accuracy of these partial sums as approximations to f on the interval.

14.5   Power Series Expansion of the Exponential Function

14.5.1   The Maclaurin Series of the Exponential Function

We recall that the exponential function exp that was introduced in Chapter 10 is a strictly increasing function on $\mathbf{R}$, that the range of the function exp is the interval $(0, \infty)$, and that for every real number x we have

$$\exp'(x) = \exp(x).$$

Since

$$\exp^{(n)}(0) = \exp(0) = 1$$

for every nonnegative integer n, we see at once that the Maclaurin series of the function exp is

$$\sum \frac{1}{n!} x^n,$$

and we see from d'Alembert's test that this series converges at every real number. In this section we shall show that the Maclaurin series of the function exp actually converges to $\exp x$ for every real number x.

14.5.2   Summing the Maclaurin Series of the Function exp

We begin by giving a name to the sum of this Maclaurin Series. We define

$$f(x) = \sum_{n=0}^{\infty} \frac{x^n}{n!}$$

for every real number x and our task is to show that the equation $f(x) = \exp x$ holds for every number x. Since the radius of convergence of the series $\sum x^n/n!$ is ∞, we know from Theorem 14.4.8 that f is differentiable on $\mathbf{R}$ and that for every real number x we have

$$f'(x) = \sum_{n=1}^{\infty} \frac{nx^{n-1}}{n!} = \sum_{n=1}^{\infty} \frac{x^{n-1}}{(n-1)!} = \sum_{n=0}^{\infty} \frac{x^n}{n!} = f(x).$$

We see also that $f(0) = 1$. We now define

$$g(x) = \frac{f(x)}{\exp(x)}$$

for every real number x. In order to show that $\exp x = f(x)$ for every number x, we need to show that g is the constant function 1. Now for every number x it follows from the quotient rule that

$$g'(x) = \frac{f'(x)\exp(x) - f(x)\exp'(x)}{(\exp(x))^2} = \frac{f(x)\exp(x) - f(x)\exp(x)}{(\exp(x))^2} = 0$$

and so g is certainly constant. Since $g(0) = 1$ we conclude that g is the constant function 1, and we conclude that

$$\exp x = \sum_{n=0}^{\infty} \frac{x^n}{n!}$$

for every number x.

14.5.3   Some Exercises on the Series Expansion of $\exp$

1. ◨ Prove that if c and x are any real numbers, then

$$\exp(x) = \sum_{n=0}^{\infty} \frac{e^c}{n!}(x-c)^n.$$

2. Prove that

$$\int_0^1 e^{-x^2}\,dx = \sum_{n=0}^{\infty} \frac{(-1)^n}{(2n+1)\,(n!)}.$$

3. Show that even before we have showed that the function f defined in the proof of Theorem 14.5.2 is the exponential function $\exp$ we could have seen from the binomial theorem and Cauchy's theorem on products of series (Theorem 12.10.4) that for all numbers x and y we have

$$f(x)f(y) = f(x+y).$$

4. (a) Prove that

$$e = \sum_{n=0}^{\infty} \frac{1}{n!}.$$

   (b) Prove that if m is any positive integer, then the number

$$(m!)\sum_{j=0}^{m} \frac{1}{j!}$$

is an integer and prove that

$$0 < (m!) \sum_{j=m+1}^{\infty} \frac{1}{j!} < 1.$$

(c) **H** Prove that if m is any positive integer, then the number $(m!)\, e$ is not an integer and deduce that the number e is irrational.

14.6   Binomial Series

14.6.1   Introduction to the Binomial Function

Given any real number α, the **binomial function** with exponent α is the function B_α defined by the equation

$$B_\alpha(x) = (1 + x)^\alpha$$

whenever the right side is defined. In the event that α is a nonnegative integer, the right side is a polynomial and is defined at every number x. For other values of α we may have to make restrictions. For example, if $\alpha < 0$, we need to avoid the possibility $x = -1$; and if α is not an integer, we may need to require $x > -1$ to ensure that $1 + x$ is positive. We also deduce from the results of Section 10.6 that whenever $x > -1$ we have

$$B'_\alpha(x) = \alpha\, (1 + x)^{\alpha - 1}$$

and, more generally, if n is any positive integer, then

$$B_\alpha^{(n)}(x) = \alpha\, (\alpha - 1)\, (\alpha - 2) \cdots (\alpha - n + 1)\, (1 + x)^{\alpha - n}\,.$$

From this observation we conclude that the Maclaurin series of the binomial function B_α is

$$\sum \frac{\alpha\, (\alpha - 1)\, (\alpha - 2) \cdots (\alpha - n + 1)}{n!} x^n$$

and, following the notation that we introduced in Example 3 of Subsection 12.7.9, we define the **binomial coefficient**

$$\binom{\alpha}{n} = \begin{cases} \dfrac{\alpha\, (\alpha - 1)\, (\alpha - 2) \cdots (\alpha - n + 1)}{n!} & \text{if } \ n \text{ is a positive integer} \\[2ex] 1 & \text{if } \ n = 0 \end{cases}$$

The Maclaurin series of the binomial function B_α is therefore

$$\sum \binom{\alpha}{n} x^n,$$

which we call the **binomial series** with exponent α, and the natural question to ask is whether the equality

$$(1 + x)^\alpha = \sum_{n=0}^{\infty} \binom{\alpha}{n} x^n$$

holds for certain values of x.

14.6.2 Binomials with Nonnegative Integer Exponents

Suppose that α is a nonnegative integer. Since the function B_α is a polynomial with degree α, we can choose numbers $a_0, a_1, \cdots, a_\alpha$ such that the equation

$$B_\alpha(x) = \sum_{j=0}^{\alpha} a_j x^j$$

holds for every number x. For each n we can differentiate both sides n times to obtain

$$\alpha (\alpha - 1) (\alpha - 2) \cdots (\alpha - n + 1) (1 + x)^{\alpha - n}$$

$$= \sum_{j=n}^{\alpha} a_j j (j - 1) \cdots (j - n + 1) x^{j-n}$$

and, putting $x = 0$, we obtain

$$a_n = \frac{\alpha (\alpha - 1) (\alpha - 2) \cdots (\alpha - n + 1)}{n!} = \binom{\alpha}{n}.$$

Thus the equation

$$(1 + x)^\alpha = \sum_{n=0}^{\alpha} \binom{\alpha}{n} x^n$$

holds for every number x. This equation is the standard algebraic form of the binomial theorem. Note, however, that, because α is a nonnegative integer, the binomial coefficient $\binom{\alpha}{n}$ must be zero whenever $n > \alpha$, because one of the factors in the product

$$\alpha (\alpha - 1) (\alpha - 2) \cdots (\alpha - n + 1)$$

must be $\alpha - \alpha$. Therefore we have the option of writing the binomial theorem in the form

$$(1 + x)^\alpha = \sum_{n=0}^{\infty} \binom{\alpha}{n} x^n$$

even though this summation is not really infinite.

14.6.3   Convergence of the Binomial Series

Suppose that α is a real number but is not a nonnegative integer. Since $\left|\binom{\alpha}{n}\right|$ is positive for every n, we can use d'Alembert's test to find the radius of convergence of the series $\sum \binom{\alpha}{n} x^n$. In fact, since

$$\lim_{n \to \infty} \frac{\left|\binom{\alpha}{n+1} x^{n+1}\right|}{\left|\binom{\alpha}{n} x^n\right|} = \lim_{n \to \infty} \frac{\left|\binom{\alpha}{n+1} x^{n+1}\right|}{\left|\binom{\alpha}{n} x^n\right|} = \lim_{n \to \infty} \frac{n - \alpha}{n + 1} |x| = |x|,$$

we deduce that the radius of convergence of this series is 1. We also know from Example 3 of Subsection 12.7.9 that the interval of convergence is as described in the following table.

Location of the number α	$\alpha > 0$	$-1 < \alpha < 0$	$\alpha \le -1$
Interval of convergence of $\sum \binom{\alpha}{n} x^n$	$[-1, 1]$	$(-1, 1]$	$(-1, 1)$

14.6.4   A Technical Fact About Binomial Coefficients

Suppose that α is any real number and that n is a nonnegative integer. Then

$$(n + 1) \binom{\alpha}{n + 1} + n \binom{\alpha}{n} = \alpha \binom{\alpha}{n}.$$

We leave the proof of this algebraic fact as a simple exercise.

14.6.5   The Behavior of the Sum of a Binomial Series

Suppose that α is any real number and that

$$f(x) = \sum_{n=0}^{\infty} \binom{\alpha}{n} x^n$$

whenever $|x| < 1$. Then whenever $|x| < 1$ we have

$$(1 + x) f'(x) = \alpha f(x).$$

Proof. Whenever $|x| < 1$ we have

$$(1+x) f'(x) = (1+x) \sum_{n=1}^{\infty} \binom{\alpha}{n} n x^{n-1} = \sum_{n=1}^{\infty} \binom{\alpha}{n} n x^{n-1} + \sum_{n=1}^{\infty} \binom{\alpha}{n} n x^n$$

$$= \sum_{n=0}^{\infty} \binom{\alpha}{n+1} (n+1) x^n + \sum_{n=1}^{\infty} \binom{\alpha}{n} n x^n$$

$$= \binom{\alpha}{1} + \sum_{n=1}^{\infty} \left((n+1) \binom{\alpha}{n+1} + n \binom{\alpha}{n} \right) x^n$$

$$= \alpha \binom{\alpha}{0} + \sum_{n=1}^{\infty} \alpha \binom{\alpha}{n} x^n = \alpha \sum_{n=0}^{\infty} \binom{\alpha}{n} x^n = \alpha f(x). \quad \blacksquare$$

14.6.6   The Sum of a Binomial Series

Suppose that α is any real number. Then at any number x at which the binomial series $\sum \binom{\alpha}{n} x^n$ converges we have

$$(1+x)^\alpha = \sum_{n=0}^{\infty} \binom{\alpha}{n} x^n.$$

Proof. We define

$$f(x) = \sum_{n=0}^{\infty} \binom{\alpha}{n} x^n$$

and we define

$$g(x) = \frac{f(x)}{(1+x)^\alpha}$$

whenever $|x| < 1$. We shall show that g is the constant function 1. Since

$$g'(x) = \frac{f'(x)(1+x)^\alpha - f(x)\alpha(1+x)^{\alpha-1}}{(1+x)^{2\alpha}}$$

$$= \frac{f'(x)(1+x)(1+x)^{\alpha-1} - f(x)\alpha(1+x)^{\alpha-1}}{(1+x)^{2\alpha}}$$

$$= \frac{\alpha f(x)(1+x)^{\alpha-1} - f(x)\alpha(1+x)^{\alpha-1}}{(1+x)^{2\alpha}} = 0$$

for each x, we know that g is constant on $(-1, 1)$, and, since $g(0) = 1$, this constant must be 1. We conclude that the equation

$$f(x) = \sum_{n=0}^{\infty} \binom{\alpha}{n} x^n$$

holds whenever $|x| < 1$, and the extension of this equation to any endpoint of the interval $(-1, 1)$ at which the series converges follows from Abel's theorem (Theorem 12.7.7). ∎

14.6.7   Some Exercises on Binomial Series

1. ⊞ Given that $\alpha > -1$, prove that

$$2^{\alpha} = \sum_{n=0}^{\infty} \binom{\alpha}{n}.$$

2. Given that $\alpha > 0$, prove that

$$\sum_{n=0}^{\infty} (-1)^n \binom{\alpha}{n} = 0.$$

3. Given that α and β are any real numbers and $|x| < 1$, apply Cauchy's theorem on products of series (Theorem 12.10.4) to the Maclaurin expansions of $(1 + x)^{\alpha}$ and $(1 + x)^{\beta}$ to deduce that the equation

$$\sum_{j=0}^{n} \binom{\alpha}{n-j} \binom{\beta}{j} = \binom{\alpha + \beta}{n}$$

holds for every positive integer n.

4. In this exercise we define

$$A_n^{\alpha} = \frac{(\alpha + 1)(\alpha + 2) \cdots (\alpha + n)}{n!}$$

whenever $\alpha > -1$ and n is a positive integer. We also define $A_0^{\alpha} = 1$.

(a) Prove that if $\alpha > -1$ and n is a nonnegative integer, we have

$$\binom{-\alpha - 1}{n} = (-1)^n A_n^{\alpha}.$$

(b) Use Exercise 3 to prove that if α and β are greater than -1 and n is a nonnegative integer, then

$$\sum_{j=0}^{n} A_{n-j}^{\alpha} A_{j}^{\beta} = A_{n}^{\alpha+\beta+1}.$$

(c) Prove that if α and n are nonnegative integers, then

$$A_{n}^{\alpha} = \frac{(n+1)(n+2)\cdots(n+\alpha)}{\alpha!},$$

where the numerator of the right side is understood to be 1 when $\alpha = 0$.

(d) Prove that if α, β, and n are nonnegative integers, we have

$$\sum_{j=0}^{n} \left(\frac{(n-j+1)\cdots(n-j+\alpha)}{\alpha!} \right) \left(\frac{(j+1)\cdots(j+\beta)}{\beta!} \right)$$

$$= \frac{(n+1)\cdots(n+\alpha+\beta+1)}{(\alpha+\beta+1)!}.$$

5. (a) ▐H▌ Prove that if

$$f(x) = \sqrt{1-x}$$

whenever $0 \le x \le 1$, then there exists a sequence of polynomials that converges uniformly to f on the interval $[0, 1]$.

(b) Prove that if f is the function defined in part a and if

$$g(x) = f\left(1 - x^2\right)$$

whenever $-1 \le x \le 1$, then there exists a sequence of polynomials that converges uniformly to g on the interval $[-1, 1]$.

(c) Prove that if $g(x) = |x|$ for all $x \in [-1, 1]$, then there exists a sequence of polynomials that converges uniformly to g on the interval $[-1, 1]$.

(d) (N) Use *Scientific Notebook* to calculate some nth Maclaurin polynomials of the function f defined in part a. For each chosen value of n, if f_n is the nth Maclaurin polynomial, and $h_n(x) = f_n\left(1 - x^2\right)$ for each x, ask *Scientific Notebook* to sketch the graph of the function h together with the graph of the absolute value function and observe graphically that the sequence (h_n) is converging uniformly to the absolute value function on the interval $[-1, 1]$. The case $n = 35$ is illustrated in Figure 14.15.

This exercise is of considerable importance because it may be used as the starting point for a major theorem known as the **Stone-Weierstrass theorem**.

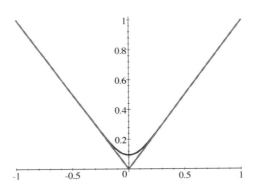

Figure 14.15

You can find an elementary presentation of the Stone-Weierstrass theorem in Rudin [26], starting with Corollary 7.27.

14.7   The Trigonometric Functions

14.7.1   The Precalculus Approach to the Trigonometric Functions

In an elementary mathematics course, the trigonometric functions cos and sin are introduced as the coordinates of a point at the end of a "terminal line" of a given "angle". The typical approach is to say that, if θ is any given real number and if $P(\theta)$ is defined to be the point on the unit circle to which one would arrive by starting at the point $(1, 0)$ and traveling around the circle through a total distance of θ, counterclockwise if θ is positive and clockwise if θ is negative, then $\cos \theta$ and $\sin \theta$ are the coordinates of the point $P(\theta)$. The approach is illustrated in Figure 14.16.

Ideally, we would have liked to base the presentation of the trigonometric functions in this book on the elementary approach. By doing so, we would demonstrate that the elementary approach can be made precise. But, unfortunately, the process of writing the elementary approach precisely presents us with several technical hurdles. The problem lies in the definition of the point $P(\theta)$, because even after we have sorted out what we ought to mean by "counterclockwise", we still have to describe the process of winding around the circle until the arclength we have travelled is θ. And this process is complicated, even if θ is small and positive. From the perspective of elementary calculus, the arclength

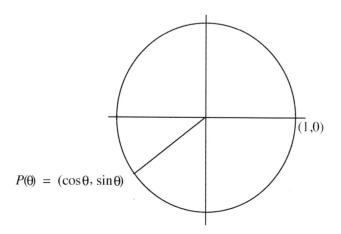

Figure 14.16

along the semicircle

$$x = \sqrt{1 - y^2}$$

in a counterclockwise direction from the point $(1, 0)$ to a given point (x_1, y_1) in the first quadrant, as depicted in Figure 14.17, is given by the integral

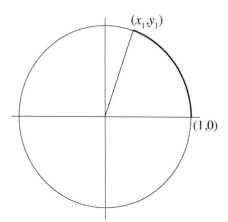

Figure 14.17

$$\int_0^{y_1} \sqrt{1 + \left(\frac{dx}{dy}\right)^2} \, dy = \int_0^{y_1} \sqrt{1 + \left(\frac{-y}{\sqrt{1 - y^2}}\right)^2} \, dy = \int_0^{y_1} \frac{1}{\sqrt{1 - y^2}} dy,$$

which, from the perspective of elementary calculus, comes out as $\arcsin y_1$. Therefore, if we want to approach the trigonometric functions by giving a precise description of the elementary definition, we must begin by defining the function arcsin as an integral.

We see therefore that a precise description of the elementary approach to trigonometry can be rather involved, and so we shall not try to give one. Instead, we shall keep in mind the two following fundamental properties of the functions sin and cos.

- For every number x we have $\sin' x = \cos x$ and $\cos' x = -\sin x$.
- $\sin 0 = 0$ and $\cos 0 = 1$.

We shall use an infinite series approach to provide us with two functions that have these fundamental properties and we shall show that there can be only one such pair of functions. We begin with a simple lemma that will help us to show that these functions are unique.

14.7.2   Summing the Squares

Suppose that f and g are functions defined on the set $\mathbf{R}$ and that the equations $f'(x) = g(x)$ and $g'(x) = -f(x)$ hold for every number x. Then the function $f^2 + g^2$ is constant.

Proof. We define

$$h(x) = (f(x))^2 + (g(x))^2$$

for every number x and we observe that if x is any real number, then

$$h'(x) = 2f(x)f'(x) + 2g(x)g'(x) = 2f(x)g(x) - 2f(x)g(x) = 0.$$

Therefore the function h must be constant. ∎

14.7.3   Trigonometric Pairs of Functions

We shall say that a pair of functions s and c form a **trigonometric pair** if the following conditions hold:

1. The functions s and c are differentiable on the entire number line $\mathbf{R}$ and for every real number x we have $s'(x) = c(x)$ and $c'(x) = -s(x)$.
2. $s(0) = 0$ and $c(0) = 1$.

14.7.4   Uniqueness of the Trigonometric Pair

There cannot exist more than one trigonometric pair of functions.

Proof. Suppose that s_1 and c_1 form a trigonometric pair and that s_2 and c_2 also form a trigonometric pair. Since

$$(s_1 - s_2)' = c_1 - c_2 \quad \text{and} \quad (c_1 - c_2)' = -(s_1 - s_2),$$

we know that the function

$$(s_1 - s_2)^2 + (c_1 - c_2)^2$$

must be constant. Therefore, if x is any real number, we have

$$(s_1(x) - s_2(x))^2 + (c_1(x) - c_2(x))^2 = (s_1(0) - s_2(0))^2 + (c_1(0) - c_2(0))^2 = 0,$$

and we deduce that $s_1 = s_2$ and $c_1 = c_2$. ∎

14.7.5 The Existence of a Trigonometric Pair

In this subsection we shall define two functions s and c that form a trigonometric pair. In order to motivate the definition, suppose for the moment that we already have two such functions.

We observe that

$$
\begin{array}{l|l}
s(0) = 0 & c(0) = 1 \\
s'(0) = c(0) = 1 & c'(0) = -s(0) = 0 \\
s''(0) = -s(0) = 0 & c''(0) = -c(0) = -1 \\
s^{(3)}(0) = -c(0) = -1 & c^{(3)}(0) = s(0) = 0 \\
s^{(4)}(0) = s(0) = 0 & c^{(4)}(0) = c(0) = 1
\end{array}
$$

after which the cycle is repeated. Thus the Maclaurin series for s is

$$\sum \frac{s^{(n)}(0)}{n!} x^n = x - \frac{x^3}{3!} + \frac{x^5}{5!} - \cdots = \sum \frac{(-1)^n x^{2n+1}}{(2n+1)!},$$

and the Maclaurin series for c is

$$\sum \frac{c^{(n)}(0)}{n!} x^n = 1 - \frac{x^2}{2!} + \frac{x^4}{4!} - \cdots = \sum \frac{(-1)^n x^{2n}}{(2n)!}.$$

Note that each of these series is convergent for every real number x. With these observations in mind we define

$$\cos x = \sum_{n=0}^{\infty} \frac{(-1)^n x^{2n}}{(2n)!}$$

and

$$\sin x = \sum_{n=0}^{\infty} \frac{(-1)^n x^{2n+1}}{(2n+1)!}$$

for every real number x. From the theorem on differentiation power series term by term we see easily that these functions form a trigonometric pair.

14.7.6   Some Facts About $\cos$ and $\sin$

1. Given any number x, we have $\cos(-x) = \cos x$ and $\sin(-x) = -\sin x$.
2. Given any number x, we have $\cos^2 x + \sin^2 x = 1$.
3. Given any numbers a and b, we have

$$
\begin{aligned}
\cos(a-b) &= \cos a \cos b + \sin a \sin b \\
\sin(a-b) &= \sin a \cos b - \cos a \sin b \\
\cos(a+b) &= \cos a \cos b - \sin a \sin b \\
\sin(a+b) &= \sin a \cos b + \cos a \sin b.
\end{aligned}
$$

Proof. Parts 1 and 2 follow at once from the preceding observations. To prove the first two identities of part 3 we suppose that b is a given real number and for every number x we define

$$f(x) = \sin(x-b) - (\sin x \cos b - \cos x \sin b)$$

and

$$g(x) = \cos(x-b) - (\cos x \cos b + \sin x \sin b).$$

Since $f' = g$ and $g' = -f$, we know that $f^2 + g^2$ is constant. Therefore, for every number x we have

$$(f(x))^2 + (g(x))^2 = (f(0))^2 + (g(0))^2 = 0,$$

and we conclude that each of the functions f and g is the constant 0. The other two identities now follow from the usual methods. ∎

14.7.7   Zeros of the Cosine Function

In this subsection we shall show that there is a positive number x such that $\cos x = 0$. Since $\cos 0 = 1$, the desired result will follow from Bolzano's intermediate value theorem (Theorem 8.10.2) if we can show that $\cos 2 < 0$. We observe that

$$\cos 2 = 1 - \frac{2^2}{2!} + \sum_{n=2}^{\infty} \frac{(-1)^n\, 2^{2n}}{(2n)!} = -1 + \sum_{n=2}^{\infty} \frac{(-1)^n\, 2^{2n}}{(2n)!},$$

and the fact that this number is negative follows from Theorem 12.7.6, which may be used to show that

$$\sum_{n=2}^{\infty} \frac{(-1)^n \, 2^{2n}}{(2n)!} \le \frac{2}{3}.$$

14.7.8   The Number π

Since the set

$$\{x > 0 \mid \cos x = 0\}$$

is nonempty, closed, and bounded below, it has a least member. We define π to be twice the least member of this set. Thus $\pi/2$ is the least positive number at which the function cos takes the value 0.

14.7.9   The Periodic Behavior of the Functions sin and cos

We begin this discussion by looking at the behavior of the functions cos and sin on the interval $[0, \pi/2]$. Since $\cos 0 = 1$ and $\cos x$ is never zero when $x \in [0, \pi/2)$, it follows from the continuity of cos that $\cos x > 0$ for every $x \in [0, \pi/2)$. From the fact that $\sin' = \cos$, we deduce that sin is strictly increasing on the interval $[0, \pi/2]$ and that $\sin x > 0$ whenever $x \in (0, \pi/2]$. From the fact that

$$\sin^2 \frac{\pi}{2} = 1 - \cos^2 \frac{\pi}{2} = 1$$

we conclude that $\sin \pi/2 = 1$. Finally, since $\cos' = -\sin$, the function cos is strictly decreasing on the interval $[0, \pi/2]$. Figure 14.18 illustrates the behavior of cos and sin on the interval $[0, \pi/2]$.

We now discuss the behavior of cos and sin on the interval $[\pi/2, \pi]$, and for this purpose we shall use the identities

$$\cos\left(\frac{\pi}{2} + x\right) = -\sin x \quad \text{and} \quad \sin\left(\frac{\pi}{2} + x\right) = \cos x,$$

which follow directly from Theorem 14.7.6. From the first of these two identities and the fact that sin increases from 0 to 1 on the interval $[0, \pi/2]$ we see that cos decreases from 0 to -1 on the interval $[\pi/2, \pi]$. In the same way, since cos decreases from 1 to 0 on the interval $[0, \pi/2]$, it follows from the second identity that sin decreases from 1 to 0 on the interval $[\pi/2, \pi]$. Figure 14.19 illustrates the behavior of cos and sin on the interval $[\pi/2, \pi]$.

We can now turn our attention to the interval $[\pi, 2\pi]$, using the identities

$$\sin(\pi + x) = -\sin x \quad \text{and} \quad \cos(\pi + x) = -\cos x,$$

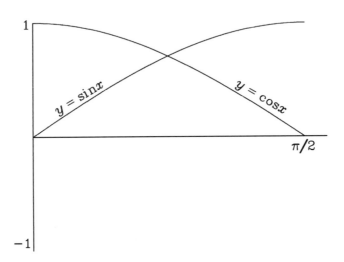

Figure 14.18

and, continuing in this way, one may verify all of the usual relationships between the functions cos and sin and the number π. In particular, one may see that cos and sin are periodic functions with period 2π. It would be a good exercise for you to write a careful and complete argument showing how all of the familiar properties of cos and sin can be made to follow from their definitions.

14.7.10   Exercises on the Trigonometric Functions

1.   Given any real number x, prove that $\sin x = 0$ if and only if x is an integer multiple of π. Prove that $\cos x = 0$ if and only if x is an odd multiple of $\pi/2$. Prove that if n is any integer, then $\cos n\pi = (-1)^n$.

2.   Prove that if α is any real number, then the equation

$$\sin (x + \alpha) = \sin x$$

holds for every real number x if and only if α is an even multiple of π.

3.   Prove that the restriction of the function sin to the interval $[-\pi/2, \pi/2]$ is a strictly increasing function from $[-\pi/2, \pi/2]$ onto the interval $[-1, 1]$. Prove that if the function arcsin is now defined to be the inverse function of this restriction of sin, then for every number $u \in (-1, 1)$ we have

$$\arcsin u = \int_0^u \frac{1}{\sqrt{1 - t^2}} dt.$$

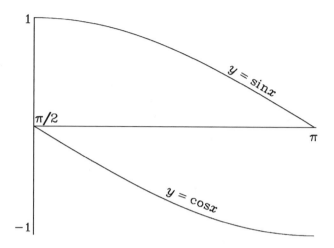

Figure 14.19

4. By analogy with the preceding exercise, give a definition of the function arctan and deduce that if u is any real number, then

$$\arctan u = \int_0^u \frac{1}{1+t^2} dt.$$

5. Prove that the restriction of the function cos to the interval $[0, \pi]$ is a strictly decreasing function from the interval $[0, \pi]$ onto the interval $[-1, 1]$. Prove that if the function arccos is now defined to be the inverse function of this restriction of cos, then for every number $u \in (-1, 1)$ we have

$$\arccos u = \frac{\pi}{2} - \int_0^u \frac{1}{\sqrt{1-t^2}} dt.$$

6. Prove that if x and y are real numbers that are not both zero and if

$$\alpha = \arccos\left(\frac{x}{\sqrt{x^2+y^2}}\right),$$

then

$$\sin \alpha = \pm \frac{y}{\sqrt{x^2+y^2}}.$$

Deduce that if x and y are real numbers that are not both zero, then there exists a positive number r and a real number $\theta \in [0, 2\pi)$ such that

$x = r\cos\theta$ and $y = r\sin\theta$.

14.8   Analytic Functions of a Real Variable

As we saw in Example 6 of Subsection 14.4.11, a function that has derivatives of all orders in a neighborhood of a number c cannot always be expressed as the sum of a power series centered at c. We mentioned earlier that this sort of pathological behavior is in sharp contrast to the way functions behave when we are working in the complex number system. The functions that do not exhibit this sort of pathological behavior are said to be **analytic.** More precisely, a function f is said to be analytic on an open set U if, for every number $c \in U$, there exists a number $\delta > 0$ and a sequence (a_n) of numbers such that the equation

$$f(x) = \sum_{n=0}^{\infty} a_n (x - c)^n$$

holds whenever $x \in (c - \delta, c + \delta)$.

From the observations that we made in Subsection 14.4.9, we see that a function f is analytic in an open set U if and only if for every number $c \in U$ there exists a number $\delta > 0$ such that the equation

$$f(x) = \sum_{n=0}^{\infty} \frac{f^{(n)}(c)}{n!} (x - c)^n$$

holds whenever $x \in (c - \delta, c + \delta)$.

In this optional section we explore some of the main facts about analytic functions, and we demonstrate that many of the standard functions that we have encountered are analytic. If you would like to read this material, you can access it from the on-screen version of this book by clicking on the icon 🏭 .

14.9   The Inadequacy of Riemann Integration

We end this chapter with some optional reading that is designed to convey some understanding of the inadequacy of Riemann integration as an integration concept in modern analysis.

Every theory of integration begins with the integration of step functions and is then extended to other types of functions by some sort of limit process. We saw one such process in Chapter 11, where we encountered the notion of a Riemann integral. As we saw there, a bounded function f is Riemann integrable if it can be approximated above and below by step functions whose integrals are close to each other. That approach was the mainstay of nineteenth-century mathematics, but it did not extend into the twentieth century because the method of approximating functions by step functions in the Riemann theory is too restrictive. During the twentieth century new and more powerful theories of integration arose, beginning in 1901 with the theory developed by Henri Lebesgue. Such theories are beyond our scope, but you can find a short discussion in the on-screen version of the book that motivates the need for something better than the Riemann theory. To reach that discussion, click on the icon .

Chapter 15
Calculus of a Complex Variable
(Optional)

This optional chapter will introduce you to some of the elementary properties of sequences, limits, continuity, and derivatives in the system C of ▮ complex numbers. In order to read this chapter you should be comfortable with the notion of limits and continuity in the Euclidean plane $\mathbf{R}^2$, which is, after all, what the system C really is. You will certainly have the needed familiarity with limits in $\mathbf{R}^2$ if you have read the chapter on ▮ metric spaces and the chapter on ▮ sequences in metric spaces, although those chapters may supply more than you need here. To access this chapter from the on-screen version of the book, click on the icon ▮ .

Chapter 16
Integration of Functions
of Two Variables

16.1   The Purpose of This Chapter

In this chapter we study integrals of the form

$$\int_a^b f(x,y)dx.$$

In an expression of this type, the function f is defined on a set of points (x, y) in the Euclidean plane $\mathbf{R}^2$ and we are integrating with respect to x with y held constant. For example,

$$\int_1^2 ye^{xy}dx = e^{xy}\Big|_{x=1}^{x=2} = e^{2y} - e^y.$$

Since this expression involves only the variable y, we can differentiate or integrate it with respect to y. For example,

$$\frac{d}{dy}\int_1^2 ye^{xy}dx = \frac{d}{dy}\left(e^{2y} - e^y\right) = 2e^{2y} - e^y$$

and

$$\int_0^1 \int_1^2 ye^{xy}dxdy = \int_0^1 \left(e^{2y} - e^y\right)dy = \frac{1}{2}e^2 - e + \frac{1}{2},$$

and so a natural question that we can ask is whether the operations with respect to x and y can be interchanged. In other words, we could ask whether

$$\frac{d}{dy}\int_1^2 ye^{xy}dx = \int_1^2 \frac{\partial}{\partial y}ye^{xy}dx$$

and

$$\int_0^1 \int_1^2 ye^{xy}dxdy = \int_1^2 \int_0^1 ye^{xy}dydx,$$

and, if you care to evaluate these expressions, you will see that the latter two equations are true. Notice that in each of the four expressions that appear in these

459

equations, the inside operation takes place with respect to one of the variables x and y with the other held constant.

In this chapter we shall introduce the notation precisely and we shall show that identities of the type

$$\frac{d}{dy} \int_a^b f(x,y)dx = \int_a^b \frac{\partial}{\partial y} f(x,y)dx$$

and

$$\int_c^d \int_a^b f(x,y)dxdy = \int_a^b \int_c^d f(x,y)dydx$$

hold for a wide variety of functions. We shall see that such identities sometimes hold even when the integrals are improper.

16.2   Functions of Two Variables

16.2.1   Vertical and Horizontal Sections of a Plane Set

As we said in Subsection 4.2.8, the **Euclidean plane**, which we write as $\mathbf{R}^2$, is defined to be the set of all ordered pairs (x, y) of real numbers.

Now suppose that S is a subset of $\mathbf{R}^2$. Given any real number x, the **vertical** x-**section** of S is the set

$$S_x = \{y \in \mathbf{R} \mid (x,y) \in S\},$$

and, given any real number y, the **horizontal** y-**section** of S is the set

$$S^y = \{x \in \mathbf{R} \mid (x,y) \in S\}.$$

Vertical and horizontal sections are illustrated in Figure 16.1.

16.2.2   Some Examples of Sections of a Set

1. Suppose that

$$S = \{(x,y) \mid 0 \le x \le 1 \text{ and } 0 \le y \le x\}.$$

This set is illustrated in Figure 16.2. Given any number x we have

$$S_x = \begin{cases} [0,x] & \text{if } x \in [0,1] \\ \emptyset & \text{if } x \in \mathbf{R} \setminus [0,1], \end{cases}$$

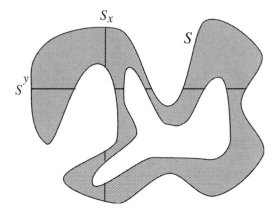

Figure 16.1

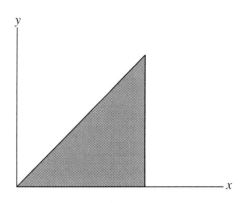

Figure 16.2

and given any number y we have

$$S^y = \begin{cases} [y, 1] & \text{if} \quad y \in [0, 1] \\ \emptyset & \text{if} \quad y \in \mathbf{R} \setminus [0, 1] . \end{cases}$$

2. Suppose that S is the circular sector illustrated in Figure 16.3. Given any number x we have

$$S_x = \begin{cases} \left[\frac{3}{4}x, \sqrt{25 - x^2}\right] & \text{if} \quad x \in [0, 4] \\ \emptyset & \text{if} \quad x \in \mathbf{R} \setminus [0, 4] , \end{cases}$$

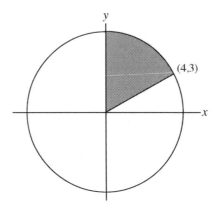

Figure 16.3

and given any number y we have

$$S^y = \begin{cases} [0, \frac{4}{3}y] & \text{if } y \in [0, 3] \\ \left[0, \sqrt{25 - y^2}\right] & \text{if } y \in [3, 5] \\ \emptyset & \text{if } y \in \mathbf{R} \setminus [0, 5]. \end{cases}$$

3. Suppose that A and B are any sets of numbers and that $S = A \times B$. We see that

$$S_x = \begin{cases} B & \text{if } x \in A \\ \emptyset & \text{if } x \in \mathbf{R} \setminus A \end{cases}$$

and

$$S^y = \begin{cases} A & \text{if } y \in B \\ \emptyset & \text{if } y \in \mathbf{R} \setminus B. \end{cases}$$

16.2.3   Vertical and Horizontal Sections of a Function

Suppose that f is a function defined on a subset S of the Euclidean plane $\mathbf{R}^2$. Given any number x, the **vertical x-section** of f is the function $f(x, \cdot)$ that is defined on the set S_x by the equation

$$f(x, \cdot)(y) = f(x, y)$$

for every number $y \in S_x$. This concept of a vertical section is a precise way of saying that, if we hold x constant, then the expression $f(x, y)$ can be looked upon as a "function of y only".

Given any number y, the **horizontal y-section** is the function $f(\cdot, y)$ that is defined on the set S^y by the equation

$$f(\cdot, y)(x) = f(x, y)$$

for every number $x \in S^y$.

16.2.4   Partial Derivatives

Suppose that f is a function defined on a subset S of the Euclidean plane $\mathbf{R}^2$. Given any point (x, y) in the set S, the **partial derivative** $D_1 f(x, y)$ of the function f at the point (x, y) is defined to be the limit

$$\lim_{u \to x} \frac{f(u, y) - f(x, y)}{u - x},$$

as long as this limit exists. Another way of looking at $D_1 f(x, y)$ is to say that it is the derivative at the number x of the function $f(\cdot, y)$. The expression $D_1 f$ stands for the function that has the value $D_1 f(x, y)$ at each point (x, y) at which $D_1 f(x, y)$ exists.

In the same way, the partial derivative $D_2 f(x, y)$ of the function f at the point (x, y) is defined to be the limit

$$\lim_{v \to y} \frac{f(x, v) - f(x, y)}{v - y}$$

(as long as this limit exists), and $D_2 f(x, y)$ is the derivative at y of the function $f(x, \cdot)$. The expression $D_2 f$ stands for the function that has the value $D_2 f(x, y)$ at each point (x, y) at which $D_2 f(x, y)$ exists.

As you can see we have avoided the symbolism

$$\frac{\partial f(x, y)}{\partial x} \quad \text{and} \quad \frac{\partial f(x, y)}{\partial y}$$

because it is hard to make this sort of notation precise.

16.2.5   Partial and Iterated Riemann Integrals

Suppose that a, b, c, and d are given real numbers and that $a < b$ and $c < d$. Suppose that f is a function defined on the rectangular set $[a, b] \times [c, d]$ that is illustrated in Figure 16.1. Given any number $y \in [c, d]$, if the function $f(\cdot, y)$ is Riemann integrable on the interval $[a, b]$, then we define the **partial integral**

$$\int_a^b f(x, y) dx$$

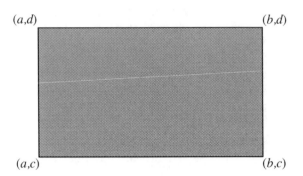

Figure 16.4

to be the integral on the interval $[a, b]$ of the function $f(\cdot, y)$. In the same way, if $x \in [a, b]$ and if the function $f(x, \cdot)$ is Riemann integrable on the interval $[c, d]$, then we define the partial integral

$$\int_c^d f(x, y)dy$$

to be the integral on the interval $[c, d]$ of the function $f(x, \cdot)$.

In the event that $f(\cdot, y)$ is integrable on $[a, b]$ for every number $y \in [c, d]$, we define the **iterated integral** (also called a **repeated integral**)

$$\int_c^d \int_a^b f(x, y)dxdy = \int_c^d \left(\int_a^b f(x, y)dx \right) dy.$$

In the same way, if $f(x, \cdot)$ is integrable on $[c, d]$ for every $x \in [a, b]$, then we define

$$\int_a^b \int_c^d f(x, y)dydx = \int_a^b \left(\int_c^d f(x, y)dy \right) dx.$$

All of these integrals have improper analogs that are defined in the obvious way. For example, we define

$$\int_c^{\to d} \int_a^{\to b} f(x, y)dxdy = \lim_{v \to d-} \int_c^v \left(\lim_{u \to b-} \int_a^u f(x, y)dx \right) dy.$$

16.3   Continuity of a Partial Integral

In this section we show that if a function f is defined on a set of the form $[a, b] \times S$, where S is a set of real numbers, then, under certain conditions, the partial integral

$$\int_a^b f(x,y)dx$$

will be as "continuous in y" as the function f itself. In fact, we shall do a bit better than this. We shall obtain our main theorem for improper integrals of the form $\int_a^{\to b}$.

16.3.1   The Continuity Theorem for Partial Riemann Integrals

Suppose that a and b are real numbers and $a < b$, that S is a set of real numbers, and that f is a bounded function defined on the set $[a,b] \times S$. Suppose that the function $f(\cdot, y)$ is Riemann integrable on $[a,b]$ for every $y \in S$ and that

$$F(y) = \int_a^b f(x,y)dx$$

for every $y \in S$. Suppose that the function $f(x, \cdot)$ is continuous on S for every $x \in [a,b]$. Then F is continuous on S.

Instead of proving the theorem in this form, we shall prove the following sharper version that applies to improper Riemann integrals.

16.3.2   The Continuity Theorem for Partial Improper Integrals

Suppose that $-\infty < a < b \le \infty$, that S is a set of real numbers, and that

$$f : [a,b) \times S \to \mathbf{R}.$$

Suppose that the function $f(\cdot, y)$ is improper Riemann integrable on $[a,b)$ for every $y \in S$ and that

$$F(y) = \int_a^{\to b} f(x,y)dx$$

for every $y \in S$. Suppose that the function $f(x, \cdot)$ is continuous on S for every $x \in [a,b)$ and suppose, finally, that there exists an improper Riemann integrable function g on $[a,b)$ such that

$$|f(x,y)| \le g(x)$$

for every point $(x,y) \in [a,b) \times S$. Then F is continuous on S.

Proof. Suppose that $y \in S$. To show that the function F is continuous at the number y we shall use Theorem 8.7.6. Suppose that (y_n) is a sequence in the set S and that $y_n \to y$ as $n \to \infty$. For every number $x \in [a,b)$ the continuity at the number y of the function $f(x, \cdot)$ guarantees that $f(x, y_n) \to f(x, y)$ as $n \to \infty$. It therefore follows from Theorem 14.3.8 that

$$\lim_{n \to \infty} \int_a^{\to b} f(x, y_n)dx = \int_a^{\to b} f(x, y)dx. \quad \blacksquare$$

16.3.3   An Example Showing How Continuity Can Fail

We define

$$F(y) = \int_{0 \leftarrow}^1 \frac{x - y}{(x + y)^3}dx.$$

We observe that

$$F(y) = -\frac{1}{(1 + y)^2}$$

whenever $y > 0$ but that $F(0) = \infty$. Thus the function F fails to be continuous at 0, in spite of the fact that the integrand is continuous in y for each $x \in (0, 1]$.

This example demonstrates the need in the continuity theorem, Theorem 16.3.2, for the existence of the function g.

16.4   Differentiation of a Partial Integral

In this section we show that if a function f is defined on a set of the form $[a, b] \times S$, where S is a set of real numbers, then, under certain conditions we have the identity that is often expressed in elementary calculus in the form

$$\frac{d}{dy} \int_a^b f(x, y)dx = \int_a^b \frac{\partial}{\partial y}f(x, y)dx.$$

In fact we shall do a bit better than this. We shall obtain our main theorem for improper integrals of the form $\int_a^{\to b}$, and then we shall derive another extension of the theorem that is known as the **Leibniz rule.**

16.4.1   Differentiation of a Partial Riemann Integral

Suppose that a and b are real numbers and $a < b$, that S is an interval, and that f is a function defined on the set $[a, b] \times S$. Suppose that the partial derivative $D_2f(x, y)$ exists for every point $(x, y) \in [a, b] \times S$, that the function D_2f is bounded, and that both of the functions $f(\cdot, y)$ and $D_2f(\cdot, y)$ are integrable on $[a, b]$ for every $y \in S$. Then, if we define

$$F(y) = \int_a^b f(x, y)dx$$

for every $y \in S$, the function F is differentiable on S and for every $y \in S$ we have

$$F'(y) = \int_a^b D_2 f(x, y) dx.$$

Instead of proving the theorem in this form we shall prove the following sharper version that applies to improper Riemann integrals.

16.4.2   Differentiation of a Partial Improper Riemann Integral

Suppose that $-\infty < a < b \leq \infty$, that S is an interval, and that f is a function defined on the set $[a, b) \times S$. Suppose that the partial derivative $D_2 f(x, y)$ exists for every point $(x, y) \in [a, b) \times S$ and that both of the functions $f(\cdot, y)$ and $D_2 f(\cdot, y)$ are improper Riemann integrable on $[a, b)$ for every $y \in S$. Suppose finally that there exists an improper Riemann integrable function g on $[a, b)$ such that

$$|D_2 f(x, y)| \leq g(x)$$

for every point $(x, y) \in [a, b) \times S$.

Then, if we define

$$F(y) = \int_a^{\to b} f(x, y) dx$$

for every $y \in S$, the function F is differentiable on S and for every $y \in S$ we have

$$F'(y) = \int_a^{\to b} D_2 f(x, y) dx.$$

Proof. Suppose that $y \in S$. To obtain $F'(y)$ we shall make use of Theorem 9.3.6. Suppose that (y_n) is a sequence in the set $S \setminus \{y\}$ and that $y_n \to y$ as $n \to \infty$. For every number $x \in [a, b)$ we know that

$$\lim_{n \to \infty} \frac{f(x, y_n) - f(x, y)}{y_n - y} = D_2 f(x, y).$$

The idea of the proof is to use Theorem 14.3.8 to obtain

$$\lim_{n \to \infty} \frac{F(y_n) - F(y)}{y_n - y} = \lim_{n \to \infty} \int_a^{\to b} \frac{f(x, y_n) - f(x, y)}{y_n - y} dx = \int_a^{\to b} D_2 f(x, y) dx,$$

and the validity of this method follows from the fact that if $x \in [a, b)$ and n is a positive integer, then we can use the mean value theorem (Theorem 9.4.3) to choose a number t between y and y_n such that

$$\frac{f(x,y_n) - f(x,y)}{y_n - y} = D_2 f(x,t)$$

from which we can deduce that

$$\left| \frac{f(x,y_n) - f(x,y)}{y_n - y} \right| = |D_2 f(x,t)| \le g(x).$$

16.4.3　The Leibniz Rule

Suppose that a and b are real numbers and $a < b$, that S is an interval, and that f is a function defined on the set $[a,b] \times S$. Suppose that the partial derivative $D_2 f(x,y)$ exists for every point $(x,y) \in [a,b] \times S$, and that the function $D_2 f$ is bounded. Suppose that, for every number $y \in S$, the function $f(\cdot,y)$ is continuous on $[a,b]$ and the function $D_2 f(\cdot,y)$ is integrable on $[a,b]$. Suppose that u and v are differentiable functions on the interval S whose ranges are included in the interval $[a,b]$. Then, if we define

$$F(y) = \int_{u(y)}^{v(y)} f(x,y)dx$$

for every $y \in S$, the function F is differentiable on S and for every $y \in S$ we have

$$F'(y) = \int_{u(y)}^{v(y)} D_2 f(x,y)dx + f(v(y),y)v'(y) - f(u(y),y)u'(y).$$

The proof of the Leibniz rule can be found in the on-screen version of the text.

16.5　Some Applications of Partial Integrals

We suppose that α is a given positive number and we define

$$F(y) = \int_0^{\to \infty} e^{-\alpha x} \frac{\sin xy}{x} dx$$

for every real number y. As we mentioned in Subsection 13.4.7, this integral is not improper at 0. Although this integral is hard to evaluate directly, we can evaluate it in the following way:

Given any $x > 0$ and any number y we define

$$f(x,y) = e^{-\alpha x}\frac{\sin xy}{x}.$$

For all such x and y we see that

$$D_2 f(x,y) = e^{-\alpha x}\cos xy.$$

In view of the inequality

$$\left|D_2 f(x,y)\right| = \left|e^{-\alpha x}\cos xy\right| \le e^{-\alpha x},$$

we can use Theorem 16.4.2 to obtain

$$F'(y) = \int_0^{\to\infty} e^{-\alpha x}\cos xy\, dx.$$

The latter integral is easy to evaluate, and, by doing so, we obtain

$$F'(y) = \frac{\alpha}{\alpha^2 + y^2}.$$

Therefore, for every number y we have

$$F(y) = F(y) - F(0) = \int_0^y \frac{\alpha}{\alpha^2 + u^2}du = \arctan\left(\frac{y}{\alpha}\right),$$

and, putting $y = 1$, we obtain

$$\int_0^{\to\infty} e^{-\alpha x}\frac{\sin x}{x}dx = \arctan\left(\frac{1}{\alpha}\right) = \frac{\pi}{2} - \arctan\alpha.$$

It now seems reasonable to allow α to approach 0 and obtain

$$\int_0^{\to\infty} \frac{\sin x}{x}dx = \lim_{\alpha\to 0+} \int_0^{\to\infty} e^{-\alpha x}\frac{\sin x}{x}dx = \frac{\pi}{2} - \arctan 0.$$

To justify the latter step we recall from Subsection 13.4.7 that the integral

$$\int_0^{\to\infty} \frac{\sin x}{x}dx$$

is convergent and we apply Abel's theorem for integrals (Theorem 13.4.6). We have therefore shown that

$$\int_0^{\to\infty} \frac{\sin x}{x}dx = \frac{\pi}{2}.$$

16.6   Interchanging Iterated Riemann Integrals

In this section we present a very beautiful theorem of Fichtenholz that says, essentially, that iterated Riemann integrals can always be interchanged. We shall state the theorem in two forms:

16.6.1   First Form of Fichtenholz's Theorem

Suppose that f is a bounded function on a rectangle $[a, b] \times [c, d]$. Then the identity

$$\int_a^b \int_c^d f(x, y) dy dx = \int_c^d \int_a^b f(x, y) dx dy$$

will hold as long as both sides exist as iterated Riemann integrals.

16.6.2   Second (Sharper) Form of Fichtenholz's Theorem

Suppose that f is a bounded function on a rectangle $[a, b] \times [c, d]$. Suppose that the integral

$$\int_a^b f(x, y) dx$$

exists for every number $y \in [c, d]$ and that the integral

$$\int_c^d f(x, y) dy$$

exists for every number $x \in [a, b]$. Then we have

$$\int_a^b \int_c^d f(x, y) dy dx = \int_c^d \int_a^b f(x, y) dx dy.$$

This second form of the theorem asserts that, as long as the inside integrals exist on both sides of the preceding identity, the outside integrals will exist automatically and, of course, the identity will hold.

16.6.3   A Note About Fichtenholz's Theorem

One of the most interesting features of Fichtenholz's theorem is that it is truly a theorem about Riemann integrals. It has no analog for the more sophisticated kinds of integral, such as the Lebesgue integral, that are used in modern mathematics. When one is working with the Lebesgue integral, it is easy to find an example of a function f defined on the rectangle $[0, 1] \times [0, 1]$ such that $0 \leq f(x, y) \leq 1$ for each point (x, y) and such that, although the repeated

Lebesgue integrals exist, we have

$$\int_0^1 \int_0^1 f(x,y)dxdy = 0 \qquad \text{and} \qquad \int_0^1 \int_0^1 f(x,y)dydx = 1.$$

This is not to say that interchange of iterated integrals is impossible in the Lebesgue theory. In the Lebesgue theory there is a theorem known as **Fubini's theorem** that tells us that if a function is integrable as a function of two variables on a rectangle $[a,b] \times [c,d]$, then

$$\iint_{[a,b] \times [c,d]} f = \int_a^b \int_c^d f(x,y)dydx = \int_c^d \int_a^b f(x,y)dxdy.$$

Fubini's theorem asserts that the two iterated integrals are equal to one another because each of them is equal to the 2-variable integral of f on the rectangle. The important difference between Fubini's theorem and Fichtenholz's theorem is that Fichtenholz's theorem makes no requirement about integrability of the function as a function of two variables. As you know, we have not studied integrals of functions of more than one variable. You can find such integrals in the optional chapter on calculus of several variables that is available in the on-screen version of this book.

Although the proof of Fichtenholz's theorem is fairly easy, it relies on the optional topic of Darboux's theorem that was introduced in Section 11.9. Because Darboux's theorem was optional, we shall make the proof of Fichtenholz's theorem optional as well. If you would like to read this proof, click on the icon ▓ for a review of Darboux's theorem and then click on the icon ▓ .

16.6.4   Failure of Fichtenholz's Theorem for Improper Integrals

The examples contained in this subsection show that iterated improper integrals cannot be interchanged as widely as ordinary Riemann integrals.

Observe that

$$\int_{0\leftarrow}^1 \int_0^1 \frac{x-y}{(x+y)^2}dxdy = -\frac{1}{2} \qquad \text{and} \qquad \int_{0\leftarrow}^1 \int_0^1 \frac{x-y}{(x+y)^2}dydx = -\frac{1}{2}.$$

You may also want to consider the integral

$$\int_1^{\to\infty} \int_1^{\to\infty} \frac{x-y}{(x+y)^2}dxdy$$

and the integral

$$\int_1^{\to\infty} \int_1^{\to\infty} \frac{x^2 - y^2}{(x^2 + y^2)^2} dxdy.$$

Although the integrals in these examples are convergent, they do not converge absolutely.

16.6.5   Fichtenholz's Theorem for Nonnegative Functions

Suppose that f is a nonnegative function on a rectangle $[a, b) \times [c, d)$. Then the identity

$$\int_a^{\to b} \int_c^{\to d} f(x, y)dydx = \int_c^{\to d} \int_a^{\to b} f(x, y)dxdy$$

will hold as long as both sides exist as iterated improper Riemann integrals.

16.6.6   Some Exercises on Iterated Riemann Integrals

1. Given that

$$f(x) = \int_0^x \exp\left(-t^2\right) dt \qquad g(x) = \int_0^1 \frac{\exp\left(-x^2\left(t^2 + 1\right)\right)}{t^2 + 1} dt$$

    and given

$$h(x) = (f(x))^2 + g(x)$$

    for every real number x, prove that the function h must be constant. What is the value of this constant? Deduce that

$$\int_0^{\to\infty} \exp\left(-x^2\right) dx = \frac{\sqrt{\pi}}{2}.$$

2. Given that

$$g(y) = \int_0^{\to\infty} \exp\left(-x^2\right) \cos 2xydx \qquad \text{and} \qquad f(y) = \exp\left(y^2\right) g(y)$$

    for every real number y, prove that the function f must be constant. What is the value of this constant?

3. Find an explicit formula for the integral

$$\int_0^{\to\infty} \exp\left(-x^2\right) \cos 2xydx.$$

4. Evaluate the integral

$$\int_0^{\to\infty} \int_0^{\to\infty} \exp\left(-x^2\right) \cos 2xy \, dx \, dy.$$

What happens in this integral if we invert the order of integration?

5. Apply Fichtenholz's theorem to the integral $\int_0^{\to\infty} \int_0^{\to\infty} f(x,y) \, dx \, dy$, where

$$f(x,y) = \begin{cases} \exp\left(-y^3\right) & \text{if } x < y^2 \\ 0 & \text{if } x \geq y^2. \end{cases}$$

Now evaluate the integral

$$\int_0^{\to\infty} \int_{\sqrt{x}}^{\to\infty} \exp\left(-y^3\right) dy \, dx.$$

6. (a) Express the integrand of the following integral in partial fractions and show that if x and y are positive numbers, then

$$\int_0^{\to\infty} \frac{1}{\left(1 + t^2x^2\right)\left(1 + t^2y^2\right)} dt = \frac{\pi}{2\left(x + y\right)}.$$

   (b) Apply Fichtenholz's theorem (more than once) to the integral

$$\int_{\leftarrow 0}^{1} \int_0^{1} \frac{\pi}{2\left(x + y\right)} dx \, dy$$

   and deduce that

$$\int_0^{\to\infty} \frac{\left(\arctan x\right)^2}{x^2} dx = \pi \log 2.$$

   (c) Evaluate the integrals

$$\int_0^{\pi/2} \frac{x^2}{\sin^2 x} dx \quad \text{and} \quad \int_0^{\pi/2} x \cot x \, dx \quad \text{and} \quad \int_{\leftarrow 0}^{\pi/2} \log \sin x \, dx.$$

7. Prove that if f is a function defined on a rectangle $[a, b) \times [c, d)$, then the identity

$$\int_a^{\to b} \int_c^{\to d} f(x,y) \, dy \, dx = \int_c^{\to d} \int_a^{\to b} f(x,y) \, dx \, dy$$

   will hold as long as both sides exist as iterated improper Riemann integrals and the left side converges absolutely. Hint: Use the fact that $f = (|f| + f) - |f|$.

8. Given that f is improper Riemann integrable on $[0, \infty)$, that $a \geq 0$, and that $g(u) = f(u - a)$ whenever $u > a$, prove that g is improper Riemann integrable on the interval $[a, \infty)$ and that

$$\int_{\leftarrow 0}^{\rightarrow \infty} f(x)dx = \int_{\leftarrow a}^{\rightarrow \infty} g(u)du.$$

9. In this exercise we suppose that f and g are nonnegative improper Riemann integrable functions on the interval $(0, \infty)$ and that the function h is defined on the set $(0, \infty) \times (0, \infty)$ by the equation

$$h(x, y) = \begin{cases} f(x - y)g(y) & \text{if } y \leq x \\ 0 & \text{if } y > x. \end{cases}$$

(a) Prove that

$$\int_{\leftarrow 0}^{\rightarrow \infty} \int_{\leftarrow 0}^{\rightarrow \infty} h(x, y)dxdy = \int_{\leftarrow 0}^{\rightarrow \infty} \int_{\leftarrow y}^{\rightarrow \infty} f(x - y)g(y)dxdy$$

$$= \left(\int_{\leftarrow 0}^{\rightarrow \infty} f \right) \left(\int_{\leftarrow 0}^{\rightarrow \infty} g \right).$$

(b) Apply Fichtenholz's theorem for improper integrals to the first integral in part a and deduce that the integral is equal to

$$\int_{\leftarrow 0}^{\rightarrow \infty} \int_{\leftarrow 0}^{\rightarrow x} f(x - y)g(y)dydx.$$

16.6.7   Some Exercises that Explore the Gamma Function

The exercises in this subsection can be used to develop most of the basic facts about the **gamma function** and the related **beta function** that were defined in Exercise 2b of Subsection 13.3.3. In these exercises we assume that α and β are given positive numbers. The expressions $\Gamma(\alpha)$ and $B(\alpha, \beta)$ are defined as follows:

$$\Gamma(\alpha) = \int_{0\leftarrow}^{\rightarrow \infty} x^{\alpha-1}e^{-x}dx$$

$$B(\alpha, \beta) = \int_{0\leftarrow}^{\rightarrow 1} (1 - t)^{\alpha-1} t^{\beta-1}dt.$$

1. Apply Exercise 9 of Subsection 16.6.6 to the functions f and g defined by

the equations

$$f(x) = x^{\alpha-1}e^{-x} \quad \text{and} \quad g(x) = x^{\beta-1}e^{-x}$$

for all $x > 0$. Deduce that

$$\Gamma(\alpha)\Gamma(\beta) = \int_{0\leftarrow}^{\rightarrow\infty}\int_{0\leftarrow}^{\rightarrow x} e^{-x}(x-y)^{\alpha-1}y^{\beta-1}dydx.$$

2. By making the substitution $y = ux$ in the inside integral in Exercise 1, deduce that

$$\Gamma(\alpha)\Gamma(\beta) = \Gamma(\alpha+\beta)B(\alpha,\beta).$$

3. Apply the method of integration by parts to the integral that defines $\Gamma(\alpha)$ and deduce that

$$\Gamma(\alpha+1) = \alpha\Gamma(\alpha).$$

4. Make the substitution $t = \sin^2 y$ in the definition of the beta function and deduce that

$$B(\alpha,\beta) = 2\int_{0\leftarrow}^{\rightarrow\pi/2} \sin^{2\alpha-1}\theta\cos^{2\beta-1}\theta d\theta.$$

5. Use Exercise 4 to evaluate $B\left(\frac{1}{2},\frac{1}{2}\right)$ and deduce that

$$\Gamma\left(\frac{1}{2}\right) = \sqrt{\pi}.$$

6. Make the substitution $t = u^2$ in the definition of $\Gamma(\alpha)$ and deduce that

$$\Gamma(\alpha) = 2\int_{0\leftarrow}^{\rightarrow\infty} u^{2\alpha-1}\exp\left(-u^2\right)du,$$

and then use Exercise 5 to find another way of showing that

$$\int_{0\leftarrow}^{\rightarrow\infty} \exp\left(-x^2\right)dx = \frac{\sqrt{\pi}}{2}.$$

Recall that we obtained this identity in Exercise 1 of Subsection 16.6.6.

7. With an eye on the exercises in Subsection 11.13.5, prove that if $p > -1$, then

$$\int_{0\leftarrow}^{\rightarrow\pi} \sin^p\theta d\theta = 2\int_{0\leftarrow}^{\rightarrow\pi/2} \sin^p\theta d\theta = \frac{\sqrt{\pi}\Gamma\left(\frac{p+1}{2}\right)}{\Gamma\left(\frac{p}{2}+1\right)}.$$

8. Prove that

$$B\left(\alpha, \alpha\right) = 2^{1-2\alpha} B\left(\alpha, \frac{1}{2}\right)$$

and deduce that

$$\sqrt{\pi}\Gamma\left(2\alpha\right) = 2^{2\alpha-1}\Gamma\left(\alpha\right)\Gamma\left(\alpha + \frac{1}{2}\right).$$

9. Prove that

$$\int_0^{\to \pi/2} \sqrt{\tan x}\, dx = \frac{\pi}{\sqrt{2}}.$$

10. In this exercise we define

$$\phi(p) = \int_0^\pi \sin^p \theta\, d\theta$$

whenever $p > 0$.

(a) Prove that ϕ is a decreasing function on the interval $(0, \infty)$.

(b) Prove that

$$\lim_{p\to\infty} \frac{\phi(2p+2)}{\phi(2p)} = 1.$$

(c) Combine the first two parts of this question and deduce that

$$\lim_{p\to\infty} \frac{\phi(2p+1)}{\phi(2p)} = 1.$$

(d) Prove that

$$\lim_{p\to\infty} \frac{4^p \Gamma^2\left(p+1\right)}{\sqrt{p}\,\Gamma\left(2p+1\right)} = \sqrt{\pi}.$$

This assertion is known as **Wallis's formula**

(e) Assuming that p is restricted to be a positive integer, rewrite Wallis's formula in terms of factorials.

11. The purpose of this exercise is to encourage you to read a proof of an interesting theorem known as **Stirling's formula**, which states that

$$\lim_{x\to\infty} \frac{\Gamma\left(x+1\right)}{x^x e^{-x}\sqrt{x}} = \sqrt{2\pi}.$$

Note that if n is a large positive integer, then Stirling's formula suggests that an approximate value for $n!$ is

$$\frac{\sqrt{2\pi n}\, n^n}{e^n}.$$

The proof of Stirling's formula that is provided with a link at this point in the on-screen version of the text is based on a proof that is provided on page 195 of Walter Rudin's classic text [26]. The proof provided here is actually a little simpler because it makes use of the improper integral form of the bounded convergence theorem (Theorem 14.3.8).

Chapter 17
Sets of Measure Zero (Optional)

This optional chapter introduces the concept of *measure zero* and uses this concept to prove some deeper theorems on Riemann integration that cannot be proved in Chapter 11. To reach this chapter from the on-screen version of the book, click on the icon .

Chapter 18
Calculus of Several Variables (Optional)

This optional chapter introduces differentiation and integration of functions of several variables and goes on to present the inverse function and implicit function theorems for several variables and the change of variable theorem for multiple integrals. To reach this chapter from the on-screen version of the book, click on the icon .

Bibliography

[1] Tom Apostol. *Mathematical Analysis*, 2nd ed. Reading, MA.: Addison-Wesley, 1974.

[2] Kendall E. Atkinson. *An Introduction to Numerical Analysis*. New York: Wiley, 1978.

[3] Anatole Beck. *Continuous Flows in the Plane*. New York: Springer-Verlag, 1974.

[4] E. T. Bell. *Men of Mathematics,* London: Victor Gollancz, 1937.

[5] George Berkeley. *The Analyst, or a Discourse Addressed to an Infidel Mathematician*. Vol. III of *The Works of George Berkeley*. Edited by A.C. Fraser. Oxford, 1901.

[6] R. Boas. *A Primer of Real Functions*. Carus Mathematical Monographs, No. 13. New York: Wiley, 1960.

[7] N. Dunford and J. Schwartz. *Linear Operators Vol II*. New York: Interscience, 1963

[8] Howard Eves. *Foundations and Fundamental Concepts of Mathematics,* 3rd ed. Boston: PWS-Kent, 1990.

[9] Gottlob Frege. *The Basic Laws of Arithmetic*. Translated and edited by Montgomery Furth. Berkeley: University of California Press, 1964.

[10] B. R. Gelbaum and J. M. H. Olmsted. *Counterexamples in Analysis*. San Francisco: Holden-Day, 1964.

[11] Paul R. Halmos. "Has Progress in Mathematics Slowed Down?" *American Mathematical Monthly* 97(7), 1990, pp. 561–588.

[12] G. H. Hardy. *Divergent Series*. Oxford: Oxford University Press, 1949.

[13] Patrick Hughes and George Brecht. *Vicious Circles and Infinity*. Garden City, NY: Doubleday and Company Inc., 1975.

[14] Konrad Jacobs. *Invitation to Mathematics*. Princeton, NJ: Princeton University Press, 1992.

[15] John L. Kelley. *General Topology*. Princeton, NJ: Van Nostrand 1955.

[16] Morris Kline. *Mathematical Thought from Ancient to Modern Times*. New York: Oxford University Press, 1972.

[17] Knight. "A Strong Inverse Function Theorem." *American Mathematical Monthly* 95(7), 1988, pp. 648–651.

[18] Jonathan Lewin. "Automatic Continuity of Measurable Group Homomorphisms." *Proc. A.M.S.* 87(1), 1983 pp. 78–82.

[19] Jonathan Lewin. "A Truly Elementary Proof of the Bounded Convergence Theorem." *American Mathematical Monthly*, 93(5), 1986, pp. 395–397.

[20] Jonathan Lewin. "Some Applications of the Bounded Convergence Theorem for an Introductory Course in Analysis." *American Mathematical Monthly*, 94(10), 1987, pp. 988–993.

[21] Jonathan Lewin. "A Simple Proof of Zorn's Lemma." *American Mathematical Monthly* 98(4), 1991, pp. 353–354.

[22] Jonathan Lewin. *Precalculus with Scientific Notebook*. Dubuque IA: Kendall/Hunt, 1997.

[23] W.A.J. Luxemburg. "The Abstract Riemann Integral and a Theorem of G. Fichtenholz on Equality of Repeated Riemann Integrals." IA and IB. *Proc. Ned. Akad. Wetensch.* Ser. A 64 1961 pp. 516–545; *Indag. Math.* 23, 1961.

[24] Ivan Niven. *Numbers, Rational And Irrational,* New York: Random House, 1962.

[25] John M. H. Olmsted. *Advanced Calculus*. Englewood Cliffs, NJ: Prentice-Hall, 1961.

[26] Walter Rudin. *Principles of Mathematical Analysis,* 3rd ed. New York: McGraw-Hill, 1976.

[27] Walter Rudin. *Real and Complex Analysis,* 3rd ed. New York: McGraw-Hill, 1987.

[28] W. Sierpinski. "Sur un problème concernant les ensembles mesurables superficiellement." *Fundamenta Mathematica*, 1, 1920, pp. 112–115.

[29] George F. Simmons. *Introduction to Topology and Modern Analysis*. New York: McGraw-Hill, 1963.

[30] George B. Thomas and Ross L. Finney, *Calculus and Analytic Geometry,* 7th ed. Reading, MA: Addison-Wesley, 1988.

[31] Richard J. Trudeau. *The Non-Euclidean Revolution*. Boston: Birkhauser, 1987.

[32] F. M. A. Voltaire. *Letters Concerning the English Nation*. London, 1733.

Index of Symbols and General Index

On-Screen Index Entries

Both the printed version and the on-screen version of this book are supplied with indices. Because the on-screen book is more extensive than the printed book, its index is correspondingly larger. In addition, the on-screen index contains entries, such as links to the World Wide Web for bigraphical information, that cannot be provided in a printed index.

I have, however, included a variety of entries in the printed index that refer to material that exists only in the on-screen book. Their purpose is to make it possible to see, at a glance, what sort of material is covered in the on-screen book. All entries in the printed index that target material in the on-screen book point to page 482, which is the page you are reading now. To reach the targets of these index entries, go to the on-screen version of the book and click on them from there.

Index for the Printed Book

Printed in the United States
By Bookmasters